Infosys Science Foundation Series

Infosys Science Foundation Series in Mathematical Sciences

The **Infosys Science Foundation Series in Mathematical Sciences**, a Scopus-indexed book series, focuses on high-quality content in the domain of pure and applied mathematics, biomathematics, financial mathematics, operations research and theoretical computer science. With this series, Springer hopes to provide readers with monographs, textbooks, edited volume, handbooks and professional books of the highest academic quality on current topics in relevant disciplines. Literature in this series will appeal to a wide audience of researchers, students, educators and professionals across the world.

Aref Jeribi · Najib Kaddachi

Topological Fixed-Point Theory in Suitable Banach Algebras with Applications

Springer

Aref Jeribi
Department of Mathematics
University of Sfax
Sfax, Tunisia

Najib Kaddachi
Department of Mathematics
University of Sfax
Sfax, Tunisia

ISSN 2363-6149 ISSN 2363-6157 (electronic)
Infosys Science Foundation Series
ISSN 2364-4036 ISSN 2364-4044 (electronic)
Infosys Science Foundation Series in Mathematical Sciences
ISBN 978-981-96-5541-0 ISBN 978-981-96-5542-7 (eBook)
https://doi.org/10.1007/978-981-96-5542-7

Mathematics Subject Classification: 47H10, 34A60, 37C25, 39B42, 47A08, 47H09

This Springer imprint is published by the registered company Springer Nature Singapore Pte Ltd.
The registered company address is: 152 Beach Road, #21-01/04 Gateway East, Singapore 189721, Singapore

To
my mother Sania,
my father Ali,
my wife Fadoua,
my children Adam and Rahma,
my brothers Sofien and Mohamed Amin,
my sister Elhem, and
all members of my extended family

—Aref Jeribi

To
my mother Torkia,
my father Dhaou,
To my wife Mouna,
my children Jouri and Joud,
To my brothers Walid, Hassan and Yassin,
To my sister Naziha

—Najib Kaddachi

Preface

The realm of fixed point theory for single-valued mapping and multi-valued mappings in Banach algebras has evolved into a pivotal domain, intricately interwoven with both pure and applied mathematics. The principal aim of this book resides in its aspiration to disseminate the fundamental techniques and outcomes of this theory. This text focuses specifically on the fixed point theory concerning block operator matrices with multi-valued inputs nestled within appropriate Banach algebras and explores its applications across a diverse spectrum of equations, including nonlinear biological models [195] and two-dimensional boundary value problems associated with growing cell populations and functional systems of differential and integral inclusions.

Across these domains, one encounters conundrums such as the diminishing compactness of mappings and the topological structure of their underlying domains. To address these challenges, we find it expedient to emphasize fixed point results in the context of the weak topology, while concurrently introducing the realm of topological fixed point theory for diverse classes of both single-valued and multi-valued maps.

In the process of curating the material, our focus has been limited to the exploration of topological methods within the fixed point theory of multi-valued mappings and their applications, predominantly pertaining to differential inclusions, integral inclusions, integral equations, and differential equations.

This book commences its journey with an exploration of contractions, offering a proof of this fundamental result, and subsequently delves into an extensive survey of fixed point theory, encompassing both single-valued and multi-valued mappings. Chapter 1 further delves into the generalization of Banach's fixed point theorem, see Theorem 1.2.4, exemplified by Theorem 1.2.26 and Theorem 1.4.38. The Banach contraction principle, hinging on an iterative procedure, possesses computational applicability in ascertaining the fixed point of a contractive map. We bestow particular attention upon the contraction Banach principle, weak compactness, notably the Eberlein-S̃mulian theorems, and the venerable results of fixed point theory. The closing section is dedicated to the classical outcomes of fixed point theory within generalized Banach algebras.

By establishing the fixed point theorem within the framework of suitably structured Banach algebras, Chap. 2 lays the cornerstone for subsequent exploration and analysis of operator equations in this milieu. This chapter turns its focus toward the challenge of verifying in a Banach algebra, the existence of solutions to equation

$$x = Ax \cdot Bx + Cx,$$

where A, B and C are nonlinear multi-valued mapping, within Banach algebras meeting a specific sequential criterion. Our attention converges on the examination of fixed points entailing weakly sequentially continuous operators. Furthermore, Perov's fixed point theorem, a potent tool in the realm of fixed point theory, is deployed to corroborate the existence of fixed points within suitable generalized Banach algebras. The Perov's fixed point theorem serving as a versatile framework, accommodates diverse mapping types. In the closing section of this chapter, we endeavor to define the concept of relatively weakly compact Banach algebras, abbreviated as RWC-Banach algebras, as originally introduced by J. Banas and L. Olszowy in [61].

In Chap. 3, our focus shifts to the investigation of fixed point theory for block operator matrices. Based on the previous extension established under the weak topology setting, we are planning in this chapter to extend these results by proving the existence of a fixed point in $S \times S$ of the 2×2 block operator matrix

$$\begin{pmatrix} A & B \cdot B' \\ C & D \end{pmatrix},$$

where S is nonempty, closed, and convex subset of a Banach spaces or a Banach algebras and A, B, B', C, and D are nonlinear operators. The chapter's initiation concerns single-valued inputs under the influence of weak topology. The subsequent section is dedicated to the quest for fixed points, covering weakly asymptotically regular mappings concerning the operator $(I-D)^{-1}C$, pseudo-contractive mappings, k-set-weakly contractive mappings, and countably β-contractive mappings within nonseparable Banach spaces. Another crucial facet broached in this chapter relates to the critical scenario where the operator

$$\frac{I-A}{B(I-D)^{-1}C}$$

is non-invertible and $B(I-D)^{-1}C$ is supposed regular i.e. $B(I-D)^{-1}C$ maps S into the set of all invertible elements of the Banach algebra X. This poses an unique challenge within the confines of traditional fixed point theory, as

$$\left(\frac{I-A}{B(I-D)^{-1}C}\right)^{-1} \tag{1}$$

does not invariably correspond to a single-valued operator. Then, the operator (1) could be seen as a multi-valued mapping. Nevertheless, the chapter manages to derive valuable results applicable to such circumstances.

The subsequent stage of inquiry pertains to block operator matrices housing multi-valued mappings as their entries, introducing additional intricacies when contrasted with the single-valued scenario. The primary objective is to ascertain the existence of a fixed point within this context under stringent conditions. In this endeavor, Boyd and Wong's theorem, in conjunction with Dhage's findings, plays a pivotal role. The specifics of these conditions, theorems and techniques employed from Boyd-Wong theorem and Dhage's fixed point results are expounded upon in the referenced work, serving as a foundation for establishing fixed point results under demanding conditions for block operator matrices with multi-valued mappings.

In an extension of the aforementioned outcomes, a methodology is presented to yield practical and applicable results within the ambit of weak topology. This involves the utilization of a novel class of multi-valued mappings termed countably $\mathscr{D}$-set-Lipschitzian class. By leveraging this class, the concluding section of this chapter endeavors to furnish practical results that find relevance in real-world scenarios, potentially involving the identification of fixed points or other pertinent solutions within the framework of weak topology.

As the narrative unfolds in subsequent chapters, (Chaps. 4–7), the book explores applications to specialized problems in fixed point theory, frequently arising within the applied sciences, underscoring the tangible significance of the theoretical framework presented.

We hope the book will be of use to graduate theoretical and applied mathematicians who work in fixed point theory, operator theory, integral inclusion, quadratic differential inclusions, and other related areas.

The authors express their gratitude to several colleagues, including K. Ben Amara, O. Fahem, and Z. Laouar, who reviewed and offered insights on the manuscript. These comments and suggestions are certain to facilitate enhancements in a forthcoming edition of this book.

Sfax, Tunisia
December 2024

Aref Jeribi
aref.jeribi@fss.rnu.tn

Najib Kaddachi
najibkadachi@fsegs.usf.tn

Introduction

In nonlinear analysis, fixed point theorems are very important because they provide an interesting way to prove that a fixed point exists or, in other words, that a solution to certain problems exists. These theorems give the conditions in which maps have fixed points. One of the most well-known and significant existence theorems in mathematics is the Brouwer fixed point theorem [1]. The first premise of Brouwer's theorem is that any continuous mapping

$$f : \overline{\mathscr{B}}_{\mathbb{R}^n} \longrightarrow \overline{\mathscr{B}}_{\mathbb{R}^n}$$

possesses a fixed point. Since the theorem and its many equivalent formulations or extensions, see for example [2] are powerful tools in showing the existence of solutions for many problems in pure and applied mathematics. The details can be found in any standard books on fixed point theory. Perhaps, the most well-known result in this theory is Banach's contraction principle. Precisely, the author S. Banach formulated and proved a theorem which focused, under appropriate conditions, on the existence and uniqueness of a fixed point in a complete metric space. It is regarded as the simplest and one of the most versatile elementary findings in fixed point theory. The Banach contraction principle relies on an iterative process and can be implemented computationally to determine the fixed point of a contraction map. It generates approximations with any desired level of accuracy and, moreover, provides the capability to determine the number of iterations required to achieve a specified level of precision. This makes the Banach contraction principle a valuable tool in practical applications, where numerical methods can be employed to compute fixed points efficiently. In 1930, J. Schauder established Schauder's fixed point theorem [3], which states that every continuous mapping defined on a nonempty, compact and convex subset of a Banach space possesses a fixed point. This theorem stands as a significant contribution to fixed point theory. The Brouwer's fixed point theorem is generalized to normed spaces by J. Schauder and to locally convex topological vector spaces by A. Tychonoff. These are all unified by M. Hukuhara in [4]. Moreover, K. Fan in [5] deduced the Tychonoff fixed point theorem. In [6], A. Tychonoff extended Schauder's result to compact and convex subsets of a locally convex topological vector spaces, see also [7]. The Tychonoff's fixed point theorem contains as a special case the earlier result in Schauder asserting the existence of a fixed point for each weakly continuous self mapping of a weakly compact and convex subset S of a Banach space.

One of the key extensions of Banach's fixed point theorem is fixed point theorems for nonexpansive mappings. For this reason, numerous writers have concentrated on establishing fixed points for nonexpansive mappings in Banach spaces, yielding numerous insightful findings (see, for instance, [8–11]). These results have been extended by several authors to the case of multi-valued mappings, see for example [12–15] and the references therein.

In 1935, the author in [6] extended Schauder's fixed point theorem to apply to locally convex topological vector spaces, by using the Kuratowski measure of non-compactness α, for more detail concerning α, we can see [16, 17].

G. Darbo in [18] generalize and improve Schauder's fixed point theorems to the class of k-set-contractive operators, that is, operator T of a Banach space into itself satisfying the following inequality

$$\alpha(T(S)) \leq k\alpha(S),$$

for some $k < 1$, where S is a nonempty bounded subset of a Banach space. Darbo's fixed point theorem is a very important generalization of Schauder's fixed point theorems and it includes the existence part of Banach's fixed point theorem. Later, B. N. Sadovskii in [19] expanded on Darbo's fixed point result by proving an extension applicable to the class of condensing operators, see also [20]. The notion of Hausdorff measure of non-compactness, alternatively known as ball measure of non-compactness, was introduced by L. S. Goldenstein, I. T. Gohberg and A. S. Markus in [21]. The measures of non-compactness have proved to be a very useful and efficient tools in nonlinear analysis and in fixed point theory and in the theory of operator equations and operator inclusions in Banach algebras.

In [22], F. S. De Blasi made an extension of the definition of the Hausdorff measure of non-compactness to encompass the weak topology. This extension was made with the intention of proving the analogous version of Sadovskii's fixed point theorem, specifically tailored for the weak topology. Due to the difficulty to prove the weak continuity condition in several cases, A. Ben Amar and M. Mnif in [23], have established new fixed point theorem for weakly sequentially continuous operators, that is, operator T of a Banach space X into itself satisfying the following condition, if $(x_n)_n$ is a sequence such that $x_n \rightharpoonup x$, then $T(x_n) \rightharpoonup T(x)$, here $\rightharpoonup$ denotes the weak convergence, who proved that every weakly sequentially continuous and condensing mapping with respect to the De Blasi measure of weak non-compactness on a nonempty, closed and convex subset of a Banach space has a fixed point.

An extension and a combination of the Banach's fixed point theorem and the Schauder's fixed point theorem to the sum of a contraction and a compact mapping were discovered by M. A. Krasnoselskii in [24], see also [25]. Indeed, many nonlinear problems in science can be mathematically recast as nonlinear equations of the form

$$x = Ax + Bx, \ x \in S, \tag{2}$$

where S is a nonempty, closed and convex subset of a Banach space X and A, $B : S \longrightarrow X$ are two nonlinear mappings. The equation (2) was initially introduced by M. A. Krasnoselskii in [24]. His result states that, if A is a contraction, B is continuous such that $B(S)$ is relatively compact and S is left invariant by the operators sum $A + B$, then equation (2) has a solution in S. This result has proven to be instrumental in solving perturbed nonlinear integro-differential equations, as well as in allied areas of mathematics, where existence theorems are established under mixed Lipschitz and compactness conditions. In [26], the author proved a new version of Krasnoselskii's fixed point theorem by using the arguments of weak topology. The strategy consists in giving sufficient conditions on the operators A and B which ensure of $I - A$, in order to deal with the mapping $(I - A)^{-1}B$. If $I - A$ is not invertible, then $(I - A)^{-1}B$ could be seen as a multi-valued mapping. This case was considered in [27]. Additional references such as [28, 29], and others provide further insights and explore applications in this area. Over time, there have been numerous extensions and generalizations of Krasnoselskii's fixed point theorem, which have significantly enriched the field.

Several generalizations of Krasnoselskii's fixed point result were proved over the course of time, see [30, 31] and the references therein. In [32], B. C. Dhage has proved another hybrid fixed point theorems in Banach algebras similar to that of Krasnoselskii having the following form

$$x = Ax \cdot Bx + Cx \tag{3}$$

and has obtained useful results in this context, where A, B and C, are nonlinear operators defined on a nonempty closed convex subset of a Banach algebra.

However, challenges arise due to issues like the loss of compactness of mappings and the absence of appropriate geometric and topological structures in the underlying domains. To address these challenges, it becomes beneficial to focus on fixed point results within the context of the weak topology. This allows us to investigate the existence of solutions for different types of nonlinear integral equations and differential equations in Banach spaces. The weak topology serves as a natural framework for various applications, and as a result, several fixed point results concerning the operator equation (3) have been established, specifically in terms of weak sequential continuity.

In this regard, A. Ben Amar, S. Chouayekh and A. Jeribi in [33] introduced an analog version of Dhage's results and their generalization. To overcome the issue of non-convergence of the product of two weakly convergent sequences to their product limits, they introduced a class of Banach algebras satisfying a specific sequential condition known as "condition $(\mathscr{P})$". We will say that the Banach algebra X satisfies the sequential condition $(\mathscr{P})$, if for any sequences $(x_n)_n$ and $(y_n)_n$ in X such that $x_n \rightharpoonup x$ and $y_n \rightharpoonup y$, then

$$x_n \cdot y_n \rightharpoonup x \cdot y.$$

Many authors have focused on the resolution of equation (3) in Banach algebra satisfying the sequential condition ($\mathscr{P}$) and have obtained several valuable results. Several fixed point results have been proven in [22, 33–38] and the references therein. These results are formulated in terms of the weak sequential continuity and involve the operator equation (3). In [39], the authors have proved some fixed point theorems for (3) based on the notion of condensing maps with respect to the De Blasi measure of weak non-compactness [22].

The results obtained in the study of equation (3) have nice applications to nonlinear integral equations and differential equations, see for example, [24, 32, 40] and the references therein. Integro-differential equations often arise when modeling problems in physics, biology, applied sciences, ... In many cases, these equations can be transformed into finding a fixed point for a continuous map defined on a functional space. This approach is equivalent to solving the equations using standard methods. Examples of such applications include transport equations in the kinetic theory of gas, see for example [41, 42], stationary nonlinear biological models [43] and boundary value problems in growing cell populations. To tackle these problems, we can turn to comprehensive existence theories for integral equations and differential equations. In [44], the authors have provided us with a solid foundation in these areas, see also [41]. They offer a study of nonlinear and partial differential equations, covering not only classical physics but also the modeling of transport in the kinetic theory of gas.

Different extensions and generalizations of Dhage's results have also been proved in [32, 45] and the references therein. Their proofs were mainly based on the convexity of the bounded domain, in order to guarantee the compactness of the operator $\frac{I-C}{A}$. Note that if $\frac{I-C}{A}$ is not invertible, then $\left(\frac{I-C}{A}\right)^{-1}$ could be seen as a multi-valued mapping. Many results in fixed point theory for single-valued mappings were extended to the case of multi-valued mapping, since the last has many useful applications in different fields of applied sciences, for instance, in game theory and optimization theory, see for example [13]. Of course, these extensions reaches the class of β-nonexpansive mappings, i.e., operator T of a Banach space into itself satisfying the following inequality

$$\beta(T(M)) \leq \beta(M),$$

for any bounded subset M of a Banach space with $\beta(M) > 0$ such that $T(M)$ is bounded, where β is the De Blasi measure of weak non-compactness. Because the convex-power condensing operators improve condensing operators, that is,

$$\beta(T(M)) < \beta(M), \tag{4}$$

for any bounded subset M a Banach space with $\beta(M) > 0$, it's interesting to do a continuation some previous studies around this concept in some section of this book. In [46], D. Guo, V. Lakshmikantham and X. Liu obtained a generalization of the Sadovskii's fixed point theorem on a bounded domain, which assert that if S is a nonempty closed, bounded and convex subset of a Banach space X and if an operator $T : S \longrightarrow S$ is convex-power condensing, that is, $T(S)$ is bounded and there exists a point $x_0 \in S$ and a strictly positive integer n_0 such that, for any bounded subset M of S with $\beta(M) > 0$, we have

$$\beta(T^{(n_0,x_0)}(M)) < \beta(M),$$

where

$$\begin{cases} T^{(1,x_0)}(M) = T(M) \\ T^{(n,x_0)}(M) = T\big(\overline{co}\{T^{(n-1,x_0)}(M) \bigcup \{x_0\}\}\big), \end{cases}$$

here $\overline{co}(M)$ denote the closed convex hull of M. Then, T has a fixed point in S. In 2012, R. P. Agarwal, D. O'Regan and M. A. Taoudi in [47] proved the following extension.

Theorem 1 *Let S be a nonempty, closed and convex subset of a Banach space X and let* $T : S \longrightarrow S$ *be a weakly sequentially continuous operator such that T(S) is bounded. If T is convex-power condensing with respect to the De Blasi measure, then T has, at least, one fixed point.*

The study of fixed points for multi-valued mappings using the Hausdorff metric was first studied in [48, 49]. In 1970, H. Covitz and S. B. Nadler jr. in [50] have introduced the concept of multi-valued contractions and established the fact that a contraction multi-valued with closed values in a complete metric space possesses a fixed point. Many authors have devoted themselves to study the validity of the multi-valued version of the classical fixed point theorems for some mappings. The specific formulation and analysis depend on the characteristics of the problem at hand, and often require the use of numerical methods and mathematical techniques for their solution. Indeed, the method of analogies can be employed to reformulate problems that may initially be expressed as integral inclusions and differential inclusions into fixed point problems involving multi-valued mappings. Thus, numerous authors have focussed to the existence of solutions for the nonlinear inclusion of type

$$x \in Ax \cdot Bx + Cx, \quad x \in S, \tag{5}$$

under some suitable conditions on the multi-valued mapping A, B and C, where S is a nonempty, closed, convex and bounded subset of a Banach algebra. In a series of papers [51, 52], B. C. Dhage has focused on the existence of a solution for the equation (5) and obtained several valuable results in Banach algebras, see also [45, 53, 54] and the references therein. They were mainly based on the convexity of the bounded domain, as well as some properties of the multi-valued mapping A, B and C (such as lower semi-continuous, compact, k-set contractive,...) and the potential tool of the axiomatic measures of non-compactness, were used, see for example [51] and the reference therein. Recently, A. Ben Amar, M. Boumiza and D. O'Regan in [55] improved and extended these theorems to the case of sequentially weakly upper semi-compact multi-valued mapping, that is, a multi-valued mapping Q is sequentially weakly upper semi-compact if, for any weakly convergent sequence $(x_n)_n$ of elements in S and for any arbitrary $y_n \in Q(x_n)$, the sequence $(y_n)_n$ has a weakly convergent subsequence. These studies were mainly based on the convexity of the unbounded domain, as well as some conditions on A, B and C, in order to guarantee the weak compactness of the multi-valued mapping $\left(\frac{I-C}{A}\right)^{-1}B$.

In [56], A. I. Perov and A. V. Kibenko have extended the Banach fixed point theorem to spaces equipped with vector-valued metrics, see also [57]. Perov's type of fixed point theorem has played a crucial role in solving various problems in different areas of natural science, such as nonlinear analysis [58, 59] systems of ordinary differential equations [60–64]. Several authors have further developed and extended Perov's results in different directions, emphasizing the properties of vector-valued metrics [65, 66, 56]. Subsequent progress was made in [61–64], where the focus shifted towards studying Krasnoselskii-type of fixed point theorems for both single-valued and multi-valued mapping. Extensions and generalizations of the Bohnenblust-Karlin's fixed point theorems and Krasnoselskii's fixed point theorems have been established over time, with notable contributions by J. J. Nieto, A. Ouahab and R. Rodriguez-Lopez in [61]. These results often require conditions such as lower semi-continuity, or upper semi-continuity of the underlying multi-valued mapping.

By utilizing new fixed point results under the weak topology setting, we can effectively address the existence of solutions for various types of nonlinear integral equations and differential equations in Banach algebra satisfying the sequential condition $(\mathscr{P})$, thereby overcoming the limitations posed by the lack of compactness and appropriate geometric and topological structures.

One of the difficulties arising, when dealing with such situations, is that a Banach algebra X does not necessarily satisfy condition $(\mathscr{P})$. That is why, J. Banas and L. Olszowy [67] have introduced a large class than those satisfying condition $(\mathscr{P})$ and it so-called the relatively weakly compact Banach algebra (RWC-Banach algebras, in shorts), which are preserved the relatively weak compactness structure of the product of two relatively weakly compact subsets. Collaborating with K. Ben Amara and A. Jeribi, in [68], we proved a sequential characterization of RWC-Banach algebras, so called condition $(\overline{\mathscr{P}})$, we managed to overcome the problem of the use of condition $(\mathscr{P})$. We will say that a Banach algebra X satisfies condition $(\overline{\mathscr{P}})$, if

$$(\overline{\mathscr{P}})\begin{cases}\text{For any sequences } (x_n)_n \text{ and } (y_n)_n \text{ of } X \text{ such that } x_n \rightharpoonup x \text{ and } y_n \rightharpoonup y,\\ \text{then the sequence } (x_n \cdot y_n)_n \text{ has a weakly convergent subsequence.}\end{cases}$$

Such an approach enables us to draw new and meaningful results about fixed points which work in more general Banach algebras than those which are satisfying condition $(\mathscr{P})$. This provides an efficient tool to develop new extensions of the fixed point theory. In this book, we use the theoretical results to prove the existence of solutions for some nonlinear functional integral equations defined on RWC-Banach algebras. In particular, we illustrate how the weak compactness structure of Banach algebras and the sequential condition $(\mathscr{P})$ requirements used in the literature for some class of nonlinear integral equations may be relaxed. The conditions were formulated in terms of strong continuity and weak sequential continuity for three nonlinear operators A, B and C.

Theorem 2 *Let S be a nonempty, closed and convex subset of a RWC-Banach algebra X and let $A, B : S \longrightarrow X$ be two strongly continuous operators such that*

(i) *A and B are countably $\mathscr{D}$-set-Lipschitzians with $\mathscr{D}$-functions Φ and Ψ, respectively,*
(ii) *$A(S)$ and $B(S)$ are bounded,*
(iii) *$Ax \cdot Bx \in S$, for all $x \in S$.*

Then, the operator equation $x = Ax \cdot Bx$ has a fixed point in S provided that

$$\|B(S)\|\Phi(r) + \|A(S)\|\Psi(r) + \Phi(r)\Psi(r) < r, \text{ for } r > 0.$$

Motivated by some results concerning the operator equation (3) and the nonlinear inclusion (5), we have the idea to extend the obtained results by proving, in this book, that under certain conditions on its nonlinear entries that the following 2×2 block operator matrix.

$$\begin{pmatrix} A & B \cdot B' \\ C & D \end{pmatrix}, \tag{6}$$

has, at least, one fixed point with both single-valued entrees and multi-valued entrees acting in $S \times S'$, where S and S' constitute two nonempty, closed and convex subsets of a Banach algebra satisfying the sequential condition $(\mathscr{P})$. This studies is based on the presence or absence of the invertibility of the following operator

$$\frac{I - A}{B(I - D)^{-1}C}. \tag{7}$$

Another crucial facet broached relates to the critical scenario, where the operator (7) is non-invertible, where $B(I-D)^{-1}C$ is supposed regular i.e. $B(I-D)^{-1}C$ maps S into the set of all invertible elements of a Banach algebra. This poses an unique challenge within the confines of traditional fixed point theory, as

$$\left(\frac{I - A}{B(I - D)^{-1}C}\right)^{-1}, \tag{8}$$

does not invariably correspond to a single-valued operator. Then, the operator (8) could be seen as a multi-valued mapping. In this case, several fixed point results of Chap. 3 enable us to get new results for a block operator matrix involving multi-valued operators such as, e.g., $\mathscr{D}$-set-Lipschitzian, weakly sequentially closed graph,

..., acting on a Banach algebra satisfying certain sequential conditions. Let us notice that A. Jeribi, N. Kaddachi and B. Krichen in [69] have initiated, in 2013 the study of the existence of a fixed point of block operator matrix (6). The theoretical tools were based on the fact that the mapping N, which is defined on a nonempty, closed and convex subset S of a Banach algebra X by $N(y) = z$, where z is the unique solution for the operator equation

$$z = A(x) + B(I - D)^{-1}C(x) \cdot B(I - D)^{-1}C(y),$$

is continuous and totally bounded. In this context, we proved the following theorem.

Theorem 3 *Let S be a nonempty, bounded, closed and convex subset of a Banach algebra X and let A, B, B', C, $D : S \longrightarrow X$ be five operators such that*

(i) *A, B and C are Lipschitzians with constants q_1, q_2 and q_3, respectively,*
(ii) *$(I - D)^{-1}$ is Lipschitzian with a Lipschitz constant δ on $C(S)$,*
(iii) *B' is continuous and C is compact,*
(iv) *$A(x) + B(I - D)^{-1}C(x) \cdot B'(I - D)^{-1}C(y) \in S$ for all $x, y \in S$.*

Then, the block operator matrix (6) has a fixed point in $S \times S$ provided that

$$q_1 + q_2 q_3 \delta M < 1,$$

where $M = \|B'(I - D)^{-1}C(S)\|$.

In [70], we continued the study, by proving fixed point results, in a Banach algebra satisfying the sequential condition $(\mathscr{P})$, for the block operator matrix (6) Our assumptions are as follows: A, C maps a nonempty, bounded, closed and convex subset S of a Banach algebra X satisfying the sequential condition $(\mathscr{P})$ into X, B, B', D from X into X.

Theorem 4 *Let S be a nonempty, bounded, closed and convex subset of a Banach algebra X satisfying the sequential condition $(\mathscr{P})$ and let A, $C : S \longrightarrow X$ and B, B', $D : X \longrightarrow X$ be five weakly sequentially continuous operators such that*

(i) *A is condensing with respect to β,*
(ii) *C is weakly compact,*
(iii) *D is a nonlinear contraction with $\mathscr{D}$-function Φ_T,*
(iv) *$(I - D)^{-1}C(S)$ is bounded,*
(v) *$A(x) + B(I - D)^{-1}C(x) \cdot B'(I - D)^{-1}C(x) \in S$, for all $x \in S$.*

Then, the block operator matrix (6) has, at least, one fixed point in $S \times X$.

As an application of this result, we discuss the following system of nonlinear functional integral equations occurring in some quantic physical problems.

$$\begin{cases} x(t) = f(t, x(t)) + \left[a(t)y(t)\right] \cdot \left[\left(\int_0^{\sigma_1(t)} k(t, s) f_1(s, y(\eta(s)))ds\right)u\right] \\ y(t) = \left[\left(q(t) + \int_0^{\sigma_2(t)} p(t, s, x(s), x(\lambda s))ds\right)v\right] + g(t, y(t)), \end{cases} \tag{9}$$

where $u, v \in X\backslash\{0\}$, $\lambda \in [0, 1]$, $a, q, \sigma_1, \sigma_2, \eta$ are five continuous real functions on $[0, \delta]$, $0 < \delta < \infty$, and f_1 and g are supposed weakly sequentially continuous on $[0, \delta] \times X$. Note that the problem (9) can be written in the following form

$$\begin{pmatrix} x(t) \\ y(t) \end{pmatrix} = \begin{pmatrix} A & B \cdot B' \\ C & D \end{pmatrix} \begin{pmatrix} x(t) \\ y(t) \end{pmatrix},$$

where

$$\begin{cases} (Ax)(t) = f(t, x(t)), \\ (Bx)(t) = a(t)x(t), \\ (Cx)(t) = \left(q(t) + \int_0^{\sigma_2(t)} p(t, s, x(s), x(\lambda s))ds\right)v, \\ (Dx)(t) = g(t, x(t)), \\ (B'x)(t) = \left(\int_0^{\sigma_1(t)} k(t, s)f_1(s, x(\eta(s)))ds\right)u. \end{cases}$$

In [71], we prove some new fixed point theorems for the operator (6) in the case when $B'x = 1_X$, by using the concepts of countably condensing mappings with respect to β, i.e., the condition (4) holds only for countable bounded sets, or countably $\mathscr{D}$-set-contraction, which generalize, extend and improve well-known results in [37, 70, 72, 73]. Our assumptions are as follows: A map a nonempty, closed and convex subset S of a Banach space X into $\mathscr{P}(X)$, where $\mathscr{P}(X)$ denote the family of nonempty subsets of X, B from X into $\mathscr{P}(X)$, C from S into X and D from X into X. Let us denote by $\Xi^{(A,B,C,D)}(M)$ the following set

$$\Xi^{(A,B,C,D)}(M) = \left\{(I - A)(x) \bigcap B(I - D)^{-1}C(y), (x, y) \in M \times M\right\}.$$

Theorem 5 *Let S be a nonempty, closed and convex subset of a Banach space X and let $A : S \longrightarrow \mathscr{P}(X)$ and $B : X \longrightarrow \mathscr{P}(X)$ be two multi-valued mappings and $C : S \longrightarrow X$, $D : X \longrightarrow X$ be two nonlinear mapping such that*

(i) *A is countably $\mathscr{D}$-set-contraction with $\mathscr{D}$-function ϕ_A,*
(ii) *$C(S) \subset (I - D)(S)$,*
(iii) *A and $B(I - D)^{-1}C$ have a weakly sequentially closed graph,*
(iv) *the set $\Xi^{(A,B,C,D)}(S)$ is relatively weakly compact,*
(v) *for any $x \in S$, the set $(I - A)^{-1}B(I - D)^{-1}C(x)$ is convex in S.*

Then, the block operator matrix (3.3.5) *has a fixed point in $S \times S$.*

Note that Theorem 5 is applied to establish the existence of solutions, in $C(J, X)$, the space of all continuous functions, defined on the interval $J = [0, 1]$ into X, where X is a Banach space, for the following two-dimensional problems for nonlinear integral inclusion given by

$$\begin{cases} x(t) \in \int_0^{\sigma_1(t)} k(t,s)F(s,x(\tau(s)))ds + \int_0^{\sigma_2(t)} G(s,y(s))ds \\ y(t) = \Phi\left(t, \int_0^{\sigma_3(t)} f(s,x(\nu(s)))ds\right) + g(t,y(t)), t \in J, \end{cases} \tag{10}$$

where σ_1, σ_2, σ_3, τ, ν are continuous real functions on J, k is a continuous real function on $J \times J$, the single-valued maps $f, g, \Phi : J \times X \longrightarrow X$ are supposed to be weakly sequentially continuous with respect to the second variables and the multi-valued maps $F, G : J \times X \longrightarrow \mathscr{P}(X)$ are supposed to have weakly sequentially closed graphs.

Note that the problem (10) can be written as the following fixed point

$$\begin{pmatrix} x(t) \\ y(t) \end{pmatrix} \in \begin{pmatrix} A & B \\ C & D \end{pmatrix} \begin{pmatrix} x(t) \\ y(t) \end{pmatrix},$$

where

$$\begin{cases} (Ax)(t) = \left\{ \int_0^{\sigma_1(t)} k(t,s)v(s)ds : v(s) \in F(s,x(\tau(s)))\ s \in J \right\} \\ (By)(t) = \left\{ \int_0^{\sigma_2(t)} v(s)ds : v(s) \in G(s,y(s))\ s \in J \right\} \\ (Cx)(t) = \Phi\left(t, \int_0^{\sigma_3(t)} f(s,x(\eta(s)))ds\right) \\ (Dy)(t) = g(t,y(t)). \end{cases}$$

The theory of fixed point theorem of block operator matrices opens up a new line of attack of these problems. It guarantees the existence of solutions, enables the study of nonlinear systems, connects abstract mathematics with concrete applications, supports numerical methods, and facilitates the analysis of optimization problems. Its importance lies in its ability to provide rigorous theoretical foundations and practical tools for solving problems involving block operator matrices. In this contexts, many problems arising in mathematical physics, biology, etc. may be described, in a first formulation, using systems of partial or ordinary differential inclusions. In [74], the authors have established the existence of fixed point for (6), in order to obtain the existence of solutions for a system of transport equations arising in biology. A. Ben Amar, A. Jeribi and B. Krichen in [75] have presented some new fixed point theorems for the block operator matrix (6) in Banach spaces when $B' = 1$ and they have applied their results to a two-dimensional boundary problem in the Banach space $L^p \times L^p$ for $p \in]1, +\infty[$, on the basis on a Schauder's fixed point theorem and Krasnoselskii's fixed point theorems. Their analysis was carried out via the arguments of weak topology and particularly the measure of weak non-compactness.

In a series of papers [38,76], N. Kaddachi has proved some existence results for the block operator matrix (6) under weak topology with multi-valued inputs acting on a Banach algebra satisfying the sequential condition $(\mathscr{P})$.

Theorem 6 *Let S be a nonempty, closed and convex subset of a Banach algebra X satisfying the sequential condition $\mathcal{P}$ and let A, B, $C : S \longrightarrow X$, $D : X \longrightarrow X$ and $B' : S \longrightarrow \mathcal{P}_{cv}(X)$ be five multi-valued mappings such that*

(i) A, B and C are $\mathcal{D}$-Lipschitzians with $\mathcal{D}$-functions ϕ_A, ϕ_B and ϕ_C, respectively,
(ii) A, B, C and D are weakly sequentially continuous,
(iii) $A(S)$, $B(S)$ and $B'(S)$ are bounded and $C(S) \subseteq (I - D)(S)$,
(iv) D is a contraction mapping with a constant k,
(v) B' is weakly completely continuous.

Then, the operator matrix (6) has a fixed point in $S \times S$ provided that

$$\|B'(S)\|\phi_B \circ \left(\frac{1}{1-k}\phi_C\right)(r) + \phi_A(r) < r.$$

As an application of Theorem 6, we discuss the existence of solutions, on $C([0, \delta], X)$, where X is a Banach algebra satisfying the sequential condition $(\mathcal{P})$, for the following two-dimensional quadratic problems of ordinary differential inclusion occurring in some physical and biological problems

$$\begin{cases} \frac{\partial}{\partial t}\left(\frac{x(t)-k(t,x(t))}{f(t,y(t))}\right) \in G(t, y(t)) \\ y(t) - f_1(t, y(t)) = \left[\left(\lambda + \int_0^{\sigma(t)} p(s, x(\lambda s))ds\right)v\right] \\ (x(0), y(0)) = (x_0, y_0) \in \mathbb{R}^2, \end{cases}$$

where f, k and f_1 are supposed to be weakly sequentially continuous single-valued mapping on $[0, \delta] \times X$, $0 < \delta < \infty$, with respect to the second variable and the multi-valued mapping $G : [0, \delta] \times X \longrightarrow \mathcal{P}(X)$ is supposed to be weakly completely continuous with respect to the second variable.

Differential inclusion systems indeed play a crucial role in various applied sciences due to their ability to capture the behavior of dynamic systems with uncertain dynamics. These systems involve sets of differential equations with multiple possible solutions, allowing for a more realistic modeling of complex phenomena. Block operator matrices serve as efficient tools for solving differential inclusion systems. By organizing the operators involved in the system into a matrix format, we can analyze the collective behavior of the system and study its properties. This matrix representation allows us to apply various mathematical techniques and computational methods to analyze the system and find solutions.

The use of block operator matrices provides several advantages in solving differential inclusion systems. Firstly, it allows for a systematic and structured approach to studying the system, simplifying the analysis and facilitating the development of numerical algorithms. By leveraging block operator matrix with multi-valued inputs, we can gain insights into the existence of solutions to differential inclusion systems. This enhances our understanding of complex phenomena in applied sciences and enables us to make more informed decisions based on the obtained solutions.

Like as the fixed point theory for operator equation, the extension and analogues of fixed point results for block operator matrices with multi-valued inputs to the weak topology setting have attracted the attention of numerous authors. This research direction offers a refined and rigorous framework for studying the existence of solutions of such systems and has led to advancements in both theoretical understanding and computational methods for solving systems of nonlinear integral inclusions.

Overall, Chap. 1 serves as an introductory chapter that lays the groundwork for the subsequent chapters of the book. It familiarizes readers with the necessary definitions, concepts and tools of nonlinear functional analysis and provides classic fixed point results that will be utilized in the study of nonlinear problems in Banach spaces and Banach algebras throughout the book.

By establishing the fixed point theorem in the context of suitable Banach algebras under the strong topology, Chap. 2 provides a foundation for further exploration and analysis of operator equations in this setting. The theorem and its proof contribute to the understanding of the behavior and properties of solutions to such equations and offer a valuable tool for solving related problems in mathematical analysis and other applied sciences. Secondly, the book turns its attention to the problem of establishing the existence of fixed points for equation (3) in Banach algebras that satisfy the sequential condition $(\mathscr{P})$. The focus is on studying the existence of fixed points involving weakly sequentially continuous operators. In addition, the Perov's fixed point theorem, which is a powerful tool in fixed point theory, is then employed to prove the existence of fixed points in generalized Banach algebras satisfying the $(\mathscr{P})$ condition. This theorem provides a general framework for establishing fixed point results and allows for the consideration of various types of mappings. In the last section of Chap. 2, we aim to characterize the notion of RWC-Banach algebras, which was introduced by J. Banas and L. Olszowy in the paper [67]. We observe that the condition $(\mathscr{P})$, which is typically used to guarantee the existence of fixed point results for different hybrid operator equations, may not be satisfied for a large class of Banach algebras. Therefore, we propose a more general condition, referred to as condition $(\overline{\mathscr{P}})$. This condition is designed to extend several fixed point results to the class of RWC-Banach algebras. By introducing this condition, we aim to overcome the limitations posed by condition $(\mathscr{P})$ and establish a framework in which these fixed point results can be applied more broadly.

In Chap. 3, the focus is on investigating a fixed point theory for block operator matrices. The chapter begins by addressing the problem in the context of single-valued inputs. This means that we consider a block operator matrix with a single-valued mapping as its entries. One important aspect discussed in Chap. 3 is the critical case. However, despite this critical scenario, the chapter establishes useful results that are applicable in such cases.

In the subsequent step, the focus shifts to a block operator matrix with a multi-valued mapping as its entries. This introduces additional complexity compared to the single-valued case. The main objective is to investigate the existence of a fixed point for this scenario under strong conditions. To tackle this problem, the chapter employs the Boyd-Wong's fixed point theorem and Dhage's results. The details of the specific conditions, theorems and techniques employed from the Boyd-Wong's fixed point theorem and Dhage's fixed point result would be elaborated upon in the referenced work. These results would aid in establishing the existence of fixed points under strong conditions for block operator matrices with multi-valued mappings as their entries.

In an extension of the aforementioned results, a method is proposed to obtain practical and useful outcomes for the problem in the context of weak topology. This involves the utilization of a new class of multi-valued mappings known as countably $\mathscr{D}$-set-Lipschitzian class. By employing the concepts of the countably $\mathscr{D}$-set-Lipschitzian class, the chapter aims to derive practical results that are applicable to real-world situations. These results likely involve the establishment of fixed points or other relevant solutions within the weak topology framework.

In Part II, the theoretical results obtained throughout previous chapters are applied to solve various integro-differential equations and integro-differential inclusions. These types of equations and inclusions arise in several problems of applied sciences, indicating the practical relevance of the developed theory.

Integro-differential equations combine both differential and integral operators, which makes them more complex and challenging to solve compared to ordinary differential equations. By utilizing the theoretical findings from earlier chapters, the book aims to provide methods and techniques to effectively tackle these integro-differential equations. The specific integro-differential equations and inclusions addressed in Part II, as well as the techniques used to solve them, would be described in detail within the referenced book. These applications in applied sciences provide concrete examples, where the developed theory can be directly applied, demonstrating the practical significance of the research.

References

[1] Brouwer L.E.J.: Uber Abbildungen von Mannigfaltigkeiten. Math. Ann. **71**(1912) 97–115

[2] Goebel, K., Kirk, W.A.: Topics in Metric Fixed Point Theory, Cambridge Studies in Advanced Mathematics. vol. 28, Cambridge University Press, Cambridge, 1990

[3] Schauder J.: Der fixpunktsatz in funktionalrumen. Studia Math. **2**(1930) 171–180

[4] Hukuhara M.: Sur I'existence des points invariants d'une transformation dans I'espace fonctionnel. Jap. J. Math. **20**(1950) 1–4

[5] Fan, K.: A generalization of Tychonoff's fixed point theorem. Math. Ann. **142**(1961) 305–310

[6] Tychonoff A.: Ein Fixpunktsatz. (German) Math. Ann. **111**(1935) no. 1, 767–776

[7] Singbal, B.V.: Generalized form of Schauder-Tychnoff fixed point principle. In: Bonsall, F.F. (ed.) Lectures on Some fixed point Theorems of Functional Analysis. Mimeographed Notes. TFIR, Bombay, 1962

[8] Browder, F.E.: nonexpansive nonlinear operators in a Banach space. Proc. Nat. Acad. Sci. USA. **54**(1965) 1041–1044

[9] Göhde, D.: Zum Prinzip der Kontraktiven Abbildung. Math. Nachr. **30**(1965) 251–258

[10] Ishikawa, S.: Fixed points by a new iteration method. Proc. Am. Math. Soc. **44**(1974), 147–150

[11] Kirk, W.A.: A fixed point theorem for mappings which do not increase distance. Am. Math. Monthly, **72**(1965) 1004–1006

[12] Agarwal, R.P., Meehan, M., O'Regan, D.: Fixed Point Theory and Applications. Cambridge Tracts in Mathematics, 141. Cambridge University Press, Cambridge, 2001

[13] Ben Amar, A., O'Regan, D.: Topological Fixed Point Theory for single-valued and multi-valued Mappings and Applications. Springer, [Cham], 2016. x+194 pp. ISBN: 978-3-319-31947-6, 978-3-319-31948-3

[14] Husain, T., Tarafdar, E.: Fixed Point Theorems for Multi-Valued Mappings of nonexpansive Type. Yokohama Math. J., **28**(1980) 1–6

[15] Husain, T., Latif, A.: Fixed Points of multi-valued nonexpansive Maps. Math. Japonica, **33**(1988) 385–391

[16] Bana´s, J., Mursaleen, M.: Sequence Spaces and Measures of non-compactness with Applications to Differential and Integral equations. Springer, New Delhi, 2014

[17] Hu, S., Papageorgiou, N.S.: Handbook of multi-valued analysis. Vol. I. Theory. Mathematics and its Applications. Kluwer Academic Publishers, Dordrecht, **419**(1997) 7923–4682

[18] Darbo, G.: Punti uniti in trasformazioni a codominio non compatto. (Italian) Rend. Sem. Mat. Univ. Padova **24**(1955) 84–92

[19] Akhmerov, P.P., Kamenskii, M.I., Potapov, A.S., Rodkina, A.E., Sadovskii, B.N.: Measures of non-compactness and Condensing Operators. Birkhäuser, Basel. Boston. Berlin. 1992.

[20] Sadovskii, B.N.: On a fixed point theorem. Funktsional. Anal. i Prilozhen. **1**(1967) 74–76

[21] Goldenštein, L.S., Gohberg, I.T., Markus, A.S.: Investigations of some properties of bounded linear operators with their q-norms. U'cen. Zap. Kishinevsk. Univ. **29**(1957) 29–36

[22] De Blasi, F.S.: On a property of the unit sphere in a Banach space. Bull. Math. Soc. Sci. Math. R. S. Roumanie (N.S.) **21**(69) (1977) 259–262

[23] Ben Amar, A., Mnif, M.: Leray-Schauder alternatives for weakly sequentially continuous mappings and application to transport equation. Math. Methods Appl. Sci., **33** (2010), 80–90

[24] Krasnoselskii, M.A.: Topological Methods in The Theory of Nonlinear Integral Equations. Perg- amon Press 1964

[25] Jeribi, A.: Théorie des points fixes pour la topologie faible : exercices et problèmes corrigés, édition Cepadues, Toulouse, France (2023)

[26] Barroso, C.S.: Krasnoselskii's fixed point theorem for weakly continuous maps. Nonlinear Anal., **55**(2003) 25–31

[27] Liu, Y., Li, Z.: Schaefer type theorem and periodic solutions of evolution equations. J. Math. Anal. Appl. **316**(2006) 237–255

[28] Ben amar, A.: Krasnoselskii type fixed point theorems for multi-valued mappings with weakly sequentially closed graph. Ann. Univ. Ferrara Sez. VII Sci. Mat. **58**(2012) 1–10

[29] Ben Amar, A., Jeribi, A., Moalla, R.: Leray-Schauder alternatives in Banach algebra involving three operators with applications. Fixed Point Theory **15**(2014), no. 2, 359–372

[30] Burton, T.A.: A fixed point theorem of Krasnoselskii. Appl. Math. Lett. **11**(1998) 85–88

[31] O'Regan, D.: New fixed point results for 1-set contractive set valued maps. Comput. Math. Appl. **35**(1998) 27–34

[32] Dhage, B.C.: On a fixed point theorem in Banach algebras with applications. Appl. Math. Lett. **18**(2005) 273–280

[33] Ben Amar, A., Chouayekh, S., Jeribi, A.: New fixed point theorems in Banach algebras under weak topology features and applications to nonlinear integral equations. J. Funct. Anal. **259**(2010) 2215–2237

[34] Bana´s, J., Rivero, J.: On measures of weak non-compactness. Ann. Mat. Pura Appl. (4) **151**(1988) 213–224

[35] Bana´s, J., Taoudi, M.A.: Fixed points and solutions of operator equations for the weak topology in Banach algebras. Taiwanese J. Math., **18**(2014), No. 3, 871–893

[36] Garcia-Falset, J.: Existence of fixed points and measures of weak non-compactness. Nonlinear Anal. **71**(2009) 2625–2633

[37] Jeribi, A., Krichen, B., Mefteh, B.: Fixed point theory in WC-Banach algebras. Turkish J. Math., **40**(2016) 283–291

[38] Kaddachi, N.: Existence Results of Ordinary Differential Inclusions in Banach Algebra under Weak Topology. J. Phys. Math. **9**(2018).

[39] Ben Amar, A., Chouayekh, S., Jeribi, A.: Fixed point theory in a new class of Banach algebras and application. Afr. Mat. **24**(2013) 705–724

[40] Zeidler, E.: Nonlinear Functional Analysis and its Applications. I. fixed point theorems. Translated from the German by Peter R. Wadsack. Springer-Verlag, New York, 1986

[41] Dautray, R.: Lions, J.L.: Analyse Mathématiques et Calcul Numerique. Masson. Paris, Tome **9**(1988)

[42] Djebali, S., Górniewicz, L., Ouahab, A.: Solution Sets for Differential Equations and Inclusions, De Gruyter Series in Nonlinear Analysis and Applications. 18.Walter de Gruyter, Berlin, 2013

[43] B. Perthame, Transport Equations in Biology. Frontiers in Mathematics. Birkhauser Verlag, Basel (2007).

[44] O'Regan, D., Meehan, M.: Existence Theory for Nonlinear Integral and Integro-differential Equations. Kluwer Academic Publishers, Dordrecht (1998)

[45] Dhage, B.C.: Multi-valued operators and fixed point theorems in Banach algebras. Comput. Math. Appl. **48**(2004) 1461–1476

[46] Guo, D., Lakshmikantham, V., Liu, X.: Nonlinear Integral Equations in Abstract Spaces. Springer Science, Business Media, 2013

[47] Agarwal, R.P., O'Regan, D., Taoudi, M.A.: Fixed point theorems for convex-power condensing operators relative to the weak topology and applications to Volterra integral equations. J. Integral Equations Appl. **24**(2012) 167–181
[48] Markin, J.T.: Continuous dependence of fixed point set. Proc. Amer. Math. Soc. **38**(1973) 545–547
[49] Nadler, S.B.: Multi-valued contraction mappings. Pacific J. Math., **30**(1969) 475–488. https://doi.org/10.2140/pjm.1969.30.475
[50] Covitz, H., Nadler, jr.S.B.: Multi-valued contraction mappings in generalized metric spaces. Israel J. Math. **8**(1970) 5–11
[51] Dhage, B.C.: On solvability of operator inclusions x ε AxBx + Cx in Banach algebras and differential inclusions. Commun. Appl. Anal. **14**(2010) 567–596
[52] Dhage, B.C.: Multi-valued condensing random operators and functional random integral inclusions. Opusc. Math. **31** (2011) 27–48
[53] Dhage, B.C.: multi-valued mappings and fixed points II. Tamkang J. Math. **37**(2006) 27–46
[54] Krichen, B., O'Regan, D.: On the class of relatively weakly demi-compact nonlinear operators. Fixed Point Theory, **19**(2018) 625–630
[55] Garcia-Falset, J., Guran, L., Llorens-Fuster, E.: Fixed points for multi-valued contractions with respect to a w-distance. Sci. Math. Jpn. **71**(2010) 83–91
[56] Perov, A.I., Kibenko, A.V.: On a certain general method for investigation of boundary value problems. Izv. Akad. Nauk SSSR Ser. Mat. **30**(1966) 249–264, (in Russian)
[57] Perov, A.I.: On the Cauchy problem for a system of ordinary differential equation. (Russian) Pribliz. Metod. Resen. Differ. Uravn., **2**(1964) 115–134
[58] Perov, A.I.: On the Cauchy problem for a system of ordinary differential equation. (Russian) Pribliz. Metod. Resen. Differ. Uravn., **2**(1964) 115–134
[59] Browder, F.E.: Nonlinear operators and nonlinear equations of evolution in Banach spaces. Nonlinear functional analysis (Proc. Sympos. Pure Math., Vol. XVIII, Part 2, Chicago, Ill., 1968), Amer. Math. Soc., Providence, R. I., (1976) 1–308
[60] Conway, J.B.: A course in functional analysis. Second edition. Springer-Verlag, Berlin 1990
[61] Graef, J.R., Henderson, J., Ouahab, A.: Topological methods for differential equations andinclusions. Monographs and Research Notes in Mathematics, CRC Press, 2019
[62] Ouahab, A.: Some Perov's and Krasnosel'skii type fixed point results and application. Comm. Appl. Nonlinear Anal, (2015) 623–642
[63] Petre, I.R.: A multi-valued version of Krasnosel'skii's theorem in generalized Banach spaces. An. ¸ Stiin¸t. Univ. "Ovidius" Constan¸ta Ser. Mat., **22**(2014) 177–192
[64] Petre, I.R., Petrusel, A.: Krasnoselskii's theorem in generalized Banach spaces and application. Elec. J. Qual. Theo. of Diff. Equa., **85**(2012) 1–20
[65] O'Regan, D., Precup, R.: Continuation theory for contractions on spaces with two vectorvalued metrics. Applicable Anal. **82**(2003) 131–144

[66] O'Regan, D., Shahzad, N., Agarwal, R.P.: Fixed point theory for generalized contractive maps on spaces with vector-valued metrics. Fixed point theory and applications, **6**(2007) 143–149

[67] Chandrasekhar, S.: Radiative Transfer. Dover Publications, New York, 1960

[68] Ben Amara, K., Kaddachi, N., Jeribi, A.: Equivalence of some properties in the theory of Banach algebras and applications. J. Math. Anal. Appl. **520**(2023) 126865

[69] Kaddachi, N., Jeribi, A., Krichen, B.: Fixed point theorems of block operator matrices on Banach algebras and an application to functional integral equations. Math. Methods Appl. Sci. **36**(2013) 659–673

[71] Jeribi, A., Kaddachi, N., Krichen, B.: Fixed point Theorems for multi-valued Operator Matrix Under Weak Topology with an Application. Bull. Malays. Math. Sci. Soc. 43 (2020), 1047–1067

[72] K. Ben Amara, N. Kaddachi, A. Jeribi, New fixed point theorems for countably condensing maps with an application to functional integral inclusions. Math. Slovaca **71**(2021) 1487–1510

[73] Alsheikh, A.A., Ben Amar, A., O'Regan, D.: Fixed point theorems for the sum of two multivalued mappings and an application to an integral inclusion. Bull. Malays. Math. Sci. Soc. **40**(2017) no. 3, 1307–1320

[74] Jeribi, A., Krichen, B.: Nonlinear Functional Analysis in Banach Spaces and Banach Algebras Fixed Point Theory under Weak Topology for Nonlinear Operators and Block Operator Matrices with Applications. Monographs and Research Notes in Mathematics. Boca Raton, FL: CRC Press, (2016)

[75] Ben Amar, A., Jeribi, A., Krichen, B.: Fixed point theorems for block operator matrix and an application to a structured problem under boundary conditions of Rotenberg's model type. Math. Slovaca. **64**(2014) 155–174

[76] Kaddachi, N.: Generalized form of fixed point theorems in Banach algebras under weak topology with an application. Filomat **33**(2019) 4281–4296

Contents

Symbol Description

$\mathbb{N}$	The set of natural numbers
$\mathbb{R}$	The fields of real numbers
$\mathbb{C}$	The fields of complex numbers
$\mathbb{R}^n$	The n-dimensional real Euclidean space
l^∞	The space of bounded sequences, equipped with the norm $\|\cdot\|_\infty$
$\mathbb{M}_{n,m}(\mathbb{C})$	Matrices of size $n \times m$
$\mathscr{B}_X(x, r)$	The open ball in a space X of radius r about x
$\mathscr{B}_X(r)$	The open ball in a space X centered at origin with radius r
$\overline{\mathscr{B}}_X(x, r)$	The closed ball in a space X centered at x with radius r
$\overline{\mathscr{B}}_X(r)$	The closed ball in a space X centered at origin with radius r
$\overline{\mathscr{B}}_X$	The closed unit ball in X
$\overline{S}_X$	The closed unit sphere in X
$\mathscr{B}^p(x, r)$	The open p-ball centered at x with radius r, where p is a semi-norm
$\overline{S}$	The closure of the set S
$diam(S)$	The diameter of the set S
$\overline{S}^w$	The weak closure of the set S
$\max(S)$	Maximum of S
$\min(S)$	Minimum of S
$\inf(S)$	Infimum, greatest lower bound of S
$\sup(S)$	Supremum, least upper bound of S
∂S	The boundary of the set S
$\partial_\Omega S$	The boundary in Ω of the set S
$A \Rightarrow B$	A implies B
$A \Leftrightarrow B$	A and B are equivalent
$x \in S$	x is a member of the set S
$A \subset B$	A is a subset of B
(a, b)	Open interval $\{x, a < x < b\}$
$\text{dist}(x, y)$	The distance between x and y
dist_H	The Hausdorff metric induced by $dist(\cdot, \cdot)$
$diam(A)$	The diameter of a set A
$\|x\|$	The norm of x

X^*	Dual space of a Banach space X
$\mathscr{B}(X)$	The family of all nonempty bounded subset of a space X
$\mathscr{W}(X)$	The family of all relatively weakly compact subset of X
$C(J, X)$	The space of all continuous functions defined of the compact interval $J \subset \mathbb{R}$ into X
$C(K, X)$	The space of all continuous functions of a compact Hausdorff space K into X
$L^1(J, \mathbb{R})$	The space of Lebesgue integrable real-valued functions on a closed interval J
$L^1(J, X)$	Space of all Bochner integrable functions on a closed interval J
(X, Σ, ν)	Finite measure space
$L_1(X, \nu)$	The family of all complex-valued measurable and integrable functions on X
ℓ^2	Space of sequence $x = (x_n)_n$ of real numbers with $\|x\| = \sqrt{\sum_{n=0}^{\infty} x_n^2} < \infty$
$C(X)$	The space of all continuous functions on X with values in $\mathbb{C}$
$co(S)$	The convex hull of S
$\overline{co}(S)$	The closed convex hull of S
$Gr(A)$	The graph of an operator or multi-valued A
$x_n \to x$	Sequence $(x_n)_n$ converges to x
$x_n \rightharpoonup x$	Sequence $(x_n)_n$ converges weakly to x
$\mathscr{P}(X)$	The family of nonempty subsets of X
$\mathscr{P}_p(X)$	The class of all nonempty subsets of X with property p
$\mathscr{P}_{cl}(X)$	The family of nonempty and closed subsets of X
$\mathscr{P}_{cv}(X)$	The family of nonempty and convex subsets of X
$\mathscr{P}_{bd}(X)$	The family of nonempty and bounded subsets of X
$\mathscr{P}_{cp}(X)$	The family of nonempty and compact subsets of X
$\mathscr{P}_{cl,cv}(X)$	The family of nonempty, closed and convex subsets of X
$\mathscr{P}_{cl,bd}(X)$	The family of nonempty, closed and bounded subsets of X
$\mathscr{P}_{bd,cv}(X)$	The family of nonempty, bounded and convex subsets of X
$\mathscr{P}_{cp,cv}(X)$	The family of nonempty, compact and convex subsets of X
$\mathscr{P}_{cv,rcp}(X)$	The family of nonempty, convex and relatively compact subsets of X
$\mathscr{P}_{bd,cl,cv}(X)$	The family of nonempty, bounded, closed and convex subsets of X
μ	The Hausdorff measure of non-compactness
α	The Kuratowski measure of non-compactness
β	The De Blasi measure of weak non-compactness
β_G	The generalized De Blasi measure of weak non-compactness

Part I
Fixed Point Theory in Suitable Banach Algebras

Chapter 1
Topological Structure

The aim of this chapter is to introduce the basic concepts and elementary results that are used throughout the book. A survey of the basic fixed point theory for both single-valued and multi-valued mappings will be presented. Starting from the basics of Banach's fixed point theorem, most of the main results and techniques are developed and fixed point results are established for several classes of maps. The results in this chapter may be found in most standard books dealing with functional analysis and fixed point theory. For further studies connected with the material of this chapter, see [1–14] and the references therein.

1.1 Basic Tools

Let X be a nonempty set. A function $\text{dist} : X \times X \longrightarrow \mathbb{R}_+$ is called to be a metric distance if the following properties are satisfied

- $\text{dist}(x, y) = 0$ if, and only, if $x = y$, for all $x, y \in X$.
- $\text{dist}(x, y) = \text{dist}(y, x)$, for all $x, y \in X$.
- $\text{dist}(x, y) \leq \text{dist}(x, z) + \text{dist}(z, y)$, for all $x, y, z \in X$.

The pair (X, dist) is called metric space.

A nonempty set X is called to be an algebra if it is a vector space endowed with an internal composition law noted by

$$\begin{aligned}(\cdot) : X \times X &\longrightarrow X \\ (x, y) &\longrightarrow x \cdot y,\end{aligned}$$

which is associative and bilinear. Assume that X is a vector space over $\mathbb{K}$, $\mathbb{K} = \mathbb{R}$ or $\mathbb{C}$. A function $\| \cdot \| : X \longrightarrow \mathbb{R}^+$ is called to be a norm, provided that the following conditions are satisfied

A. Jeribi and N. Kaddachi, *Topological Fixed-Point Theory in Suitable Banach Algebras with Applications*, Infosys Science Foundation Series,
https://doi.org/10.1007/978-981-96-5542-7_1

- $\|x\| = 0$ implies $x = 0$
- $\|\lambda x\| = |\lambda|\|x\|$, for all $\lambda \in \mathbb{K}$ and $x \in X$,
- $\|x + y\| \leq \|x\| + \|y\|$, for all $x, y \in X$.

If X is a vector space and $\| \cdot \|$ is a norm on X, then X is called to be a normed vector space. If X is a vector space and $\| \cdot \|$ is a norm on X, then X becomes a metric space, if we define the metric distance $\text{dist}(\cdot, \cdot)$ by

$$\text{dist}(x, y) = \|x - y\|, \quad \text{for all } x, \ y \in X. \tag{1.1.1}$$

Notice that a normed vector space, which is a complete metric space with respect to the distance (1.1.1), is called to be a Banach space. A normed algebra is an algebra endowed with a norm satisfying the following property, for all $x, y \in X$, we have

$$\|x \cdot y\| \leq \|x\|\|y\|.$$

A complete normed algebra is called to be a Banach algebra.

Remark 1.1.1 A closed linear subspace of a Banach space may always be regarded as a Banach space, since a closed subspace of a complete space is complete with respect to the metric induced by the norm.

Remark 1.1.2 In a normed algebra, the multiplication is both left and right continuous with respect to the algebra norm, that is, if $(x_n)_n$ is a convergent sequence contained in X to a point x, then

$$x_n \cdot y \to x \cdot y \ \text{ and } \ y \cdot x_n \to y \cdot x \ \text{ as } \ n \to \infty,$$

for all $y \in X$. Moreover, the multiplication is jointly continuous, that is,

$$x_n \cdot y_n \to x \cdot y \ \text{ as } \ n \to \infty,$$

which follows from the equality

$$x_n \cdot y_n - x \cdot y = (x_n - x) \cdot y_n + x \cdot (y_n - y).$$

Example 1.1.1 Every finite-dimensional normed space is a Banach algebra.

Example 1.1.2 If we take $X = \mathbb{C}$. Then, with respect to the usual addition, multiplication of complex numbers and the modulus, X becomes as a commutative Banach algebra.

Proposition 1.1.3 *Let X be a Banach algebra and let K be a compact Hausdorff space. Then, the set $C(K, X)$, the space of all continuous functions of K into X, with respect to the pointwise multiplication of functions is a commutative algebra, and with the norm*

$$\|f\|_\infty = \sup_{x \in K} |f(x)|$$

is a Banach algebra. In the case, where K is not compact, the set $C(K, X)$ is not, in general, a Banach space with respect to the uniform norm, since it may contain unbounded functions.

Remark 1.1.4 If we take $K = [0, 1]$ and $X = \mathbb{R}$, then the space $C([0, 1], \mathbb{R})$ equipped with the norm $\|\cdot\|_1$ given by

$$\|f\|_1 = \int_0^1 |f(t)|dt$$

is not a Banach algebra. In fact, consider the sequence of continuous functions

$$f_n(x) = \begin{cases} 0 & \text{if } x \in \left[0, \frac{1}{2}\right] \\ n(x - \frac{1}{2}) & \text{if } x \in \left[\frac{1}{2}, \frac{1}{2} + \frac{1}{n}\right] \\ 1 & \text{if } x \in [\frac{1}{2} + \frac{1}{n}, 1], \end{cases}$$

Then, $(f_n)_n$ is a Cauchy sequence, since

$$\|f_n - f_m\|_1 = \frac{1}{2}\left|\frac{1}{n} - \frac{1}{m}\right|.$$

But it is not converge in $C([0, 1], \mathbb{R})$. Suppose that, there exists a function f such that

$$\lim_{n \to +\infty} \|f_n - f\|_1 = 0.$$

In other words, we get

$$\lim_{n \to +\infty} \int_0^1 |f_n(t) - f(t)|dt = 0.$$

After simple calculations, we can verify that

$$f(x) = \begin{cases} 0 & \text{if } x \in \left[0, \frac{1}{2}\right] \\ 1 & \text{if } x \in \left]\frac{1}{2}, 1\right]. \end{cases}$$

This means, in particular, that the sequence $(f_n)_n$ converge to the discontinuous function f and so, the space $C([0, 1], \mathbb{R})$ is not a Banach algebra.

Example 1.1.3 The space

$$l^1(\mathbb{Z}) = \left\{ f : \mathbb{Z} \longrightarrow \mathbb{C} \text{ such that } \sum_{n=-\infty}^{+\infty} |f(n)| < \infty \right\},$$

with pointwise addition, scalar multiplication, and the norm

$$\|f\|_1 = \sum_{n=-\infty}^{+\infty} |f(n)| < \infty$$

is a Banach space. Moreover $l^1(\mathbb{Z})$ can be transformed into a Banach algebra in a non-obvious manner. In fact, for arbitrary $f, g \in l^1(\mathbb{Z})$, define the convolution product

$$(f * g)(x) = \sum_{k=-\infty}^{+\infty} f(n-k)g(k).$$

Then, $f * g \in l^1(\mathbb{Z})$ and,

$$\begin{aligned} \|f * g\|_1 &= \sum_{n=-\infty}^{+\infty} |(f * g)(n)| \\ &\leq \sum_{n=-\infty}^{+\infty} \left| \sum_{k=-\infty}^{+\infty} f(n-k)g(k) \right| \\ &\leq \sum_{k=-\infty}^{+\infty} |g(k)| \sum_{n=-\infty}^{+\infty} |f(n-k)| \\ &\leq \|f\|_1 \sum_{k=-\infty}^{+\infty} |g(k)| \\ &\leq \|f\|_1 \|g\|_1. \end{aligned}$$

This implies that

$$\|f * g\|_1 \leq \|f\|_1 \|g\|_1.$$

It is easy to verify that, this multiplication is associative and commutative. Assuming this, we obtain the space $l^1(\mathbb{Z})$ is a commutative Banach algebra.

Example 1.1.4 The Banach sequence space consisting of all real sequence converging to zero, equipped with the standard supremum norm and with the multiplication of two sequences $x = (x_n)_n$ and $y = (y_n)_n$ in the classical way

$$x \cdot y = (x_n y_n)_n$$

is a Banach algebra.

Example 1.1.5 The algebra of all bounded continuous real or complex-valued functions on some locally compact spaces (again with pointwise operations and sup-norm) is a Banach algebra.

Example 1.1.6 The set of $n \times n$ matrices $M_n(\mathbb{C})$, $n \geq 2$, with matrix addition and with matrix multiplication, becomes as a non-commutative Banach algebra, if we equip it with Frobenius norm defined by

$$\|A\| = \left(\sum_{i,j=1}^{n} |a_{ij}|^2 \right)^{\frac{1}{2}}.$$

1.2 Elementary Fixed Point Theorems

As part of his Ph.D. thesis in 1922, Polish mathematician Stefan Banach addressed a significant fixed point theorem known as the Banach contraction principle. This theorem is regarded as the primary source of metric fixed point theory and it is one of the most significant analytical results, see [15]. It is the most widely applied fixed point result in many branches of mathematics, because it requires the structure of complete metric space with contractive condition on the map, which is easy to test in this setting. The Banach contraction principle has been generalized in many different directions.

1.2.1 Contraction Mapping Principle

Banach's fixed point theorem concerns certain mappings of a complete metric space into itself. It states conditions sufficient for the existence and uniqueness of fixed point. Before stating it, we need the following definition.

Definition 1.2.1 Let (X, dist) be a metric space. A mapping $T : X \longrightarrow X$ is said to be Lipschitzian, if there exists a constant $k \geq 0$ such that

$$\text{dist}(Tx, Ty) \leq k\, \text{dist}(x, y), \quad \text{for all } x,\ y \in X. \tag{1.2.1}$$

In this case, it is readily seen that there exists the smallest value k, for which the inequality (1.2.1) is satisfied, called to be the Lipschitz constant of T. In particular, if $0 \leq k < 1$, we say that T is a contraction mapping with constant k or a k-contraction mapping. Sometimes, a map T satisfying the inequality (1.2.1) with $k = 1$ is called to be a nonexpansive mapping.

Remark 1.2.2 Notice that a Lipschitzian mapping is necessarily continuous (one can take $\delta = \varepsilon$ in the definition of continuity). For notational purposes, we define T^n, $n \in \mathbb{N}$ inductively by $T^0 = T$ and $T^{n+1} = T \circ T^n$.

Remark 1.2.3 If T is a contraction mapping on a metric space into itself with constant k, then T^n is also a contraction with a constant k^n, but the converse may not be true. Let $T : [0, 2] \longrightarrow [0, 2]$ be defined by the formula

$$T(x) = \begin{cases} 0 & \text{if } x \in [0, 1] \\ 1 & \text{if } x \in \,]1, 2]. \end{cases}$$

Then, $T^2(x) = 0$, for all $x \in [0, 2]$, and so T^2 is a contraction mapping. Note that T is not continuous, and thus it is not a contraction map.

We have the following result.

Theorem 1.2.4 (Banach's fixed point theorem) *Let $(X, dist)$ be a complete metric space and let $T : X \longrightarrow X$ be a contraction mapping with a constant k. Then, T has an unique fixed point $x^* \in X$.*

Proof The proof is constructive, meaning that we will construct a sequence converging to the fixed point of T. Let us choose any $x_0 \in X$ and define the iterative sequence $(x_n)_n$ by

$$x_{n+1} = T(x_n).$$

To simplify the notation, we denote the n-th iterate of T by T^n, so that

$$x_n = T^n x_0.$$

Our proof strategy will be to show that $(x_n)_n$ is a Cauchy sequence. Making use of the triangle inequality, together with the summation formula for a geometric series, enables us to have

$$\begin{aligned} \mathrm{dist}(T^n x_0, T^m x_0) &= \mathrm{dist}(x_n, x_m) \\ &\leq \mathrm{dist}(x_n, x_{n+1}) + \mathrm{dist}(x_{n+1}, x_{n+2}) + \cdots + \mathrm{dist}(x_{m-1}, x_m) \\ &\leq \sum_{i=n}^{m-1} k^i \mathrm{dist}(x_0, x_1) \\ &\leq k^n \frac{1 - k^{m-n}}{1 - k} \mathrm{dist}(x_0, x_1), \end{aligned}$$

whenever $n < m$. Since $0 \leq k < 1$, we obtain

$$\mathrm{dist}(x_n, x_m) \leq \frac{k^n}{1 - k} \mathrm{dist}(x_0, x_1).$$

This shows that $(x_n)_n$ is a Cauchy sequence. Since X is complete, it follows that there exists a point $x^* \in X$ with

$$\lim_{n \to +\infty} x_n = x^*.$$

Moreover, the continuity of T guarantees that

$$\begin{aligned} x^* &= \lim_{n\to+\infty} x_n \\ &= \lim_{n\to+\infty} T(x_{n-1}) \\ &= T\left(\lim_{n\to+\infty} x_{n-1}\right) \\ &= Tx^*. \end{aligned}$$

Then, x^* is a fixed point of T. Finally, suppose that there are two distinct points u, $v \in X$ such that $u = Tu$ and $v = Tv$. Then,

$$\begin{aligned} \operatorname{dist}(u, v) &= \operatorname{dist}(Tu, Tv) \\ &\leq k \operatorname{dist}(u, v), \end{aligned}$$

but this contradicts the fact that $k < 1$ and so, T has an unique fixed point. Q.E.D.

Remark 1.2.5 Theorem 1.2.4 requires that $k < 1$. Indeed, if we take $k = 1$, then Theorem 1.2.4 is not always applicable as shown the example $Tx = x + 1$, for $x \in \mathbb{R}$. Another natural attempt to extend Theorem 1.2.4 would be to suppose that

$$\operatorname{dist}(Tx, Ty) < \operatorname{dist}(x, y),$$

for all pair of distinct points $x, y \in X$. Again, T need not have a fixed point as the example $Tx = \ln(1 + e^x)$ shows. However, the following positive version can be found in [9], and it has been improved by Edelstein and O'Brien [16].

Remark 1.2.6 A classical application of Theorem 1.2.4 is Newton's method for funding roots of equations. Starting with a differentiable function f and an initial guess x_0 for a root of f, Newton's method suggests

$$\begin{aligned} x_{n+1} &= T(x_n) \\ &= x_n - \frac{f(x_n)}{f'(x_n)}, \end{aligned}$$

as a sequence of successively better approximations to the true root of f. Then, the equation $f(x) = 0$ may be solved by the iterative process $x_{n+1} = T(x_n)$.

Example 1.2.1 Let $f(x) = x^2 - 3$ and setting

$$\begin{aligned} Tx &= x - \frac{f(x)}{f'(x)} \\ &= \frac{1}{2}\left(x + \frac{3}{x}\right). \end{aligned}$$

After simple calculations, we can verify that $T : [\sqrt{3}, +\infty[\longrightarrow [\sqrt{3}, +\infty[$ is a contraction mapping with a constant $\frac{1}{2}$. Moreover, a point x^* is a fixed point of T if, and only, if $f(x^*) = 0$.

An easy implication of the Banach contraction principle is the following theorem. For more details, the reader may refer to [17].

Corollary 1.2.7 *Let $(X, dist)$ be a complete metric space and $T : X \longrightarrow X$ be a mapping, for which T^p is a contraction mapping with a constant k, for some positive integer $p \in \mathbb{N}^\star$. Then, T has an unique fixed point.*

Proof It follows from Theorem 1.2.4, that there exists an unique point $x^* \in X$ such that $T^p(x^*) = x^*$. Then,

$$\begin{aligned} T(T^p x^*) &= T^p(Tx^*) \\ &= Tx^*. \end{aligned}$$

Hence, x^* is the unique fixed point of T. Q.E.D.

Theorem 1.2.8 *Let $(X, dist)$ be a complete metric space and let $T : X \longrightarrow X$ be a mapping such that the following inequality*

$$dist(Tx, Ty) < dist(x, y)$$

is satisfied, for each pair of distinct points $x, y \in X$. Then, T has a fixed point in X.

Proof Notice that the function $x \longrightarrow \text{dist}(x, Tx)$ reaches its minimum, say at x^*. Then, $Tx^* = x^*$, since otherwise the following inequality

$$\text{dist}(T(Tx^*), Tx^*) < \text{dist}(Tx^*, x^*),$$

leading to a contradiction. Q.E.D.

Definition 1.2.9 Let (X, dist) be a metric space and let $T : X \longrightarrow X$ be a mapping. T is said to be an expansive mapping if, there exists a constant $h > 1$ such that the following inequality

$$\text{dist}(Tx, Ty) \geq h \, \text{dist}(x, y)$$

is satisfied, for all pair of distinct points x and y in X.

The following results, concerning the expansive mapping, are very much useful in the study of fixed point theory, the details appear in [18] and the reference therein.

Lemma 1.2.10 *Let S be a nonempty and closed subset of a metric space X and let $T : S \longrightarrow X$ be an expansive mapping with a constant $h > 1$ such that $S \subset T(S)$. Then, there exists exactly one point $x^* \in S$ such that $Tx^* = x^*$.*

Proof After simple calculations, we can verify that the inverse of the operator T exists on $T(S)$, and also $T^{-1} : S \subset T(S) \longrightarrow S$ is a contraction mapping with a constant $\frac{1}{h}$. Using the continuity of T^{-1}, we obtain $T(S)$ is a closed subset of X. Then, on account of Theorem 1.2.4, there exists an unique point $x^* \in S$ such that $T^{-1}x^* = x^*$ and so, $x^* = Tx^*$. Q.E.D.

Proposition 1.2.11 *Let S be a nonempty and closed subset of a Banach space X and let $T : S \longrightarrow X$ be an expansive mapping with constant $h > 1$. Then, the inverse of the operator $I - T$ exists on S and it is Lipschitzian.*

Proof For each pair of distinct points x and y in S, we have

$$\begin{aligned}\|(I-T)x-(I-T)y\| &= \|Tx-Ty-(x-y)\| \\ &\geq (h-1)\|x-y\|,\end{aligned}$$

which prove that $I - T$ is one-to-one and then, the operator inverse $(I - T)^{-1}$ exists and it is Lipschitzian with a Lipschitz constant $\frac{1}{h-1}$ on $(I - T)(S)$. Q.E.D.

Numerous applications of Banach's fixed point theorem have been found in analysis, one of the most recent is the work [19], which guarantees the presence of fixed points for nonexpansive mappings in Banach spaces (under appropriate conditions).

Definition 1.2.12 Let X be a Banach space and let $T : X \longrightarrow X$ be a mapping. We say that T is a nonexpansive operator if

$$\|Tx - Ty\| \leq \|x - y\|,$$

for all pair of distinct points x and y in X.

Remark 1.2.13 In Banach space, a nonexpansive operator does not always require a fixed point, as demonstrated by the translation $Tx = x + a$, for any $a \in X\backslash\{0\}$. Furthermore, the identity map shows that, generally speaking, nonexpansive maps do not satisfy uniqueness of fixed point.

Theorem 1.2.14 *Let S be a nonempty, closed, bounded, and convex subset of a Banach space X and let $T : S \longrightarrow S$ be a nonexpansive mapping. Then, for any $\varepsilon > 0$, there exists a point $x_\varepsilon \in S$ such that the following inequality*

$$\|x_\varepsilon - Tx_\varepsilon\| < \varepsilon$$

is satisfied. In this case, the sequence $(x_\varepsilon)_\varepsilon$ is called to be an approximating fixed point sequence of the mapping T.

Proof We may assume that $S = \overline{\mathscr{B}}_X(R)$, the closed ball in X centered at origin with radius R, where $R = diam(S)$, here $diam(S)$ denote the diameter of the set S. For any $0 < r < 1$, it is easy to verify that the mapping rT is a contraction, then it admits an unique fixed point $x_r \in S$ and so,

$$0 \leq \|Tx_r - rTx_r\| \leq (1 - r)R.$$

The result follows on passing to the limit as $r \to 1^-$. Q.E.D.

In the study of fixed point theory of nonexpansive mappings, the approximate fixed point sequence is essential, see [20–22] for further detail.

Theorem 1.2.15 *Let S be a nonempty and closed subset of a Banach space X and let $T : S \longrightarrow S$ be a continuous mapping such that $T(S)$ is compact. If T has an approximating fixed point sequence, then T has a fixed point.*

Proof Let us consider a sequence $(x_n)_n$ in S such that

$$\lim_{n\to\infty} \|x_n - Tx_n\| = 0.$$

Taking into account the compactness of $T(S)$, we obtain the existence of a subsequence, say $(x_{n_k})_k$, such that

$$\lim_{k\to\infty} T(x_{n_k}) = y \in T(S).$$

The continuity of T guaranties that

$$\begin{aligned} y &= \lim_{n\to\infty} x_{n_k+1} \\ &= \lim_{k\to\infty} T(x_{n_k}) \\ &= T\left(\lim_{k\to\infty} x_{n_k}\right) \\ &= T(y). \end{aligned}$$

This completes the proof. Q.E.D.

As a consequence of Theorem 1.2.15, we get the following.

Corollary 1.2.16 *Let S be a nonempty, compact, and convex subset of a Banach space X. Then, every nonexpansive mapping $T : S \longrightarrow S$ has a fixed point.*

Theorem 1.2.17 *Let S be a nonempty, closed and bounded subset of a Banach space X and let $T : S \longrightarrow S$ be a nonexpansive mapping such that $(I - T)(S)$ is closed. Then, T has a fixed point.*

Proof Assume that $0 \in S$. Let $R > 0$ be such that $\|x\| \leq R$, for all $x \in S$. Consider a convergent sequence $(\lambda_n)_n$ contained in the set $[0, 1]$ to 1 and define the mapping T_n on S by

$$T_n(x) = \lambda_n T(x).$$

After simple calculations, we can verify that T_n is a contraction mapping, and when we apply Theorem 1.2.4, we reach that T_n admits an unique fixed point $x_n \in S$. Then,

$$\begin{aligned}\|T(x_n) - x_n\| &= \|T(x_n) - T_n(x_n)\| \\ &\leq \|T(x_n) - \lambda_n T(x_n)\| \\ &\leq (1 - \lambda_n)R.\end{aligned}$$

This implies that

$$\lim_{n \to +\infty} \|T(x_n) - x_n\| = 0.$$

It follows from $(I - T)(S)$ is closed, that $0 \in (I - T)(S)$ and hence, T has a fixed point. Q.E.D.

In [23], the notion of asymptotic regularity was presented in relation to the analysis of fixed points of nonexpansive mappings.

Definition 1.2.18 A nonlinear mapping T of a Banach space X into itself is called to be asymptotically regular if, for each $x \in X$, we have

$$\lim_{n \to \infty} \|T^n x - T^{n+1} x\| = 0.$$

Remark 1.2.19 The asymptotic regularity of an operator implies the existence of an approximating fixed point sequence of that mapping, but the converse is not true. In fact, the operator $T : \mathbb{R} \longrightarrow \mathbb{R}$ defined by $Tx = -x$ is nonexpansive, but it is not asymptotically regular.

In [24], the authors proved the following theorem with no restriction on the geometry of the Banach space involved.

Theorem 1.2.20 *Let S be a nonempty, bounded, closed, and convex subset of a Banach space X and let $T : S \longrightarrow S$ be a nonexpansive mapping. Then, for each $s \in (0, 1)$, the general averaged mapping $T_s : S \longrightarrow S$ given by*

$$T_s(x) = (s - 1)x + sT(x)$$

is asymptotically regular.

Definition 1.2.21 A space X is said to be uniformly convex if, it satisfies the following geometric condition: for all $\varepsilon > 0$, there exists $\rho > 0$ such that

$$\|x - y\| \geq \varepsilon, \ \|x\| \leq 1, \ \|y\| \leq 1 \Rightarrow \left\| \frac{x + y}{2} \right\| \leq 1 - \rho,$$

for all pair of distinct points x and y in X.

Note that uniform convexity is a geometric property of the norm, an equivalent norm need not be uniformly convex.

Example 1.2.2 The space $(\mathbb{R}^2, \|\cdot\|_2)$, where

$$\|x\|_2 = \sqrt{x_1^2 + x_2^2},$$

whenever $x_1, x_2 \in X$, is uniformly convex. Here the curvature of the unit sphere is positive. But $(\mathbb{R}^2, \|\cdot\|_1)$ and $(\mathbb{R}^2, \|\cdot\|_\infty)$ are not uniformly convex, where

$$\|x\|_1 = |x_1| + |x_2| \text{ and } \|x\|_\infty = \max(|x_1|, |x_2|).$$

They both have a flat surface.

Example 1.2.3 The Lebesgue spaces $L^p(\Omega)$ is uniformly convex, for $1 < p < \infty$ and Hilbert spaces is uniformly convex. For more detail see [25].

Example 1.2.4 Any uniformly convex Banach space is reflexive and has the fixed point property. For a proof, the reader can see [15, 26, 27].

Definition 1.2.22 A partially ordered set is a pair $(X, \leq)$, where X is a set and $\leq$ is a binary relation on X (in other words $\leq$ is a subset of $X \times X$) satisfying the following properties

(i) $\leq$ is reflexive,
(ii) If $x \leq y$ and $y \leq x$, then $x = y$,
(iii) If $x \leq y$ and $y \leq z$, then $x \leq z$.
The relation $\leq$ is then called to be a partial order on X.

The following result describes the situation of nonexpansive maps in uniformly convex spaces. A proof may be found in [28] and also, in [2, 29, 30] in the case of Hilbert space, and extended to uniformly convex Banach spaces in [24, 31–33].

Theorem 1.2.23 *Let S be a nonempty, bounded, closed and convex subset of an uniformly convex space X. Then, each nonexpansive mapping $T : S \longrightarrow S$ has a fixed point.*

Proof Denote by $\bigwedge$ the set

$$\bigwedge = \{M \subset S,\ M \neq \emptyset,\ M \text{ is a closed and convex subset with } T(M) \subset M\}.$$

Notice that $\bigwedge \neq \emptyset$, since $S \in \bigwedge$. If we order $\bigwedge$ by defining

$$M_1 \preceq M_2, \text{ if } M_1 \subset M_2,$$

then $(\bigwedge, \preceq)$ becomes as a partially ordered set, which is inductive, since the intersection of the elements M_ρ of a linearly ordered subfamily of $\bigwedge$ is also a closed and convex subset M' of $\bigwedge$, which is invariant under T, and which is nonempty since all the sets M_ρ are weakly compact subsets of the reflexive Banach space X. Hence, the set $\bigwedge$ has a minimal element M_0. We remark that M_0 is the convex closure of $T(M_0)$, since if M_1 is this convex closure, M_1 is a closed and convex subset of M_0, which is nonempty and invariant under T and by the minimality of M_0 in $\bigwedge$, we get $M_0 = M_1$. To complete the proof, it suffices to show that M_0 has exactly one element.

Since M_0 is nonempty, it suffices to show that M_0 does not have two distinct points in it. Suppose it does not, let k_0 be the diameter of M_0 and choose two distinct points $x_1, x_2 \in M_0$ such that

$$\|x_1 - x_2\| \geq \frac{k_0}{2}.$$

Let x be the midpoint of the segment joining x_1 and x_2, so that $x \in M_0$. For any element $y \in M_0$, we remark that $x - y$ is the midpoint of the segment joining $x_1 - y$ and $x_2 - y$ and

$$\|x_1 - y\| \leq k_0 \quad \text{and} \quad \|x_2 - y\| \leq k_0.$$

By the uniform convexity of the space X, there exists a constant $q \in (0, 1)$ such that

$$\begin{aligned} \|x - y\| &\leq (1 - q)k_0 \\ &< k_0. \end{aligned}$$

Let $k_1 = (1 - q)k_0$ and set

$$M_2 = \bigcap_{y \in M_0} \{u \in M_0 \text{ such that } \|u - y\| \leq k_1\}.$$

Notice that M_2 is a nonempty, closed, and convex subset of M_0, since it is the intersection of closed and convex sets, and is nonempty, since x lies in M_0. The set M_2 is bounded in M_0, since k_1 is less than the diameter of M_0 and

$$T(M_2) \subset M_2.$$

For this purpose, let $u \in M_2$ and $y \in M_0$, then for any $\varepsilon > 0$, we can find a convex linear combination of Tz_j, for $z_j \in M_0$ such that

$$\left\| y - \sum_{j \geq 1} \lambda_j T z_j \right\| < \varepsilon, \text{ with } 0 \leq \lambda_j \leq 1 \text{ and } \sum_{j \geq 1} \lambda_j = 1.$$

This implies that

$$\begin{aligned} \|Tu - y\| &\leq \left\| Tu - \sum_{j \geq 1} \lambda_j T z_j \right\| + \varepsilon \\ &\leq \sum_{j \geq 1} \lambda_j \left\| Tu - T z_j \right\| + \varepsilon. \end{aligned}$$

Since T is nonexpansive and u lies in M_2, it follows that

$$\begin{aligned}\|Tu - y\| &\leq \sum_{j\geq 1} \lambda_j \left\|u - z_j\right\| + \varepsilon \\ &\leq \sum_{j\geq 1} \lambda_j k_1 + \varepsilon \\ &= k_1 + \varepsilon,\end{aligned}$$

which implies that

$$\|Tu - y\| \leq k_1,$$

for all $y \in M_0$, that is, Tu lies in M_2. However, M_0 is a minimal element of $\bigwedge$, so that, we have reached a contradiction from the assumption that M_0 has, at least, two points. Hence, $M_0 = \{x_0\}$ and so $Tx_0 = x_0$. Q.E.D.

1.2.2 Some Generalizations of the Banach Contraction

The concept of $\mathscr{D}$-Lipschitzian mapping playing an important role in fixed point theory, see [34].

Definition 1.2.24 Let X be a Banach space and let $T : X \longrightarrow X$ be an operator. If there exists a continuous and nondecreasing function $\Phi_T : \mathbb{R}^+ \longrightarrow \mathbb{R}^+$ such that

$$\|Tx - Ty\| \leq \Phi_T\left(\|x - y\|\right)$$

for all $x, y \in X$, where $\Phi_T(0) = 0$, then T is called to be a $\mathscr{D}$-Lipschitzian mapping. The function Φ_T is sometimes called to be the $\mathscr{D}$-function of the mapping T on X. Moreover, if $\Phi_T(r) < r$, for all $r > 0$, then T is called to be a nonlinear contraction mapping on X. In particular, if $\Phi_T(r) = kr$, for some constant $0 \leq k < 1$, then T is a contraction.

Remark 1.2.25 Every Lipschitzian mapping is $\mathscr{D}$-Lipschitzian, but the converse may not be true. For example, take $T(x) = \sqrt{|x|}$ and consider on $\mathbb{R}_+$ the function Φ_T by putting

$$\Phi_T(r) = \sqrt{r}.$$

After simple calculations, we can prove that the function Φ_T is continuous and nondecreasing. Moreover, for all $x, y \in \mathbb{R}$, we have

$$\begin{aligned}\left(\sqrt{|x+y|}\right)^2 &\leq |x| + |y| \\ &\leq \left(\sqrt{|x|} + \sqrt{|y|}\right)^2.\end{aligned}$$

This shows that T is sub-additive and consequently,

$$\begin{aligned}|Tx - Ty| &\leq T(x-y) \\ &\leq \Phi_T(|x-y|),\end{aligned}$$

whenever $x, y \in \mathbb{R}$. Suppose that T is a Lipschitzian mapping with a Lipschitz constant $k \geq 0$ then, for all $x \in \mathbb{R}$, we have

$$Tx \leq k|x|.$$

If $x \neq 0$, we get

$$\frac{1}{\sqrt{|x|}} \leq k.$$

Letting x to zero, we obtain a contradiction.

Boyd and Wong [4] obtained the following result, which is one of the more helpful modifications of Banach's fixed point principle, in which the Lipschitz constant k is replaced by some real-valued control function.

Theorem 1.2.26 (Boyd's fixed point theorem) *Let X be a Banach space and let $T : X \longrightarrow X$ be a nonlinear contraction with $\mathscr{D}$-function Φ_T. Then, T has an unique fixed point x^*. Moreover, for any $x \in X$, the sequence $x_n = T^n(x)$ converges to x^*.*

Proof Let $x \in X$ be fixed and define the iterative sequence

$$a_n = \|T^n x - T^{n-1}x\|.$$

By induction, one has

$$a_n \leq \Phi_T^n(\|Tx - x\|).$$

Since T is nonlinear contraction with $\mathscr{D}$-function Φ_T, it follows that $(a_n)_n$ is a decreasing sequence and has a limit c. If $c > 0$, we have $a_{n+1} \leq \Phi_T(a_n)$ and so,

$$c \leq \limsup_{t \to c^+} \Phi_T(t) \leq \Phi_T(c),$$

which is a contradiction. We prove that $(T^n x)_n$ is a Cauchy sequence, for each $x \in X$. If not the case, there exists an element $\varepsilon > 0$ and there exists a sequence of integer $\{m_k, n_k\}$ with $m_k > n_k > k$ such that

$$\|T^{m_k}x - T^{n_k}x\| \geq \varepsilon, \quad \text{for } k = 1, 2, \ldots. \tag{1.2.2}$$

We may assume that

$$\|T^{m_k - 1}x - T^{n_k}x\| < \varepsilon$$

and choosing m_k to be the smallest number exceeding n_k, for which the inequality (1.2.2) is satisfied. By using the definition of the sequence $(a_n)_n$, we get

$$\begin{aligned} \|T^{m_k}x - T^{n_k}x\| &\leq \left\|T^{m_k}x - T^{m_k-1}x\right\| + \left\|T^{m_k-1}x - T^{n_k}x\right\| \\ &< a_{m_k} + \varepsilon. \end{aligned}$$

Thus,

$$\left\|T^{m_k}x - T^{n_k}x\right\| \to \varepsilon^+, \quad \text{as} \quad k \to +\infty.$$

On the other hand, we have

$$\begin{aligned} \left\|T^{m_k}x - T^{n_k}x\right\| &\leq \left\|T^{m_k+1}x - T^{m_k}x\right\| + \left\|T^{m_k+1}x - T^{n_k+1}x\right\| + \left\|T^{n_k+1}x - T^{n_k}x\right\| \\ &\leq a_{m_k+1} + \Phi\left(\left\|T^{m_k}x - T^{n_k}x\right\|\right) + a_{n_k+1} \\ &\leq 2a_{n_k+1} + \Phi\left(\left\|T^{m_k}x - T^{n_k}x\right\|\right). \end{aligned}$$

This means that $\varepsilon \leq \Phi(\varepsilon)$, as $k \to +\infty$, which is a contradiction. Q.E.D.

Remark 1.2.27 The mapping $T : [0, 1] \longrightarrow [0, 1]$ given by

$$T(x) = x - x^2$$

does not satisfy the assumptions of Theorem 1.2.4 and has an unique fixed point $x = 0$, whereas it satisfies those of Theorem 1.2.26 with $\varphi(r) = r - r^2$.

A well-known result of Brouwer [14], which says that any continuous map of a closed ball in $\mathbb{R}^n$ into itself has a fixed point, is generalized by several fixed point theorems in infinite-dimensional spaces in mathematics. It has interesting applications to game theory, mathematical economics, and numerical mathematics. The Hahn–Banach theorem and the Weierstrass existence theorem for minima are two more crucial existence theorems in mathematics. Brouwer's fixed point theorem is extended by the Schauder fixed point theorem [1, 15]. In order to formulate Schauder's fixed point theorem, the following definitions are necessary.

Definition 1.2.28 Let X be a Banach space and let $T : X \longrightarrow X$ be a mapping. We say that T is compact if, $T(S)$ is a compact subset of X, for any bounded subset S of X. Similarly, T is totally bounded if, T maps a bounded subset of X into totally bounded subsets of X. Finally, the mapping $T : X \longrightarrow X$ is completely continuous if, it is continuous and totally bounded on X.

Remark 1.2.29 Every compact operator is totally bounded, but the converse may not be true, however, two notions are equivalent on bounded subsets of a Banach space. For example, the operator defined on $\mathbb{R}$ by $T(x) = x$ is totally bounded but it is not compact.

Definition 1.2.30 Let X be a vector space. The convex hull of any subset S of X is the intersection of all convex sets that contain S, or more intuitively, the smallest convex set enclosing S. Following our book's notation, we will denote this $co(S)$.

Proposition 1.2.31 *Let S be a nonempty subset of a vector space X. Then, the following assertions are satisfied*
(i) $co(S)$ is convex and $co(S) \subseteq S$,
(ii) If S is convex and $S_1 \subset S$, then $co(S_1) \subseteq S$,

The Schauder fixed point theorem states in one version.

Theorem 1.2.32 *Let S be a nonempty, compact, and convex subset of a Banach space X. Any continuous operator $T : S \longrightarrow S$ has, at least, one fixed point in S.*

Theorem 1.2.32 provides the following result, since a continuous operator on a compact set is always a compact operator.

Theorem 1.2.33 *Let S be a nonempty, bounded, closed, and convex subset of a Banach space X and let $T : S \longrightarrow S$ be a mapping. If T is compact, then T has, at least, a fixed point in S.*

Proof For every $n = 1, 2, \ldots$, there exists a finite-dimensional subspace X_n of X and there exists a continuous operator $T_n : S \longrightarrow X_n$ with $T_n(S) \subset co(T(S))$ such that

$$\|Tu - T_n u\| \leq \frac{1}{n},$$

for all $u \in S$. Let $S_n = X_n \bigcap S$. Keeping in mind the construction of T_n, and using the convexity of S, we obtain

$$\begin{aligned} T_n(S_n) &= T_n(X_n \textstyle\bigcap S) \\ &\subset T_n(S) \\ &\subset co(T(S)) \\ &\subset co(S) \\ &\subset S. \end{aligned}$$

On the other hand, we have

$$T_n(S_n) \subset T_n(S) \subset X_n.$$

From the above discussion, we get $T_n : S_n \longrightarrow S_n$ is well defined. Now, we show that S_n is a compact and convex subset of X. Since X_n is a finite-dimensional subspace of X and $S_n \subset X_n$ is closed and bounded ($S_n \subset S$), then S_n must be a compact subset of S. Besides, S_n is convex, since S and X_n are convex. Accordingly, the operator $T_n : S_n \longrightarrow S_n$ has, at least, one fixed point u_n, in view of Brouwer's fixed point theorem. Next, we prove that the sequence $(u_n)_n$ as well as $(T(u_n))_n$ has a convergent subsequence and the limit is equal to the fixed point of T. If we consider the compactness of the operator T and deploy the inclusion $S_n \subset S$, we get there exists a subsequence, say $(T(u_{n_k}))_k$, of $(T(u_n))_n$ such that

$$\lim_{k \to +\infty} T(u_{n_k}) = v.$$

Since S is closed, then $v \in S$. Moreover,

$$\|v - u_{n_k}\| \leq \|v - T(u_{n_k})\| + \|T(u_{n_k}) - T_{n_k}(u_{n_k})\|.$$

Thus, $u_{n_k} \rightarrow v$ and consequently, $T(u_{n_k}) \rightarrow Tv$, since T is continuous. Q.E.D.

Dhage [35] has established a fixed point theorem involving three operators, by combining Schauder's fixed point theorem with Banach's fixed point theorem.

Theorem 1.2.34 *Let S be a nonempty, closed, convex and bounded subset of a Banach algebra X and let A, $C : X \longrightarrow X$ and $B : S \longrightarrow X$ be three operators satisfying the following conditions*
(i) A and C are Lipschitzians with constants q_1 and q_2, respectively,
(ii) B is completely continuous,
(iii) If $x = Ax \cdot By + Cx, \quad y \in S$, then $x \in S$.
Then, the operator equation (0.0.3) has, at least, one solution in S provided that

$$q_1 M + q_2 < 1,$$

where $M = \|B(S)\|$.

A. N. Tychonoff expanded on Schauder's fixed point result in [36] to include compact and convex subsets of a locally convex topological vector spaces, for more information, see [37]. As a special case, the earlier result in Schauder stating that there exists a fixed point for any weakly continuous self mapping of a weakly compact and convex subset S of a Banach space is contained in Tychonoff's fixed point theorem. First, the following definitions are required, in order to demonstrate the Schauder-Tychonoff fixed point theorem.

Definition 1.2.35 Let (X, τ) be a topological space and $x \in X$. A subset U of X is called to be a neighborhood at x if, it contains an open set containing the point x.

Definition 1.2.36 (Hausdorff Topology) A Topological space (X, τ) is Hausdorff if, for each pair of distinct points $x, y \in X$, there exist two disjoint open neighborhoods U and V of x and y, respectively.

Definition 1.2.37 Let $O_i, i = 1, \ldots, n$ be an open subsets of a Hausdorff Topological space X and let K be a compact subset of X such that

$$K \subset \bigcup_{i=1}^{n} O_i.$$

A family of continuous functions $\varphi_j : X \longrightarrow [0, 1]$, for $j = 1, \ldots, n$, is said to be a partition of the unity for K subordinate to the open cover $\{O_1, \ldots, O_n\}$, if each φ_j is supported on O_j and

$$\varphi_1(x) + \cdots + \varphi_n(x) = 1, \quad \text{for all } x \in K.$$

Definition 1.2.38 A subset K of a topological space X is said to be sequentially compact if, any sequence in K possesses a convergent subsequence with limit in K.

Definition 1.2.39 Let (X, τ) be a topological space. A family $U(x)$ is called to be a basis of neighborhood at a point $x \in X$ if, for any neighborhood $V \subset X$ of x, there exists $W \in U(x)$ such that $W \subset V$.

Definition 1.2.40 A locally convex space is a topological vector space (X, τ) admitting a neighborhood basis at 0 formed by convex sets.

Remark 1.2.1 Every point in a locally convex space X admits a neighborhood basis formed of convex sets and there exists a neighborhood basis at 0 formed by open convex symmetric sets.

Remark 1.2.2 If K is a compact subset of a Hausdorff space X such that

$$K \subset \bigcup_{i=1}^{n} O_i,$$

where O_i is an open subset of X, for $i = 1, \ldots, n$, then there exists a partition of the unity $\varphi_1, \ldots, \varphi_n$ for K subordinate to O_j. For more detail, the reader may consult [17].

When talking about linear topological spaces, the concept of the semi-norm function is crucial.

Definition 1.2.41 A real-valued function $p : X \longrightarrow \mathbb{R}_+$ is called to be a semi-norm on X, if the following conditions are satisfied
(i) $p(x + y) \leq p(x) + p(y)$,
(ii) $p(\lambda x) = |\lambda| p(x)$.

Let P be a family of semi-norms on a vector space X and let S be a convex subset of X. Let $\varepsilon > 0$ and suppose that there exists $z^1, \ldots, z^n \in S$ such that

$$S \subset \bigcup_{i=1}^{n} \mathscr{B}^p(z^i, \varepsilon),$$

where the notation $\mathscr{B}^p(z^i, \varepsilon)$ is the open p-ball given by

$$\mathscr{B}^p(z^i, \varepsilon) = \{x \in X, \ \forall\, p \in P, \ p(x - z^i) < r\}, \ \text{for all } r > 0.$$

Define the functions $g^i = g^i_{p,\varepsilon}$, $w = w_{p,\varepsilon}$ and $w^i = w^i_{p,\varepsilon}$ by putting

$$\begin{cases} g^i(x) = \max\left\{\varepsilon - p(x - z^i), 0\right\} \\ w(x) = \sum_{i=1}^{n} g^i(x) \\ w^i(x) = \dfrac{g^i(x)}{w(x)}. \end{cases}$$

Define the mapping $\varphi = \varphi_{p,\varepsilon} : S \longrightarrow S$ by putting

$$\varphi(x) = \sum_{i=1}^{n} w^i(x) z^i.$$

The mapping $\varphi_{p,\varepsilon}$ is called to be the Schauder mapping. We may now formulate and demonstrate the Schauder-Tychonoff theorem. The reader can refer to the work of Cobzas [38].

Theorem 1.2.42 (Schauder-Tychonoff) *Let S be a nonempty, convex, and compact subset of a Hausdorff locally convex space (X, τ). Then, every continuous mapping $T : S \longrightarrow S$ has a fixed point.*

Proof Let $\mathscr{B}$ be a basis of 0-neighborhoods formed by open convex symmetric subsets of X. Then, the Minkowski functional p_B corresponding to a set $B \in \mathscr{B}$ is a continuous semi-norm on X. Using the compactness of the set S, we can deduce that there exists $z_B^1, \ldots, z_B^{n(B)} \in S$ such that

$$S \subset \{z_B^1, \ldots, z_B^{n(B)}\} + B.$$

Denote by φ_B the Schauder mapping corresponding to p_B, $\varepsilon = 1$ and define the set S_B by

$$S_B = co\{z_B^1, \ldots, z_B^{n(B)}\}.$$

After simple calculations, we can verify that the function

$$T_B = \varphi_B \circ T : S_B \longrightarrow S_B$$

is continuous. An application of Brouwer's fixed point theorem yields that, there exists a sequence $(x_B)_{B \in \mathscr{B}}$ such that

$$f_B(x_B) = x_B.$$

This implies that the sequence $(x_B)_{B\in\mathscr{B}}$ contains a convergent subsequence, say $(x_{\gamma(k)})_k$ to an element $x \in S$. Besides, the topology of the space X is separated Hausdorff. This is equivalent to

$$x - T(x) \in V, \quad \text{for all 0-neighborhoods } V.$$

For a 0-neighborhoods V, choose a subset $B \in \mathscr{B}$ such that

$$B + B \subset V.$$

Then, there exists an integer k_0 with $\gamma(k_0) \subset B$ and so, for all $k \geq k_0$, we have

$$\gamma(k) \subset \gamma(k_0) \subset B.$$

Making use of the fact that $\varphi_{\gamma(k)}(T(x_{\gamma(k)})) = x_{\gamma(k)}$, we get

$$p_{\gamma(k)}\left(\varphi_{\gamma(k)}(T(x_{\gamma(k)})) - x_{\gamma(k)}\right) < 1.$$

This implies that

$$x_{\gamma(k)} - \varphi_{\gamma(k)}(T(x_{\gamma(k)})) \in B.$$

Passing to limit for $k \geq k_0$, and using the continuity of T, one obtains

$$x - T(x) \in B + B \subset V.$$

This completes the proof. Q.E.D.

We provide an extension of Banach's fixed point theorem to the situation of independent contraction, pseudo-contractive, and accretive operators developed in [6, 39].

Definition 1.2.43 A mapping $T : X \longrightarrow X$ is called to be a separate contraction if, there exist two functions $\varphi, \psi : \mathbb{R}^+ \longrightarrow \mathbb{R}^+$ satisfying the following conditions
(i) ψ is strictly increasing with $\psi(0) = 0$,
(ii) $\|Tx - Ty\| \leq \varphi(\|x - y\|)$,
(iii) $\psi(r) + \varphi(r) \leq r$, for $r > 0$.

Remark 1.2.44 Every contraction mapping is separate contraction, but the converse may not be true. In fact, the mapping $T : C([0, 1], \mathbb{R}) \longrightarrow C([0, 1], \mathbb{R})$ defined by the formula

$$T(x) = f(t, x(t)),$$

where $f : [0, 1] \times \mathbb{R} \longrightarrow \mathbb{R}$ is given by

$$f(t, x) = x - \frac{x^4}{4} + \frac{\sin^2(t)}{4}$$

is a separate contraction mapping, but not a contraction mapping, see [40].

Example 1.2.5 The mapping $T : \left[0, \frac{1}{\sqrt{2}}\right] \longrightarrow \left[0, \frac{1}{\sqrt{2}}\right]$ defined by the formula

$$Tx = x - x^3$$

is separate contraction with function $\psi(r) = r - \varphi(r)$, where

$$\varphi(r) = \begin{cases} r - \dfrac{r^3}{4} & \text{if } r \leq 1 \\ \dfrac{3}{4}r & \text{if } r \geq 1. \end{cases}$$

But T is not a contraction mapping.

Proposition 1.2.45 *Let X be a Banach space and let $T : X \longrightarrow X$ be a separate contraction mapping. Then, $I - T$ is a homeomorphism.*

Proof Suppose that x and y are two distinct points in X. Then, there exists a function φ such that

$$\begin{aligned} \|(I - T)x - (I - T)y\| &\geq \|x - y\| - \|Tx - Ty\| \\ &\geq \|x - y\| - \varphi\left(\|x - y\|\right) \\ &\geq \psi\left(\|x - y\|\right), \end{aligned}$$

where

$$\psi(r) = r - \varphi(r), \quad \text{for } r > 0.$$

Thus, $I - T$ is bijective from X into $(I - T)(X)$. We claim that $(I - T)^{-1}$ is continuous. If this is not the case, there exists $y \in X$ such that

$$(I - T)x_n \to (I - T)y,$$

and there exists $\varepsilon_0 > 0$ and there exists a subsequence, say $(x_{n_k})_k$, of $(x_n)_n$ such that

$$\|x_{n_k} - y\| \geq \varepsilon_0.$$

Again, for each $\varepsilon > 0$, there exists $N > 0$ such that, for all $k \in \mathbb{N}$ with $n_k > N$, we have

$$\begin{aligned}
\varepsilon &\geq \|(I-T)x_{n_k} - (I-T)y\| \\
&\geq \|x_{n_k} - y\| - \|T(x_{n_k}) - Ty\| \\
&\geq \|x_{n_k} - y\| - \varphi\left(\|x_{n_k} - y\|\right) \\
&\geq \psi\left(\|x_{n_k} - y\|\right) \\
&\geq \psi\left(\varepsilon_0\right),
\end{aligned}$$

which is a contradiction occurs when $\varepsilon \to 0$, and the claim is proved. Q.E.D.

Theorem 1.2.46 *Let X be a complete metric space and let $T : X \longrightarrow X$ be a separate contraction mapping. Then, T has a fixed point.*

Proof Let $x_0 \in X$ be fixed and define the sequence $(a_n)_n$ by the formula

$$a_n = \text{dist}(T^n x_0, T^{n+1} x_0).$$

It is easy to verify that, there exists a function φ such that

$$0 \leq a_{n+1} \leq \varphi(a_n) < a_n \leq \varphi(a_{n-1}), \quad \text{for any } n \in \mathbb{N}^*.$$

From the above discussion, we see that both $(a_n)_n$ and $(\varphi(a_n))_n$ are two decreasing positive sequences and

$$\lim_{n\to+\infty} (a_n - \varphi(a_n)) = 0.$$

This implies that

$$\begin{aligned}
\psi\left(\lim_{n\to+\infty} a_n\right) &\leq \psi(a_n) \\
&\leq a_n - \varphi(a_n).
\end{aligned}$$

where, the function ψ is as described in Definition 1.2.43. Thus,

$$\lim_{n\to+\infty} a_n = 0.$$

Since $\varphi(r) < r$, without loss of generality, we assume that

$$\varepsilon^* = \sup_{0\leq r\leq\varepsilon} \varphi(r) < \varepsilon,$$

for fixed $\varepsilon > 0$. Now, we choose sufficiently large number N such that

$$a_N \leq \varepsilon - \varepsilon^*.$$

In order to show that the sequence $(x_n)_n$ defined by $x_n = T^n(x_0)$ is Cauchy, it is sufficient to prove that the set

$$M = \{x \in X \text{ such that } \text{dist}(x, x_N) \leq \varepsilon\}$$

is invariant under T. Indeed,

$$\begin{aligned} \mathrm{dist}(T(x), x_N) &= \mathrm{dist}(T(x), T(x_n)) + \mathrm{dist}(T(x_n), x_N) \\ &\le a_N + \varphi(\mathrm{dist}(x, x_N)) \\ &\le \varepsilon^* + \varepsilon - \varepsilon^*. \end{aligned}$$

This shows the set M is invariant under T and consequently, $(x_n)_n$ is a Cauchy sequence. If we take $x^* = \lim_{n \to +\infty} x_n$ then, by using the continuity of T, we have

$$\begin{aligned} T(x^*) &= \lim_{n \to +\infty} T(x_n) \\ &= \lim_{n \to +\infty} x_{n+1} \\ &= x^*. \end{aligned}$$

This completes the proof. Q.E.D.

Definition 1.2.47 Let X be a Banach space and $T : X \longrightarrow X$ be a mapping. We say that T is strongly pseudo-contractive if, there exists a positive constant k such that

$$\|x - y\| \le \|(1 + r)(x - y) - kr\,(T(x) - T(y))\|, \quad \text{for all } r > 0,$$

for each pair of distinct points x and y in X. In particular, if $k = 1$, then the operator T is called to be pseudo-contractive.

An equivalent definition is as follows. An operator T of a Hilbert space X into itself is strongly pseudo-contractive if, there exists $k > 0$ such that

$$\langle Tx - Ty, x - y \rangle \le k\|x - y\|^2.$$

for each pair of distinct points x and y in X. In particular, if $k = 1$, then the operator T is pseudo-contractive.

The pseudo-contractive mappings are easily seen to be more general than the nonexpansive mappings.

Example 1.2.6 Let $H = \mathbb{R}^2$ be the Hilbert space under the usual Euclidean inner product. For arbitrary $(a, b) \in H$, let us define

$$(a, b)^{\perp} = (b, -a) \in H.$$

Let $S = \overline{\mathscr{B}_H}$ and, set

$$S_1 = \left\{ x \in H \text{ such that } \|x\| \le \frac{1}{2} \right\}$$

and

$$S_2 = \left\{ x \in H \text{ such that } \frac{1}{2} \leq \|x\| \leq 1 \right\}.$$

Define the mapping $T : S \longrightarrow H$ by putting

$$Tx = \begin{cases} x + x^{\perp} & \text{if } x \in S_1 \\ \dfrac{x}{\|x\|} - x + x^{\perp} & \text{if } x \in S_2. \end{cases}$$

Notice that the operator T is Lipschitzian on S with a Lipschitz constant 5. We claim that T is pseudo-contractive. For this purpose, choose

$$\Gamma(x, y) = \|x - y\|^2 - \langle Tx - Ty, x - y \rangle.$$

To show T is pseudo-contractive, we need to prove that

$$\Gamma(x, y) \geq 0, \quad \text{for all } x, \ y \in S.$$

The claim regarding $x, y \in S_1$ is immediate. If $x \in S_1$ and $y \in S_2$, then

$$\Gamma(x, y) = 2\|y\|^2 - \|y\| + (1 - 2\|y\|) \frac{\langle x, y \rangle}{\|y\|}.$$

Since $1 - 2\|y\| \leq 0$, for all $y \in S_2$ and $\dfrac{\langle x, y \rangle}{\|y\|}$ has its minimum, for fixed $\|x\|$ and $\|y\|$ when $\dfrac{\langle x, y \rangle}{\|x\| \|y\|} = 1$, we conclude that

$$\Gamma(x, y) \geq 2\|y\|^2 - \|y\| + \|x\| - 2\|x\| \|y\| \geq 0.$$

If we take $x, y \in S_2$, one easily shows that

$$\Gamma(x, y) = 2\|x\|^2 + 2\|y\|^2 - \|x\| - \|y\| - (4\|x\| \|y\| - \|x\| - \|y\|) \frac{\langle x, y \rangle}{\|x\| \|y\|}.$$

Then, for fixed $\|x\|$ and $\|y\|$ when $\dfrac{\langle x, y \rangle}{\|x\| \|y\|} = 1$, the function $\Gamma(x, y)$ has a minimum. This minimum is equal to $2(\|x\| - \|y\|)^2$. This implies that $\Gamma(x, y) \geq 0$. Therefore, the operator T is Lipschitzian, continuous and pseudo-contractive, but it is not nonexpansive.

Example 1.2.7 The mapping $T : [0, 1] \longrightarrow \mathbb{R}$ given by

$$Tx = (1 - x\sqrt{x})\sqrt{1 - x\sqrt{x}}$$

is pseudo-contractive, since T is monotonically decreasing. If we choose $x = 4^{-3}$ and $y = 2^{-3}$, one easily shows that

$$|Tx - Ty| > |x - y|.$$

Hence, T is not nonexpansive.

Proposition 1.2.1 *Let H be a Hilbert space. An operator $T : H \longrightarrow H$ is pseudo-contractive if, and only, if for all $x,\ y \in H$, we have*

$$\|Tx - Ty\|^2 \leq \|x - y\|^2 + \|(I - T)x - (I - T)y\|^2.$$

Proof It follows from the assumption T is pseudo-contractive, that

$$\langle Tx - Ty, x - y\rangle \leq \|x - y\|^2, \quad \text{for all } x,\ y \in H.$$

This implies that

$$\begin{aligned} \|Tx - Ty\|^2 &= \|(I - T)x - (I - T)y - (x - y)\|^2 \\ &= \|x - y\|^2 + \|(I - T)x - (I - T)y\|^2 - 2\left(\|x - y\|^2 - \langle Tx - Ty, x - y\rangle\right) \\ &\leq \|x - y\|^2 + \|(I - T)x - (I - T)y\|^2. \end{aligned}$$

Conversely, if the inequality

$$\|Tx - Ty\|^2 \leq \|x - y\|^2 + \|(I - T)x - (I - T)y\|^2$$

is satisfied, then

$$\|Tx - Ty\|^2 \leq 2\|x - y\|^2 + \|Tx - Ty\|^2 - 2\langle Tx - Ty, x - y\rangle.$$

Consequently,

$$\langle Tx - Ty, x - y\rangle \leq \|x - y\|^2. \qquad \text{Q.E.D}$$

Definition 1.2.48 Let X be a Banach space and $T : X \longrightarrow X$ be a mapping. We say that T is accretive if, for each pair of distinct points x and y in X and $s > 0$, the following inequality

$$\|x - y\| \leq \|x - y - s(Tx - Ty)\|$$

is satisfied. A mapping T is said to be dissipative if, the operator $I - T$ is accretive. An accretive operator T is m-accretive if, $T + \lambda I$ is surjective, for every $\lambda > 0$.

Example 1.2.8 Every increasing real-valued mapping is accretive and every decreasing real-valued mapping is dissipative.

An alternative way to define an accretive mapping is the use of the duality map. Let X be a Hilbert space and let $J : X \longrightarrow \mathscr{P}(X)$ be denote the duality mapping of X defined by

$$J(x) = \left\{ f \in X^* \text{ such that } \langle f, x\rangle = \|x\|^2 \text{ and } \|f\| = \|x\| \right\}. \tag{1.2.3}$$

Proposition 1.2.2 *Let X be a Banach space and $T : X \longrightarrow X$ be a mapping. Then, T is accretive if, and only, if there exists $j(x-y) \in J(x-y)$ such that the following inequality*

$$\langle T(x) - T(y), j(x-y)\rangle \geq 0$$

is satisfied, for each pair of distinct points x and y in X.

Remark 1.2.49 Let X be a Banach space and $T : X \longrightarrow X$ be a mapping. Note that T is pseudo-contractive if, and only, if $I - T$ is accretive. Consequently, the fixed point theory for pseudo-contractive operators and the theory for accretive operators are intimately connected. You may have noticed that [6, 41] introduced and examined the ideas of accretive and m-accretive operators.

Proposition 1.2.3 *Let X be a Banach space and $T : X \longrightarrow X$ be a mapping such that $I + sT$ is expansive, for all $s > 0$. Then, the following assertions are satisfied*
(i) T is accretive,
(ii) $I + sT$ is one-to-one,
(iii) $(I + sT)^{-1}$ is a nonexpansive mapping from $(I + sT)(X)$ into X.

Definition 1.2.50 Let $\psi : \mathbb{R}^+ \longrightarrow \mathbb{R}^+$ be a continuous function such that $\psi(0) = 0$ and $\psi(r) > 0$, for all $r > 0$. A mapping $T : X \longrightarrow X$ is called to be ψ-expansive if, the following inequality

$$\|T(x) - T(y)\| \geq \psi(\|x - y\|)$$

is satisfied, for every $x, y \in X$.

The following result is required afterward. The reader can refer to [10] for further information.

Lemma 1.2.51 *Let X be a Banach space and $T : X \longrightarrow X$ be a m-accretive mapping. If T is ϕ-expansive on X, then T is surjective.*

1.3 Weak Topology

Recall that a topology on a set X is a collection τ of subsets of a Banach space X having the following assertions

1. $\emptyset, X \in \tau$
2. arbitrary unions of sets in τ is in τ,
3. finite intersections of sets in τ is in τ.

Let $(X, \|\cdot\|)$ be a Banach space and X^* be its dual. For an arbitrary function $f \in X^*$, we denote by $\varphi_f : X \longrightarrow \mathbb{R}$ the linear functional

$$\varphi_f(x) = f(x).$$

As f runs through X^*, we obtain a collection $(\varphi_f)_{f \in X^*}$ of maps from E into $\mathbb{R}$. The usual topology on X associated to $\|\cdot\|$ is now disregarded and a new topology is defined as follows on the set X.

Definition 1.3.1 The weak topology $\sigma(X, X^*)$ on X is coarsest (smallest) topology associated with the collection $(\varphi_f)_{f \in X^*}$, or equivalently the weak topology $\sigma(X, X^*)$ is the smallest topology that makes every element in X^* continuous.

Remark 1.3.2 For the usual topology, each map φ_f, $f \in X^*$ is a linear and continuous functional. Therefore, the weak topology is weaker (it contains fewer open sets) than the usual topology on X induced by the norm on X.

The topology induced by a norm on a vector space is a very strong topology in the sense that it has many open sets. This brings advantages to the functions whose domain is such a space, because for them is easy to be continuous but it brings disadvantages to compactness, because the richness of open sets makes it difficult for a set to be compact. For example, an infinite-dimensional normed space has so many open sets that its closed unit ball cannot be compact. Because of this, many facts about finite-dimensional normed spaces that are based on the Heine-Borel property cannot be immediately generalized to the infinite-dimensional case. Compactness is depended on the nature of topology. The weak topology has more compact sets, this fact motivates us to search for a topology defined on a normed space X, which is the weakest topology among all topologies, which we can define on X, so we can have the biggest class of compact sets. The topology, which gives us the desired result is the weak topology defined on X. This topology helps us to characterize properties of topological spaces with infinite dimensions and provides simple means to check their nature. To present the basics from the weak topology, we will use [42, 43]. Let X be a Banach space and X^* be its dual. For each $f \in X^*$, the function

$$\rho_f : x \longrightarrow |f(x)|$$

is a semi-norm on X. The topology generated by the family of semi-norms

$$\{\rho_f,\ f \in X^*\}$$

is an equivalent definition of the weak topology on X, see Definition 1.3.1. In practical situation, often one faces the problem of type partial differential equations in the weak topology. All results for a Hausdorff locally convex topology induced by a separating vector space of linear functionals hold, for the weak topology of a normed space.

Definition 1.3.3 Let X be a Banach space and $x_0 \in X$. A subset V of X is called to be weakly open around x_0 if, there exists an element $\varepsilon > 0$ and there exists a finite collection of linear functionals $x_1^*, \ldots, x_n^* \in X^*$ such that

$$\bigcap_{i=1}^{n} \{x \in X,\ |x_k^*(x) - x_k^*(x_0)| < \varepsilon\} \subset V.$$

Remark 1.3.4 The weak topology and the strong topology on X coincide if, and only, if X is finite dimensional.

Example 1.3.1 Let X be an infinite dimensional and let S_X be the closed unit sphere in X. Then, $\overline{S_X}^w = \{0_X\}$, where is the weak closure of the S_X. Indeed, let V be any weak neighborhood at origin 0_X. Then, there exists $\varepsilon > 0$ and there exists a finite collection of linear functionals $f_1, \ldots, f_n \in X^*$ such that

$$\{x \in X \text{ such that } |(f_i, x)| < \varepsilon\} \subset V.$$

The map $\phi : x \longrightarrow ((f_1, x), \ldots, (f_n, x))$ is linear on X and

$$\ker \phi = \bigcap_{i=1}^{n} \ker f_i.$$

By the rank-nullity theorem, $\ker \phi$ is infinite dimensional, and can certainly not be equal to $\{0_X\}$. This implies that, there exists $x \in X \backslash \{0_X\}$ such that $(f_i, x) = 0$, for all $i \in \{1, \ldots, n\}$. Then, for all $\lambda \in \mathbb{R}$, we have

$$\lambda x \in \{x \in X \text{ such that } |(f_i, x)| < \varepsilon\} \subset V.$$

If we take $\lambda = \dfrac{1}{\|x\|}$, we obtain

$$V \bigcap S_X \neq \emptyset.$$

Hence, any weakly open neighborhood at origin 0_X intersects S_X and consequently, $\overline{S_X}^w = \{0_X\}$. Thus, the weak and the strong closure of S are different.

Proposition 1.3.1 *Let $x_0 \in X$ be fixed and let $\varepsilon > 0$. Put a finite set*

$$\{f_1, \ldots, f_n\} \subset X^*.$$

Denote by V_ε the set

$$V_\varepsilon = \{x \in X \text{ such that } |\langle f_i, x - x_0\rangle| < \varepsilon, \text{ for all } i = 1, \ldots, n\}.$$

Then, V_ε is a neighborhood at x_0 for the topology $\sigma(X, X^*)$ and, by varying ε, n and f_i in X^*, we obtain a basis of neighborhood at x_0, for $\sigma(X, X^*)$.

Proof Obviously V_ε is open for $\sigma(X, X^*)$ and contains x_0, since

$$V_\varepsilon = \bigcap_{i=1}^{n} \varphi_{f_i}^{-1}\left(]a_i - \varepsilon,\ a_i + \varepsilon[\right),$$

with $a_i = \langle f_i, x_0\rangle$. Conversely, let U be a neighborhood at x. From the above discussion, we know that the set

$$W = \bigcap_{i \in J} \varphi_{f_i}^{-1}(w_i)$$

is open and containing x_0, where $J \subset I$ finite and w_i is a neighborhood in $\mathbb{R}$ of a_i. Then, there exists $\varepsilon > 0$ such that

$$]a_i - \varepsilon,\ a_i + \varepsilon[\subset w_i,$$

for every $i \in J$. Consequently, $x_0 \in V_\varepsilon \subset W \subset U$. Q.E.D.

Definition 1.3.5 A sequence $(x_n)_n$ in a normed space X converges weakly to a point $x \in X$, if for every bounded linear functional $f \in X^*$, the sequence $(f(x_n))_n$ converges strongly to $f(x)$. If a sequence $(x_n)_n$ of X converges to a point $x \in X$ in the weak topology, we shall write $x_n \rightharpoonup x$.

Remark 1.3.6 Every strongly convergent sequence is weakly convergent. This implication follows from the following inequality

$$|f(x_n) - f(x)| \le \|f\|_{X^*} \|x_n - x\|.$$

The converse is not in general true. We consider the following two counter-examples.

Example 1.3.2 Consider the sequence $e_n = (0, \ldots, 0, 1, 0, \ldots)$ with 1 in the n-th position in the space l^2. Then,

$$\|e_n - e_m\|_2 = \sqrt{2}, \quad \text{for } m \neq n.$$

So, $(e_n)_n$ is not Cauchy, then it cannot converge, and it even does not contain any convergent subsequence. Nonetheless, in some weak sense, we may say that this sequence converges to $(0, \ldots, 0)$. For instance, for every sequence $x = (x_n)_{n\in\mathbb{N}^*} \in \ell^2$, we have

$$\lim_{n\to\infty} \langle x, e_n\rangle = \lim_{n\to\infty} x_n = 0.$$

Example 1.3.3 Let $X = C([0, 1], \mathbb{R})$, the space of real-valued continuous functions on $[0, 1]$, equipped with the norm $\|\cdot\|_\infty$. Define the sequence of functions $(\varphi_n)_n$ on $[0, 1]$ by putting

$$\varphi_n(x) = \begin{cases} nx & \text{if } x \in \left[0, \dfrac{1}{n}\right] \\ 2 - nx & \text{if } x \in \left[\dfrac{1}{n}, \dfrac{2}{n}\right] \\ 0 & \text{if } x \in \left[\dfrac{2}{n}, 1\right]. \end{cases}$$

Notice that $\varphi_n \not\to 0$ strongly, since $\|\varphi_n\| = 1$. We claim that $(\varphi_n)_n$ is a weakly convergent sequence to the null element. If this is not the case, there exists a subsequence, say $(\varphi_{n_k})_k$, of $(\varphi_n)_n$, and there exists a linear function $f \in X^*$ such that

$$f(\varphi_{n_k}) \geq \gamma, \quad \text{for some } \gamma > 0.$$

Moreover, we may assume that the subsequence satisfies $2n_k \leq n_{k+1}$. If we denote

$$\psi_N = \sum_{k=1}^{N} \varphi_{n_k}$$

and $p = \max\{k, \quad n_k < \frac{1}{x}\}$, we obtain

$$\begin{aligned}\psi_N(x) &\leq \sum_{n_k<\frac{1}{x}} n_k x + \sum_{\frac{1}{x}\leq n_k<\frac{2}{x}} 2 - n_k x \\ &\leq \sum_{k\leq p} \frac{1}{2^{p-k}} + 2.\end{aligned}$$

This implies that

$$\|\psi_N\|_\infty \leq 4 \text{ and } |f(\psi_N)| \leq 4\|f\|.$$

Again,

$$f(\psi_N) = \sum_{k=1}^{N} f\left(\varphi_{n_k}\right) \geq N\gamma.$$

Thus,

$$\lim_{N \to +\infty} f(\psi_N) = +\infty,$$

which is a contradiction, and the claim is proved.

Remark 1.3.7 If $(x_n)_n$ is a weakly convergent sequence, then the sequence of norms $(\|x_n\|)_n$ is bounded. For more detail see [5].

Definition 1.3.8 Let X be a Banach space and let $T : X \longrightarrow X$ be a mapping. We say that T is weakly demi-compact if, for every bounded sequence $(x_n)_n$ such that

$$x_n - Tx_n \rightharpoonup x \in X,$$

for some x in X, there exists a weakly convergent subsequence of $(x_n)_n$.

Definition 1.3.9 Let X be a Banach space and let $T_1, \ T_2 : X \longrightarrow X$ be two mappings. We say that T_1 is weakly asymptotically regular with respect to T_2 if,

$$T_1^n x - T_1^{n-1} x + T_2 x \rightharpoonup 0,$$

for all $x \in X$.

Definition 1.3.10 A subset K of a Banach space X is said to be weakly sequentially compact if, any sequence in K possesses a weakly convergent subsequence with limit in K.

Definition 1.3.11 A space having a countable dense subset is said to be separable.

Example 1.3.4 • Finite-dimensional spaces are separable.

- Sequence spaces l^p, $1 \leq p < \infty$, normed by p-norm is separable.
- The space $C(K, X)$, of all continuous functions from K to X is also separable, where K is a compact Hausdorff space.
- L^∞ and sequences space l^∞ are not separable.

Definition 1.3.12 A countably compact subset S of a topological space X is one in which each countable open covering S has a finite sub-covering. A subset S is called to be relatively countably compact if, the closure of S in X is countably compact.

Lemma 1.3.13 *Let H be a Hilbert space and let $f, \ g \in H$ be with $\|g\| \neq 0$. Then, there exists $h \in H$ such that*

$$\langle h, g \rangle = 0 \quad \textit{and} \quad f = \frac{\langle f, g \rangle}{\|g\|^2} g + h.$$

Proof If we set $h = f - \dfrac{\langle f, g \rangle}{\|g\|^2} g$, then

$$\langle h, g \rangle = \langle f, g \rangle - \frac{\langle f, g \rangle}{\|g\|^2} \langle g, g \rangle = 0.$$

This completes the proof. Q.E.D.

Now, we prove a fundamental inequality for Hilbert spaces.

Theorem 1.3.14 (Cauchy Schwartz) *Let H be a Hilbert space and let $f,\ g \in H$. Then, the following inequality is satisfied*

$$|\langle f,\ g\rangle| \leq \|f\|\|g\|.$$

Proof We assume that $\|g\| \neq 0$. Consider the decomposition

$$f = \frac{\langle f, g\rangle}{\|g\|^2} g + h,$$

where h is orthogonal to g. This implies that

$$\begin{aligned}\|f\|^2 &= \left\| \frac{\langle f, g\rangle}{\|g\|^2} g \right\|^2 + \|h\|^2 \\ &= \frac{|\langle f, g\rangle|^2}{\|g\|^2} + \|h\|^2 \\ &\geq \frac{|\langle f,\ g\rangle|^2}{\|g\|^2}.\end{aligned}$$

Multiplying both sides of the above inequality, we obtain

$$|\langle f,\ g\rangle| \leq \|f\|\|g\|.$$

This completes the proof. Q.E.D.

Let H be a Hilbert space and let T be a bounded linear functional on H. The following theorem is called to be the Riesz representation theorem.

Theorem 1.3.15 *If T is a bounded linear functional on a Hilbert space H, then there exists an unique element $g \in H$ such that, for every $f \in H$, we have*

$$T(f) = \langle f,\ g\rangle, \quad \textit{and}\ \|T\| = \|g\|.$$

Proof Let us assume that H is separable for now. It's not really much harder to prove for any Hilbert space, but the separable case makes nice use of the ideas we developed regarding Fourier analysis, we can see [44–46]. Also, let us just work over $\mathbb{R}$. Since H is separable, we can choose an orthonormal basis $\phi_j,\ j \geq 1$ for H. Let $T : H \longrightarrow H$ be a bounded linear functional and set a sequence $a_j = T(\phi_j)$. Choose $f \in H$ and let $c_j = \langle f, \phi_j\rangle$. Define the mapping f_n by the formula

$$f_n = \sum_{j=1}^{n} c_j \phi_j.$$

Since ϕ_j form a basis, for $j = 1, \ldots, n$, we know that

$$\lim_{n\to\infty} \|f - f_n\| = 0.$$

Since T is linear, we get

$$T(f_n) = \sum_{j=1}^{n} a_j c_j.$$

Due to the fact that T is bounded, say with norm $\|T\| < \infty$, we obtain

$$\|T(f) - T(f_n)\| \le \|T\| \|f - f_n\|.$$

This implies that

$$\begin{aligned} T(f) &= \lim_{n\to\infty} T(f_n) \\ &= \sum_{j=1}^{\infty} a_j c_j. \end{aligned}$$

In fact, the sequence $(a_j)_j$ must itself be square-summable. To see this, first since $\|T(f)\| \le \|T\| \|f\|$, we have

$$\left| \sum_{j=1}^{\infty} a_j c_j \right| \le \|T\| \left(\sum_{j=1}^{\infty} c_j^2 \right)^{\frac{1}{2}}. \tag{1.3.1}$$

The inequality (1.3.1) must hold, for any square-summable sequence $(c_j)_j$, since any c_j corresponds to some element in H. Fix a positive integer N and define the sequence $c_j = a_j$, for $j \le N$ and $c_j = 0$, for $j > N$. Clearly such a sequence is square-summable and

$$\left(\sum_{j=1}^{N} a_j^2 \right)^{\frac{1}{2}} \le \|T\|.$$

Thus, a_j is square-summable, since the sequence of partial sum is bounded and consequently, the function

$$g = \sum_j a_j \phi_j$$

is well defined as an element of H and

$$T(f) = \sum_j a_j c_j$$
$$= \langle f, g \rangle.$$

From the above discussion, we conclude that

$$\|g\| \le \|T\|.$$

But from Theorem 1.3.14, we also have

$$|T(f)| = |\langle f, g \rangle|$$
$$\le \|f\| \|g\|,$$

and then

$$\|T\| \le \|g\|. \qquad \text{Q.E.D}$$

The following definition is necessary, in order to state the Hahn–Banach theorem.

Definition 1.3.16 A real-valued function p on a vector space X is positive-homogeneous if $p(tx) = tp(x)$, whenever $t > 0$, and is sub-additive if,

$$p(x + y) \le p(x) + p(y),$$

whenever $x, y \in X$. If p has both properties, then it is said to be a sublinear functional.

Theorem 1.3.17 (Hahn–Banach) *Let U be a nonempty, open and convex subset of a topological real vector space X and let F be a subspace of X such that*

$$F \bigcap U = \emptyset.$$

Then, there exists a closed subspace G of codimension 1 such that

$$F \subset G \text{ and } G \bigcap U = \emptyset.$$

Theorem 1.3.18 (Analytic form of the Hahn–Banach extension theorem) *Let X be a real vector space. Assume that $p : X \longrightarrow \mathbb{R}$ is a sublinear functional and f is a linear functional on a subspace Y in X such that*

$$f(y) \le p(y),$$

whenever $y \in Y$. Then, there exists a linear functional $g : X \longrightarrow \mathbb{R}$ such that the restriction of g to Y is f and

$$g(x) \le p(x),$$

whenever $x \in X$, that is, the linear functional f can be extended to a linear functional on X that is still dominated by p.

As an immediate consequence of Theorem 1.3.18, we obtain the following result.

Lemma 1.3.19 *Let X be a normed linear space and let $x \in X \backslash \{0\}$. Then, there exists a linear function $f \in X^*$, where X^* denote the dual space of X, with $\|f\| = 1$ such that $f(x) = \|x\|$.*

Proof Let F be the one-dimensional space spanned by x and define the mapping $g : F \longrightarrow X$ by putting

$$g(tx) = t\|x\|.$$

Then, $\|g\| = 1$ and there exists a function $f \in X^*$ such that $\|f\| = 1$ and, which extends g. Hence, $f(x) = g(x) = \|x\|$. Q.E.D.

Corollary 1.3.20 *Let X be a locally convex vector space. For each $x \in X \backslash \{0\}$, there exists a continuous linear functional $f : X \longrightarrow \mathbb{R}$ such that*

$$\langle x, f \rangle \neq 0.$$

Lemma 1.3.21 *The topology $\sigma(X, X^*)$ is Hausdorff.*

Proof Let y and z be two distinct points in X. According to Theorem 1.3.18, we can strictly separate $\{y\}$ and $\{z\}$ by some $f \in X^*$. Then, there exists $\lambda \in \mathbb{R}$ such that

$$Ref(y) < \lambda < Ref(z).$$

Set

$$U = \{x \in X \text{ such that } Ref(x) < \lambda\} \text{ and } V = \{x \in X \text{ such that } Ref(x) > \lambda\}.$$

It is easy to check that U and V are open for $\sigma(X, X^*)$ such that $y \in U$, $z \in V$ and $U \bigcap V = \emptyset$. Q.E.D.

Definition 1.3.22 A Banach space X is reflexive if J is surjective, where the mapping $J : X \longrightarrow X^{**}$ is defined by the formula

$$J(x)(f) = \varphi_x(f) = f(x).$$

Example 1.3.5 • Finite-dimensional spaces are reflexive, since $\dim E = \dim E^{**}$.
• L^p and l^p are reflexive if, $1 < p < +\infty$.
• L^1, L^∞, l^1 and l^∞ are not reflexive.

Proposition 1.3.23 *Every Hilbert space is reflexive.*

Proof Define the mapping $J : H \longrightarrow H^{**}$ by putting

$$J_u(f) = f(u).$$

It follows from Theorem 1.3.15, that there exists an unique element $v_1 \in H$ such that

$$f(u) = \langle u, v_1 \rangle \text{ and } \|f\|_{H^*} = \|v_1\|_H, \quad \text{for any } f \in H^*.$$

Then, for every $f, g \in H^*$, there exists an unique couple $(v, w) \in H \times H$ such that

$$\begin{aligned} \langle f, g \rangle_{H^*} &= \langle f_v, g_w \rangle_{H^*} \\ &= \langle w, v \rangle_H. \end{aligned}$$

This implies that H^* is a Hilbert space and $H \cong H^*$, since the following mapping $u \in H \longrightarrow f_u \in H^*$ is bijective. From the above discussion, we show that $H^* \cong H^{**}$. Then, for any $h \in H^{**}$, there exists $g \in H^*$ such that

$$h(f) = \langle f, g \rangle_{H^*}, \quad \text{for all } f \in H^*.$$

Then, on account of Theorem 1.3.15, there exist $v, w \in H$ corresponding to $f, g \in H^*$ such that

$$\begin{aligned} \langle f, g \rangle_{H^*} &= \langle w, v \rangle_H \\ &= f(v). \end{aligned}$$

Thus, $h = J(v)$ and so, the mapping $J : H \longrightarrow H^{**}$ is bijective and hence, the space H is reflexive. Q.E.D.

The following theorem is due to Kakutani [12] and give us a very useful criteria of reflexive spaces.

Theorem 1.3.24 *A Banach space is reflexive if, and only, if its closed unit ball* $\overline{\mathscr{B}_X}$ *is weakly compact.*

Proof Since X is reflexive, then the mapping $J : X \longrightarrow X^{**}$ is a homeomorphism and so, we can identify X with X^{**} and $\sigma(X^{**}, X^*)$ and $\sigma(X, X^*)$ have the same topological structure. Due to the fact that $J(\overline{\mathscr{B}_X}) = \overline{\mathscr{B}_{X^{**}}}$, we get $\overline{\mathscr{B}_X}$ is weakly compact. Q.E.D.

As a consequence, we can prove the following property, which enables us to construct many reflexive spaces.

Proposition 1.3.25 *If M is a closed vector subspace of a reflexive Banach space X, then M is reflexive.*

Proof It should be noted that M^* is the restriction of elements in X^* to M, in view of Theorem 1.3.18. So, the weak topology on M coincides with the topology on M, that is, induced by the weak topology on X. Since X is reflexive, then $\overline{\mathscr{B}_X}$ is weakly compact in X. On the other hand, since both $\overline{\mathscr{B}_X}$ and M are closed and convex, they are weakly closed. So, the closed unit ball of M,

$$\overline{\mathscr{B}_M} = M \bigcap \overline{\mathscr{B}_X}$$

is weakly closed in X. Then, $\overline{\mathscr{B}_M}$ is weakly compact in X, with respect to the weak topology on X, or equivalently, $\overline{\mathscr{B}_M}$ is weakly compact in M, with respect to the weak topology on M. Q.E.D.

The following result was proved in [5], see also [43, Theorem 2.5.11].

Theorem 1.3.26 *A linear operator T from a normed space X into a normed space Y is weakly continuous if, and only, if it is norm-to-norm continuous.*

Remark 1.3.27 Let τ be the original vector topology of X. A simple consequence of the fact that $\sigma(X, X^*) \subset \tau$, is that every strongly convergent sequence is a weakly convergent sequence.

Definition 1.3.28 A topological space is said to be metrizable if, its topology can be obtained from a metric on the underlying space.

For Banach spaces with finite dimensions, the weak topology is metrizable so it is quite easy to deal with compactness in that case. Even though the weak topology of Banach spaces with infinite dimensions is never metrizable it behaves like a metrizable topology because in the weak topology, the two types of compactness are still equivalent. This surprising result is called the Eberlein–S̃mulian theorem. In 1940, S̃mulian proved that every weakly compact subset of a Banach space is weakly sequentially compact. In 1947, Eberlein showed the converse. For the proof, we refer to [25].

Theorem 1.3.29 *A subset of a Banach space is relatively weakly compact if, and only, if it is relatively weakly sequentially compact.*

The well-known Eberlein–S̃mulian theorem is a consequence of Theorem 1.3.29. For its proof, we refer the reader to [43].

Theorem 1.3.30 (Eberlein–S̃mulian theorem) *Let S be a nonempty subset of a normed space X. Then, the following assertions are equivalent*

1. *The set S is relatively weakly compact.*
2. *The set S is relatively weakly countably compact.*
3. *The set S is relatively weakly limit point compact.*
4. *The set S is relatively weakly sequentially compact.*

Remark 1.3.31 If S is a relatively weakly compact subset of a normed space and $a \in \overline{S}^w$, then there exists a sequence in S, that converge weakly to a.

Combining Theorems 1.3.24 and 1.3.29, we obtain the following interesting result.

Theorem 1.3.32 *Let X be a Banach space. The following assertions are equivalent*

1. *X is reflexive.*
2. *$\overline{\mathscr{B}}_X$ is relatively weakly compact.*
3. *$\overline{\mathscr{B}}_X$ is relatively weakly sequentially compact.*

Definition 1.3.33 An operator $T : X \longrightarrow X$ is said to be weakly compact, if $T(B)$ is relatively weakly compact, for every nonempty and bounded subset B of X.

The Krein–S̃mulian theorem is a consequence of Theorem 1.3.29. A proof can be found in [8].

Theorem 1.3.34 *Let K be a weakly compact of a Banach space X. Then, the closed convex hull $\overline{co}(K)$ of K is a weakly compact subset of X.*

The well-known Mazur's theorem is a consequence of Theorem 1.3.34 and shows us that there exists a class of sets, namely convex sets, where the norm and the weak closure agree.

Theorem 1.3.35 ([43]) *Let S be a nonempty and convex subset of a normed linear space X. Then, the closure of S, in the norm topology coincides with the weak closure of K. In particular, a convex subset of a normed space is closed if, and only, if it is weakly closed.*

Proof We will assume that X is a real Banach space. Choose a point $x_0 \in X \backslash S$. Then, on account of Theorem 1.3.17, there exists a closed hyperplane, which separates $\{x_0\}$ and S, i.e., there exists a function $f \in X^*$ and $\lambda \in \mathbb{R}$ such that

$$f(x_0) \leq \lambda \leq f(x), \quad \text{for all } \ x \in S.$$

Then, the set

$$U = \{x \in X \ \text{ such that } \ f(x) < \lambda\}$$

is weakly open in X and $x_0 \in U$, which does not meet S. This shows that the complement of S is weakly open and hence, S is a weakly closed subset of X. Q.E.D.

Remark 1.3.36 Every closed, bounded, and convex subset of a reflexive Banach space is weakly compact. Indeed, $\overline{\mathscr{B}}_X$ is weakly compact, in view of Theorem 1.3.24. According to Theorem 1.3.35, we get every closed and convex subset of X is weakly closed. Therefore, any closed, convex and bounded subset of X is a weakly closed subset of a weakly compact set and consequently, must be weakly compact.

Corollary 1.3.37 *If S is a subset of a normed space X, then*

$$\overline{co}(S) = \overline{co}^{w}(S).$$

The following results are needed later. A proof can be found in [8].

Theorem 1.3.38 *If $\{f(\cdot)\}$ is a family of functions in $L_1(S, \Sigma, \nu)$, which is weakly sequentially compact. Then, the family $\{|f(\cdot)|\}$ is also weakly sequentially compact.*

Definition 1.3.39 Let (X, Σ, ν) be a measure space. A subset Ω of $L_1(\nu)$ is said to be uniformly integrable (equi-integrable) provided that $\sup_{f\in\Omega} \|f\|_{L^1(\nu)} < \infty$, and to each $\varepsilon > 0$, there exists $\delta > 0$ such that the following inequality

$$\sup_{f\in\Omega} \int_S |f| d\nu < \varepsilon$$

is satisfied, for any measurable set S with $\nu(S) < \delta$, where $L_1(\nu)$ denote the family of all complex-valued measurable on X functions that satisfy

$$\int_X |f| d\nu(x) < \infty.$$

For finite measure spaces, we have the following characterization of uniform integrability.

Proposition 1.3.40 *Let (X, Σ, ν) be a finite measure space. A subset A of $L_1(\nu)$ is uniformly integrable if, and only, if for each $\varepsilon > 0$, there exists $\delta > 0$ such that*

$$\sup_{f\in A} \int_{\{x\in X,\ |f(x)|\geq M\}} |f| d\nu < \varepsilon.$$

Theorem 1.3.41 (Dunford–Pettis theorem) *Let (X, Σ, ν) be a finite measure space and let $(f_n)_n$ be a bounded sequence in $L^1(\nu)$. The following assertions are equivalent*

1. *$(f_n)_n$ is uniformly integrable.*
2. *Every subsequence of $(f_n)_n$ has a further subsequence that converges weakly in $L^1(\nu)$.*

The Arzelà–Ascoli theorem was proved in [47] and plays a crucial role in the proof of the existence of solutions for functional integral equations.

Theorem 1.3.42 *Let $(X, dist)$ be a compact metric space. A subset S in $C(X)$, where $C(X)$ denote the space of all continuous functions on X with values in $\mathbb{C}$, is relatively compact if, and only, if it is equibounded and equi-continuous, that is, for every $\varepsilon > 0$, there exists $\delta > 0$ such that*

$$dist(f(x), f(y)) < \varepsilon, \quad \text{for all } f \in S,$$

whenever $dist(x, y) < \delta$ and $x, y \in X$.

Definition 1.3.43 Let X be a Banach space and let $T : X \longrightarrow X$ be a mapping. We say that T is weakly sequentially continuous if, for every sequence $(x_n)_n$ such that $x_n \rightharpoonup x$, we have

$$T(x_n) \rightharpoonup T(x).$$

While it is not always possible to show that a given mapping between Banach spaces is weakly continuous, quite often its weak sequential continuity offers no problem. This is mainly due to the fact that Lebesgue's dominated convergence theorem is valid for sequences, but not for nets. A very interesting discussion (including examples) about different types of continuity can be found in [48, 49]. So, O. Arino, S. Gautier and J.-P. Penot proved in [50] the analog of Schauder's fixed point theorem for weakly sequentially continuous mappings. The proof follows easily from the Schauder-Tychonoff theorem.

Theorem 1.3.44 (Arino–Gautier–Penot's fixed point theorem) *Let S be a nonempty, weakly compact, and convex subset of a Banach space (or, more generally, a metrizable locally convex linear topological space). Then, any weakly sequentially continuous map $T : S \longrightarrow S$ has, at least, a fixed point.*

Proof For each weakly closed subset S_1 of X, the set $T^{-1}(S_1)$ is sequentially closed in S. According to Theorem 1.3.29, we infer that $T^{-1}(S_1)$ is a weakly compact subset of X. As a result, T is weakly continuous. Q.E.D.

Remark 1.3.45 By using Fan's fixed point theorem [51] instead of the Schauder-Tychonoff theorem, we can prove that the same argument is valid if T is replaced by a multi-valued mapping $Q : S \longrightarrow \mathscr{P}_{cl,cv}(S)$, where $\mathscr{P}_{cl,cv}(X)$ denote the family of nonempty, closed, and convex subsets of X, which is upper semi-continuous with respect to the weak topology. This means that if $(x_n)_n$ converges weakly to a point x and if U is a weakly open subset of X containing $Q(x)$, then $(Qx_n)_n$ is contained in U for n large enough. In other words, for any weakly closed subset S_1 of S, the set $Q^{-1}(S_1)$ is sequentially closed for the weak topology on S.

An important fixed point theorem that has been commonly used in the theory of nonlinear differential and integral equations is the following result, see [52, 53].

Theorem 1.3.46 *Let S be a nonempty, closed, and convex subset of a Banach space X and let $T : S \longrightarrow S$ be a weakly sequentially continuous map. If $T(S)$ is relatively weakly compact, then T has a fixed point in S.*

Proof It follows from Theorem 1.3.34, that $\overline{co}(T(S))$ is a weakly compact and convex subset of S. On the other hand, it is easy to see that

$$\begin{aligned} T(\overline{co}(T(S))) &\subset T(S) \\ &\subset \overline{co}(T(S)), \end{aligned}$$

and when we apply Theorem 1.3.44, we reach the desired conclusion. Q.E.D.

Definition 1.3.47 A Banach algebra X is said to be weakly compact (WC-Banach algebra for short), if the product of two arbitrary weakly compact subsets of X is weakly compact.

Definition 1.3.48 We will say that the Banach algebra X satisfies the sequential condition $(\mathscr{P})$ if, for any sequences $(x_n)_n$ and $(y_n)_n$ contained in X such that $x_n \rightharpoonup x$ and $y_n \rightharpoonup y$, we have

$$x_n \cdot y_n \rightharpoonup x \cdot y.$$

Remark 1.3.49 Every finite-dimensional Banach algebra satisfies condition $(\mathscr{P})$.

The following theorem is a fundamental tool for the proofs of the existence of solutions for several functional integral equations. A proof can be found in [54].

Theorem 1.3.50 *Let K be a compact Hausdorff space and let X be a Banach space. Assume that $(f_n)_n$ is a bounded sequence in $C(K, X)$ and $f \in C(K, X)$. Then, $(f_n)_n$ is weakly convergent to f if, and only, if $(f_n(t))_n$ is a weakly convergent sequence to the point $f(t)$, for each $t \in K$.*

Remark 1.3.51 In the case when K is a compact interval of reals, Theorem 1.3.50 was demonstrated in [55]. The scenario where X is reflexive or separable is treated by Bogdanowicz [56]. One can simplify the general case to the situation where X is separable.

Remark 1.3.52 If X is a Banach algebra satisfying the sequential condition $(\mathscr{P})$, then the space of all continuous functions $C(K, X)$ is a Banach algebra satisfying condition $(\mathscr{P})$, where K is a compact Hausdorff space. The proof is based on Theorem 1.3.50. In fact, let $(f_n)_n$ and $(g_n)_n$ be two sequences contained in the set $C(K, X)$ such that $f_n \rightharpoonup f$ and $g_n \rightharpoonup g$. Then,

$$f_n(x) \rightharpoonup f(x) \text{ and } g_n(x) \rightharpoonup g(x), \quad \text{for all } x \in K.$$

The condition $(\mathscr{P})$ guaranties that

$$f_n(x) \cdot g_n(x) \rightharpoonup f(x) \cdot g(x), \quad \text{in } X.$$

Taking into account that $(f_n \cdot g_n)_n$ is a bounded sequence and using Theorem 1.3.50, we get

$$f_n \cdot g_n \rightharpoonup f \cdot g.$$

The Dunford–Pettis property is defined in [57].

Definition 1.3.53 We say that a Banach space X has the Dunford–Pettis property (DPp, in short) if, for every Banach space Y, each weakly compact linear operator $T : X \longrightarrow Y$ takes weakly compact sets in X into compact sets of Y.

Remark 1.3.54 Since every Banach algebra is a Banach space then, we can also consider Banach algebras with DPp. It can be shown that $C(K, X)$ and $L^1(\nu)$ have the DPp condition, where K is a compact Hausdorff space, X is a Banach algebra and ν is a finite measure. We refer to [58], for an excellent survey of DPp.

Let X be a Banach space. A homogenous continuous polynomial on X is a mapping P given by

$$P(x) = T(x, \dots, x),$$

where T is a multilinear continuous map on $X \times X \times \cdots \times X$. It was proved by Ryan [59], that in spaces with DPp, all multilinear forms are weakly sequentially continuous. We have the following result.

Theorem 1.3.55 *Every commutative Banach algebra with DPp satisfies condition* $(\mathscr{P})$.

Proof Let X be a commutative Banach algebra with DPp and let $\phi \in X^*$ be an arbitrary function. Define the mapping $T_\phi : X \times X \longrightarrow \mathbb{R}$ by putting

$$T_\phi(x, y) = \phi(x \cdot y).$$

It be shown that, T_ϕ is a continuous bilinear map. Let $(x_n)_n$ and $(y_n)_n$ be two sequences in X such that $x_n \rightharpoonup x$ and $y_n \rightharpoonup y$, for some $x, y \in X$. From the following equality

$$x_n \cdot y_n = \frac{1}{4}\left((x_n + y_n)^2 - (x_n - y_n)^2\right),$$

we obtain

$$\phi(x_n \cdot y_n) = \frac{1}{4}\left(T_\phi(x_n + y_n, x_n + y_n) - T_\phi(x_n - y_n, x_n - y_n)\right).$$

If we consider the weak sequential continuity of the mapping T_ϕ, we get

$$\phi(x_n \cdot y_n) \rightharpoonup \frac{1}{4}\left(T_\phi(x + y, x + y) - T_\phi(x - y, x - y)\right).$$

This means that

$$\phi(x_n \cdot y_n) \to \phi(x \cdot y).$$

In others words, we have $x_n \cdot y_n \rightharpoonup x \cdot y$. Q.E.D.

Remark 1.3.56 The weak compactness property is valid in finite-dimensional algebras and is also valid in the space $C(K, X)$, where K is a compact Hausdorff space, with pointwise multiplication and the supremum norm, when X have the weak compactness structure, i.e., X is a WC-Banach algebra, see [11], also in the class of commutative Banach algebras, which have the DPp, see [60].

In [11], A. Jeribi and B. Krichen posed the question of whether the sequential condition ($\mathscr{P}$) is satisfied by any arbitrary WC-Banach algebra. The writers of [61] will respond in the positive to this query. We will need a few auxiliary results given below. We cite [60, 61] for the proof.

Lemma 1.3.57 *Let $(x_n)_n$ be a weakly convergent sequence contained in a Banach algebra X to a point $x \in X$, then*

$$x_n \cdot y \rightharpoonup x \cdot y \quad \textit{and} \quad y \cdot x_n \rightharpoonup y \cdot x,$$

for every $y \in X$.

Proof Let $y \in X$ be fixed. Consider on X the left-hand multiplication operator $L_y(x) = yx$ and the right-hand multiplication operator $R_y(x) = xy$. Clearly L_y and R_y are linear and continuous. Taking into account the fact that a linear operator between Banach spaces is continuous if, and only, if it is weakly continuous, we deduce the desired conclusion. Q.E.D.

Proposition 1.3.58 *A Banach algebra X satisfies the sequential condition ($\mathscr{P}$) if, and only, if for arbitrary sequences $(x_n)_n$ and $(y_n)_n$ in X such that $x_n \rightharpoonup \theta$ and $y_n \rightharpoonup \theta$, we have that $x_n \cdot y_n \rightharpoonup \theta$.*

Proof Let $(x_n)_n$ and $(y_n)_n$ be two sequences of X such that $x_n \rightharpoonup x$ and $y_n \rightharpoonup y$ and take

$$\begin{cases} x'_n = x_n - x \\ y'_n = y_n - y. \end{cases}$$

It is easy to verify that,

$$x'_n \rightharpoonup \theta \quad \text{and} \quad y'_n \rightharpoonup \theta.$$

Keeping in mind our assumptions, we have that

$$x'_n \cdot y'_n \rightharpoonup \theta.$$

Using Lemma 1.3.57, we get

$$\begin{aligned} x_n \cdot y_n &= (x + x'_n) \cdot (y + y'_n) \\ &= x \cdot y + x \cdot y'_n + x'_n \cdot y + x'_n \cdot y'_n \rightharpoonup x \cdot y. \end{aligned}$$

The proof of the converse implication is obvious. Q.E.D.

Theorem 1.3.59 *A Banach algebra X with a multiplication operation satisfies the sequential condition ($\mathscr{P}$) if, and only, if X is the WC-Banach algebra.*

Proof It suffices to prove that if X is the WC-Banach algebra, then X satisfies the sequential condition ($\mathscr{P}$). Suppose that the contrary is true. Then, in virtue of

Proposition 1.3.58, we get there exist two sequences $(x_n)_n$ and $(y_n)_n$ in X such that $x_n \rightharpoonup \theta$ and $y_n \rightharpoonup \theta$, but the sequence $(x_n \cdot y_n)_n$ does not converge weakly to θ. Set

$$U = \{x_n, \ n \in \mathbb{N}\} \bigcup \{\theta\} \quad \text{and} \quad V = \{y_n, \ n \in \mathbb{N}\} \bigcup \{\theta\}.$$

Obviously these sets are weakly compact in X. Since X is a WC-Banach algebra, it follows that $U \cdot V$ is weakly compact. It follows from Theorem 1.3.29, that the sequence $(x_n \cdot y_n)_n$ contains a weakly convergent subsequence. If all such subsequences converge weakly to θ, then the sequence $(x_n \cdot y_n)_n$ is convergent weakly to θ, which is a consequence of Proposition 1.3.58. But this contradicts the fact that $(x_n \cdot y_n)_n$ does not converge weakly to θ. This implies that there exists a subsequence of $(x_n \cdot y_n)_n$, which converges weakly to some $z \in X \backslash \{0\}$. Without loss of generality, we may assume that

$$x_n \cdot y_n \rightharpoonup z.$$

Let $q > 0$ be an arbitrary number. Consider the sequences $(x'_n)_n$ and $(y'_n)_n$ by putting

$$x'_n = \left(q + \frac{1}{n}\right) x_n \quad \text{and} \quad y'_n = \left(q + \frac{1}{n}\right) y_n.$$

After simple calculations, we can verify that

$$x'_n \cdot y'_n = \left(q + \frac{1}{n}\right)\left(q + \frac{1}{n}\right) x_n \cdot y_n \rightharpoonup q^2 z.$$

Let us take into account the sets U' and V' defined as

$$U' = \{x'_n, \ n \in \mathbb{N}\} \bigcup \{\theta\} \quad \text{and} \quad V' = \{y'_n, \ n \in \mathbb{N}\} \bigcup \{\theta\}.$$

Since X is a WC-Banach algebra, then the set $U' \cdot V'$ is weakly compact. Taking into account that

$$\{x'_n \cdot y'_n, \ n \in \mathbb{N}\} \subset U' \cdot V'$$

and using Theorem 1.3.29, we get there exist $n, m \in \mathbb{N}$ such that

$$q^2 z = \left(q + \frac{1}{n}\right)\left(q + \frac{1}{m}\right) x_n \cdot y_m.$$

In other words, we have

$$x_n \cdot y_m = \frac{q^2}{\left(q + \frac{1}{n}\right)\left(q + \frac{1}{m}\right)} z. \tag{1.3.2}$$

Define the mapping $T : (0, \infty) \longrightarrow \mathbb{N} \times \mathbb{N}$ by putting

$$T(q) = (n, m),$$

where the pair (n, m) of natural numbers is chosen according to the equality (1.3.2). Due to the fact that the interval $(0, \infty)$ is uncountable and the set $\mathbb{N} \times \mathbb{N}$ is countable, we infer that there exist $q_1, q_2 \in (0, \infty)$ such that

$$q_1 \neq q_2 \text{ and } T(q_1) = T(q_2).$$

Then, there exist $n, m \in \mathbb{N}$ such that

$$\begin{aligned} x_n \cdot y_m &= \frac{q_1^2}{\left(q_1 + \frac{1}{n}\right)\left(q_1 + \frac{1}{m}\right)} z \\ &= \frac{q_2^2}{\left(q_2 + \frac{1}{n}\right)\left(q_2 + \frac{1}{m}\right)} z. \end{aligned}$$

This implies that

$$\frac{q_1^2}{\left(q_1 + \frac{1}{n}\right)\left(q_1 + \frac{1}{m}\right)} = \frac{q_2^2}{\left(q_2 + \frac{1}{n}\right)\left(q_2 + \frac{1}{m}\right)}. \tag{1.3.3}$$

However, the equality (1.3.3) contradicts the fact that the function

$$q \longrightarrow \frac{q^2}{\left(q + \frac{1}{n}\right)\left(q + \frac{1}{m}\right)}$$

is injective, for fixed natural numbers n and m. Q.E.D.

The writers in [61] have proposed the following idea in response to the previous discussions.

Definition 1.3.60 A Banach algebra is said to be relatively weakly compact (RWC-Banach algebra, for short) if, the product of two arbitrary relatively weakly compact subsets of X is relatively weakly compact.

Remark 1.3.61 Every Banach algebra satisfying the sequential condition ($\mathcal{P}$) is RWC-Banach algebra. But the converse may not be true as shown by the following counter-example. Consider the Banach algebra ℓ^2 with the norm

$$\|x\|_2 = \|x_n\|_2 = \left(\sum_{n\geq 1} |x_n|^2\right)^{1/2},$$

and the multiplication

$$x \cdot y = \left(\sum_{n\geq 1} x_n y_n, 0, 0, \dots\right). \tag{1.3.4}$$

It is easily seen that, for arbitrary relatively weakly compact subsets U, V of ℓ^2 the product $U \cdot V$ is a bounded subset of the 1-dimensional subspace $span(e_n)$, where $(e_n)_n$ denotes the sequence of canonical vectors in ℓ^2. This means that ℓ^2 is the RWC-Banach algebra. On the other hand, by taking $x_n = y_n = e_n$, we have that $x_n \rightharpoonup \theta$ and $y_n \rightharpoonup \theta$. But $x_n \cdot y_n = e_1 \not\rightharpoonup \theta$. Therefore, ℓ^2 with the multiplication, as is defined in (1.3.4), does not satisfy condition ($\mathcal{P}$).

The following, which were obtained in [62], will be used in the sequel.

Lemma 1.3.62 *Let X be a Banach algebra satisfying the sequential condition ($\mathcal{P}$) and let S and S' be two weakly compact subsets of X. Then, the set*

$$S \cdot S' = \{x \cdot y,\ x \in S \text{ and } x' \in S'\}$$

is weakly compact.

Proof Let $(x_n)_n$ be any sequence of S and $(y_n)_n$ be any sequence of S'. Then, there exists a subsequence, say $(x_{n_k})_k$, of $(x_n)_n$ and there exists a subsequence, say $(y_{n_k})_k$, of $(y_n)_n$ such that

$$x_{n_k} \rightharpoonup x \in S, \text{ and } y_{n_k} \rightharpoonup y \in S'.$$

The sequential condition ($\mathcal{P}$) guarantees that

$$x_{n_k} \cdot y_{n_k} \rightharpoonup x \cdot y \in S \cdot S'.$$

This implies that $S \cdot S'$ is a weakly sequentially compact subset of X. An application of Theorem 1.3.29 shows that $S \cdot S'$ is weakly compact. Q.E.D.

1.4 Hausdorff Metric

This section compiles the characteristics of multi-valued maps, also known as set-valued or multiple-valued maps. These qualities are essential for understanding fixed point theory and its applications to differential inclusions and nonlinear integral inclusions. Let X be a Banach algebra. A correspondence $Q : X \longrightarrow \mathscr{P}_p(X)$ is called to be a multi-valued operator or multi-valued mapping on X into itself, where $\mathscr{P}_p(X)$ denote the class of all nonempty subsets of X with property p. A point $x \in X$ is said to be a fixed point of Q if $x \in Qx$ and the set of all fixed points of Q in X is denoted by $\mathscr{F}_Q$. For the sake of convenience, we denote

$$Q(A) = \bigcup_{x \in A} Qx,$$

for a subset A of X. For any $A, B \in \mathscr{P}_p(X)$, let us denote

$$A \pm B = \{a \pm b,\ a \in A \text{ and } b \in B\}, \text{ and}$$

$$A \cdot B = \{a \cdot b,\ a \in A \text{ and } b \in B\}.$$

Lemma 1.4.1 *Let X and Y be two Banach spaces, $A \subset X$ and $B \subset Y$ be two nonempty subsets. Assume that $Q : X \longrightarrow \mathscr{P}(Y)$ is a multi-valued mapping. Then,*

$$A \subset Q^{-1}\left(Q(A)\right), \quad Q\left(Q^{-1}(B)\right) \subset B \quad \text{and } Q^{-1}\left(Y \backslash B\right) \subset X \backslash \{Q^{-1}(B)\},$$

Let X be a Banach algebra and let $Q_1, Q_2 : X \longrightarrow \mathscr{P}(X)$ be two multi-valued mappings. The product $Q_1 \cdot Q_2$ is defined by

$$\left(Q_1 \cdot Q_2\right)(x) = Q_1(x) \cdot Q_2(x).$$

Let A be a nonempty subset of a Banach space X and let $x \in X$. The distance between x and A is defined by

$$\operatorname{dist}(x, A) = \inf_{a \in A} \operatorname{dist}(x, a).$$

Let $a \in X$ with $\operatorname{dist}(x, A) = 0$. Then, for every $\varepsilon > 0$, there exists $x^{\varepsilon} \in A$ such that

$$\operatorname{dist}(a, x^{\varepsilon}) < \varepsilon.$$

This implies that

$$A \bigcap \mathscr{B}_X(a, \varepsilon) \neq \emptyset.$$

Consequently $a \in \overline{A}$. The inverse implication is obvious, because for all $x \in \overline{A}$, we have

$$A \bigcap \mathscr{B}_X(x, \varepsilon) \neq \emptyset,$$

for every $\varepsilon > 0$. Then, for each $\varepsilon > 0$, there exists $x^\varepsilon \in A$ such that

$$\text{dist}(a, x^\varepsilon) < \varepsilon.$$

Hence,

$$\overline{A} = \{x \in X \text{ such that } \text{dist}(x, A) = 0\}.$$

The distance between any two nonempty subsets A and B of X is defined as

$$D_1(A, B) = \sup_{a \in A} \text{dist}(a, B).$$

The distance function $D_1(\cdot, \cdot)$ is not a metric, because it is not always symmetric, and $D_1(A, B) = 0$ does not imply that $A = B$. Another notion of distance is

$$D_2(A, B) = \inf_{a \in A,\ b \in B} \text{dist}(a, b).$$

However, we would like the distance between two sets to be zero if, and only, if the two sets are similar, both in shape and position. For this purpose, we can create a metric by defining the Hausdorff distance. Let $A, B \in \mathscr{P}_{cl,bd}(X)$, where $\mathscr{P}_{cl,bd}(X)$ denote the family of nonempty, closed and bounded subsets of X. Define the mapping $\text{dist}_H(\cdot, \cdot)$ on $\mathscr{P}_{cl,bd}(X) \times \mathscr{P}_{cl,bd}(X)$ by putting

$$\text{dist}_H(A, B) = \max \{D_1(A, B), D_1(B, A)\}.$$

The function $\text{dist}_H(\cdot, \cdot)$ is said to be a Hausdorff metric induced by $\text{dist}(\cdot, \cdot)$. After simple calculations, we can verify that $\text{dist}_H(\cdot, \cdot)$ is symmetric and $\text{dist}_H(A, A) = 0$, for all $A \in \mathscr{P}_{cl,bd}(X)$ and

$$\text{dist}_H(A, C) \leq \text{dist}_H(A, B) + \text{dist}_H(B, C).$$

Hence, $\text{dist}_H(\cdot, \cdot)$ is an extended pseudo-metric on $\mathscr{P}(X) \times \mathscr{P}(X)$. Moreover, we can prove that

$$\text{dist}_H(A, B) = 0 \quad \text{if, and only, if} \quad \overline{A} = \overline{B}.$$

Indeed, if both $A, B \in \mathscr{P}_{bd}(X)$, where $\mathscr{P}_{bd}(X)$ denote the family of nonempty and bounded subsets of X, then $\text{dist}_H(A, B)$ is guaranteed to be finite and $\text{dist}_H(A, B) = 0$ implies that $D_1(A, B) = 0$ and $D_1(B, A)$ and so, $A \subset \overline{B}$ and $B \subset \overline{A}$. Then, the space $(\mathscr{P}_{cl,bd}(X), \text{dist}_H)$ becomes a metric space, where $\mathscr{P}_{cl,bd}(X)$ denote the family of nonempty, closed and bounded subsets of X. Thus, in order to meet the requirements of being complete metric, we take closed and bounded sets. Note that

$$\|A\|_{\mathscr{P}} = \text{dist}_H(A, \{0\}).$$

For more detail, the reader may consult [63].

Lemma 1.4.2 *Let $(X, dist)$ be a metric space. Then, the function $\varphi_A : X \longrightarrow \mathbb{R}$ defined by the formula*

$$\varphi_A(x) = dist(x, A)$$

is nonexpansive, for all $A \in \mathscr{P}_{cl}(X)$, where $\mathscr{P}_{cl}(X)$ denote the family of nonempty and closed subsets of X.

Proof Let $x, y \in X$ be such that $x \neq y$ and let $a \in A$ be fixed. By the definition of $\text{dist}(\cdot, \cdot)$, we get

$$\text{dist}(x, a) \leq \text{dist}(x, y) + \text{dist}(y, a).$$

Since $a \in A$ is arbitrary, then

$$\text{dist}(x, A) \leq \text{dist}(x, y) + \text{dist}(y, A),$$

and

$$\text{dist}(y, A) \leq \text{dist}(x, y) + \text{dist}(x, A).$$

Thus,

$$|\text{dist}(x, A) - \text{dist}(y, A)| \leq \text{dist}(x, y).$$

This completes the proof. Q.E.D.

Lemma 1.4.3 *Let $(X, dist)$ be a metric space. Then, the function $\varphi_x : \mathscr{P}_{cl}(X) \longrightarrow \mathbb{R}$ defined by the formula*

$$\varphi_x(A) = dist(x, A)$$

is nonexpansive, for all $x \in X$.

Proof Let $x \in X$ be fixed and let $A, B \in \mathscr{P}_{cl}(X)$. It follows from the definition of the Hausdorff distance, that for each $a \in A$ and for each $b \in B$, we obtain the following implications

$$\text{if dist}(x, A) \leq \text{dist}(x, b) + \text{dist}(b, A), \text{ then dist}(x, A) - \text{dist}(x, B) \leq \text{dist}_H(A, B),$$

and

$$\text{if dist}(x, B) \leq \text{dist}(x, a) + \text{dist}(a, B), \text{ then dist}(x, B) - \text{dist}(x, A) \leq \text{dist}_H(A, B).$$

This implies that,

$$|\text{dist}(x, A) - \text{dist}(x, B)| \leq \text{dist}_H(A, B).$$ Q.E.D

We shall require the following well-known facts due to definition of $\text{dist}_H(\cdot, \cdot)$. For more details, the reader may refer to [2, 3, 64, 65].

Lemma 1.4.4 *Let A, $B \in \mathscr{P}_{cl,bd}(X)$ be such that $dist_H(A, B) < \varepsilon$. Then, for each $a \in A$, there exists an element $b \in B$ such that $dist(a, b) \leq \varepsilon$.*

Lemma 1.4.5 *Let S_1, S_2, and S_3 be three nonempty, bounded and closed subsets of a Banach algebra X. Then, the following inequality is satisfied*

$$dist_H(S_1 \cdot S_3,\ S_2 \cdot S_3) \leq dist_H(0,\ S_3)\, dist_H(S_1,\ S_2).$$

Definition 1.4.6 A sequence of nonempty sets $(A_n)_n$ in $\mathscr{P}_p(X)$ is said to converge to a set A, called to be the limiting set, if

$$\lim_{n \to +\infty} \mathrm{dist}_H(A_n, A) = 0.$$

Lemma 1.4.7 *Let $(A_n)_n$ be a sequence contained in the set $\mathscr{P}_{cl,bd}(X)$ such that*

$$\lim_{n \to +\infty} dist_H(A_n, A) = 0,$$

for some $A \in \mathscr{P}_{cl,bd}(X)$ and let $(x_n)_n$ be a convergent sequence of X to a point x. If $x_n \in A_n$, then $x \in A$.

Remark 1.4.8 The completeness of (X, dist) implies that $(\mathscr{P}_{cl,bd}(X), \mathrm{dist}_H)$ is complete.

Definition 1.4.9 A multi-valued mapping $Q : X \longrightarrow \mathscr{P}(X)$ is said to be a contraction if, there exists a constant $k \in [0, 1)$ such that

$$\mathrm{dist}_H(Qx,\ Qy) \leq k\, \mathrm{dist}(x,\ y),$$

for each pair of distinct points x and y in X. In particular, if

$$\mathrm{dist}_H(Qx,\ Qy) \leq \mathrm{dist}(x,\ y),$$

then Q is called to be a nonexpansive multi-valued mapping.

Definition 1.4.10 A multi-valued mapping $Q : X \longrightarrow \mathscr{P}(X)$ is said to be an upper semi-continuous (u.s.c., for short) at a point $x \in X$ provided that, for every open subset V with $Q(x) \subset V$, there exists an open U with $x \in U$ such that $Q(y) \subset V$, for all $y \in U$. A multi-valued is upper semi-continuous on X, if it is upper semi-continuous at every point of X.

A straightforward equivalent definition is given by Górniewicz [66].

Definition 1.4.11 A multi-valued mapping Q is said to be an upper semi-continuous if the set $Q^+(S)$ given by

$$Q^+(S) = \{x \in X \text{ such that } Q(x) \subset S\}$$

is an open subset in X, for every open subset S of X.

Definition 1.4.12 A multi-valued mapping Q is said to be a lower semi-continuous if the set $Q^{-}(S)$ given by

$$Q^{-}(S) = \left\{x \in X \text{ such that } Q(x) \bigcap S \neq \emptyset\right\}$$

is an open subset in X, for every open subset S of X.

Definition 1.4.13 A multi-valued mapping Q is called to be continuous if it is lower as well as upper semi-continuous on X.

Example 1.4.1 The multi-valued mapping $Q : [0, \pi] \longrightarrow \mathbb{R}^2$ defined by

$$Q(t) = \left\{(x, y) \in \mathbb{R}^2 \text{ such that } x = \lambda \cos(t),\ \ y = \lambda \sin(t),\ \ \lambda \in [-1, 1]\right\}$$

is lower semi-continuous.

In terms of closed sets, we can formulate Definition 1.4.11 as follows.

Lemma 1.4.14 *A multi-valued mapping Q is upper semi-continuous if, and only, if for every closed subset S in X, the set*

$$Q^{-}(S) = \left\{x \in X \text{ such that } Q(x) \bigcap S \neq \emptyset\right\}$$

is closed in X.

Definition 1.4.15 Let $Q : \Omega \longrightarrow \mathscr{P}(X)$ be a multi-valued mapping. We say that

(*i*) Q is closed, if $Gr(Q)$ is closed in $X \times X$.
(*ii*) Q is compact, if $Q(M)$ is a relatively compact subset of X, for any bounded subset M of Ω.

Through the use of sequential conditions, the following conclusions describe the lower semi-continuity and the upper semi-continuity multi-valued mapping. The reader can consult [22, 66, 67] for additional information.

Lemma 1.4.16 *Let X and Y be two metric spaces and let $x \in X$. A multi-valued mapping $Q : X \longrightarrow \mathscr{P}(Y)$ is upper semi-continuous at x if, and only, if for every sequence $(x_n)_n$ in X, which converges to x, and for any open set $V \subset X$ such that $Q(x) \subset V$, there exists $n_0 \in \mathbb{N}$ with $Q(x_n) \subset V$, for each $n \geq n_0$.*

Proof Let $(x_n)_n$ be a convergent sequence of X to a point $x \in X$ and let V be an open subset of X such that $Q(x) \subset V$. Then, there exists an open U_x of X such that

$$Q(U_x) \subset V.$$

Since $\lim_{n\to\infty} x_n = x$, there exists $n_0 \in \mathbb{N}$ such that

$$x_n \in U_x, \quad \text{for each } n \geq n_0.$$

This implies that $Q(x_n) \subset V$, for each $n \geq n_0$. Conversely, suppose that Q is not upper semi-continuous at x. Then, there exists an open subset V of Y containing $Q(x)$ such that, for every open set U_x of X, we have

$$Q(U_x) \not\subset V.$$

Consider the family of balls

$$U_x = \left\{y \in X,\ \text{dist}(x, y) < \frac{1}{n},\ n = 1, 2, \ldots\right\}.$$

Then, there exists a sequence $(x_n)_n$ such that

$$Q(x_n) \not\subset V, \quad \text{for each } n \in \mathbb{N}^*.$$

By using definition of U_x, we get

$$\text{dist}(x_n, x) < \frac{1}{n}.$$

Thus, $\lim_{n\to\infty} x_n = x$. But this contradicts the existence of $n_0 \in \mathbb{N}$ such that

$$Q(x_n) \subset V,$$

for all $n \geq n_0$. Q.E.D.

Lemma 1.4.17 *Let X and Y be two metric spaces and let $x \in X$. A multi-valued mapping $Q : X \longrightarrow \mathscr{P}(Y)$ is lower semi-continuous at a point x if, and only, if for every sequence $(x_n)_n$ of X such that $x_n \to x$, and for any point $y \in Q(x)$, there exists a sequence $(y_n)_n$ of Y with $y_n \in Q(x_n)$ such that $y_n \to y$.*

Example 1.4.2 The multi-valued mapping $Q : \mathbb{R} \longrightarrow \mathscr{P}(\mathbb{R})$ defined by the formula

$$Q(x) = \begin{cases} [-1, 1] & \text{if } x \neq 0 \\ 0 & \text{if } x = 0 \end{cases}$$

is lower semi-continuous at zero, but not upper semi-continuous at zero.

Proposition 1.4.18 *Let X be a Banach space and let $Q : X \longrightarrow \mathscr{P}_p(X)$ be an upper semi-continuous multi-valued mapping. Then, the graph $Gr(Q)$ of Q is a closed subset of $X \times X$.*

Proof Let $x, y \in X$ be such that $y \notin Q(x)$ and choose two open neighborhood V_y and $V_{Q(x)}$ of y and $Q(x)$, respectively in X such that

$$V_y \bigcap V_{Q(x)} = \emptyset.$$

Let $U_x = Q^+(V_{Q(x)})$. Then, U_x is an open neighborhood at x in X and the set $U_x \times V_y$ is an open neighborhood at (x, y). It is easy to see that

$$\left(U_x \times V_y\right) \bigcap Gr(Q) = \emptyset.$$

Indeed, if we take $(x_1, y_1) \in U_x \times V_y$, we get $Q(x_1) \in V_{Q(x)}$ and so, $y_1 \notin V_{Q(x)}$ and $(x_1, y_1) \notin Gr(Q)$, since $V_y \bigcap V_{Q(x)} = \emptyset$. Q.E.D.

Example 1.4.3 The mapping $Q : \mathbb{R} \longrightarrow \mathbb{R}$ defined by the formula

$$Q(x) = \begin{cases} \dfrac{1}{x} & \text{if } x \neq 0 \\ 0 & \text{if } x = 0, \end{cases}$$

has a closed graph, but it is not upper semi-continuous.

Lemma 1.4.19 *Let X be a Banach space and let $Q : X \longrightarrow \mathscr{P}_{cp}(X)$, be an upper semi-continuous multi-valued mapping, where $\mathscr{P}_{cp}(X)$ denote the family of nonempty and compact subsets of X. If S is a compact subset of X, then $Q(S)$ is compact.*

Proof Let $\{V_t,\ t \in I\}$ be an open covering $Q(S)$. Since $Q(x)$ is compact, for all $x \in S$, then $Q(x) \subset W_x$, where W_x is the union of a finite sets V_t, for every $x \in S$. This implies that,

$$Q(S) \subset \bigcup_{x \in S} W_x.$$

Let $U_x = Q^-(W_x),\ x \in S$. Then, the family $\{U_x,\ \ x \in S\}$ is an open covering of S in X. Since S is compact, there exists a finite subcover $U_{x_1}, \ldots, U_{x_n}$ of this covering. Consequently, the sets $W_{x_1}, \ldots, W_{x_n}$ cover $Q(S)$ and since every W_{x_i} is a finite union of sets in $\{V_t\}$, we obtain a finite subcover $V_{t_1}, \ldots, V_{t_k}$ of the covering $\{V_t\}$ and the proof is completed. Q.E.D.

Example 1.4.4 The multi-valued mapping $Q : [0, \pi] \longrightarrow \mathscr{P}_{cp,cv}(\mathbb{R})$ defined by

$$Q(x) = \begin{cases} [\tan(x), 1 + \tan(x)] & \text{if } x \neq \dfrac{\pi}{2} \\ \{0\} & \text{if } x = \dfrac{\pi}{2}, \end{cases}$$

where $\mathscr{P}_{cp,cv}(\mathbb{R})$ denote the family of nonempty, compact and convex subsets of $\mathbb{R}$, is neither upper semi-continuous nor lower semi-continuous, but it has a closed graph, see [68].

Lemma 1.4.20 (see [66]) *Let X be a Banach space and let $Q_1, Q_2 : X \longrightarrow \mathscr{P}_{cp}(X)$ be two upper semi-continuous multi-valued mappings. Then, the composition $Q_1 \circ Q_2$ is also an upper semi-continuous multi-valued mapping with compact values.*

Remark 1.4.21 The composition of two closed multi-valued mapping need not be closed as shown in the following example.

Example 1.4.5 Define the two mapping $Q_1, Q_2 : \mathbb{R} \longrightarrow \mathscr{P}_{cp,cv}(\mathbb{R})$ by putting

$$Q_1(x) = \begin{cases} \left\{\dfrac{1}{x}\right\} & \text{if } x \neq 0 \\ \{0\} & \text{if } x = 0 \end{cases}$$

and

$$Q_2(x) = \begin{cases} \left\{\dfrac{1}{x}\right\} & \text{if } x \neq 0 \\ \{1\} & \text{if } x = 0. \end{cases}$$

Notice that both Q_1 and Q_2 are closed, while the composition

$$Q_2 \circ Q_1(x) = \begin{cases} \{x\} & \text{if } x \neq 0 \\ \{1\} & \text{if } x = 0. \end{cases}$$

is not closed.

Here we now introduce Schauder's fixed point theorem in a multi-valued form. The Kakutani or Bohnenblust–Karlin fixed point theorem are other names for it. In fact, this theorem was proved by Kakutani [12] in the space $\mathbb{R}^n$, and it was extended to normed linear spaces by Bohnenblust and Karlin [69].

Theorem 1.4.22 *Let S be a nonempty, convex, and bounded subset of a normed linear space X and let $Q : S \longrightarrow \mathscr{P}_{cl,cv}(S)$ be an upper semi-continuous multi-valued mapping. Then, Q has a fixed point.*

The following result is an equivalence of Theorem 1.4.22.

Theorem 1.4.23 *Let S be a nonempty, convex, and compact subset of a normed linear space X and let $Q : S \longrightarrow \mathscr{P}_{cp,cv}(S)$ be an upper semi-continuous multi-valued mapping. If $Q(S)$ is compact, then Q has a fixed point.*

Proof Let $\mathscr{K} = \overline{co}(Q(S))$. It follows from Theorem 1.3.34, that $\mathscr{K}$ is a convex and compact subset of X. Then, the multi-valued mapping

$$Q : \mathscr{K} \longrightarrow \mathscr{P}_{cp,cv}(\mathscr{K})$$

is upper semi-continuous and when we apply Theorem 1.4.22, we reach the desired conclusion. Q.E.D.

We need the following definitions in the sequel.

Definition 1.4.24 Let $a, b \in \mathbb{R}$. A multi-valued mapping $Q : [a, b] \longrightarrow \mathscr{P}_{cp}(\mathbb{R})$ is said to be measurable if, for any $y \in X$, the function

$$t \longrightarrow D(y, Q(t)) := \inf \{|y - x|,\ x \in Q(t)\}$$

is measurable.

Let us recall some facts concerning the superposition operators required below.

Definition 1.4.25 Let $a,\ b \in \mathbb{R}$. A multi-valued mapping $Q : [a, b] \times \mathbb{R} \longrightarrow \mathscr{P}_{cp}(\mathbb{R})$ is called to be a Carathéodory if the following assertions are satisfied
$(i)\ t \longrightarrow Q(t, x)$ is measurable, for each $x \in \mathbb{R}$, and
$(ii)\ x \longrightarrow Q(t, x)$ is an upper semi-continuous almost everywhere, for $t \in [a, b]$.

Definition 1.4.26 Let $a,\ b \in \mathbb{R}$. A multi-valued mapping $Q : [a, b] \times \mathbb{R} \longrightarrow \mathscr{P}_{cp}(\mathbb{R})$ is called to be a L^1-Carathéodory if, it is a Carathéodory multi-valued mapping and, for each real number $r > 0$, there exists a function $h_r \in L^1([a, b], \mathbb{R})$ such that

$$\|Q(t, x)\|_{\mathscr{P}} \leq h_r(t)\ a.e.\ t \in [a, b],$$

for all $x \in \mathbb{R}$ with $|x| \leq r$, where $L^1([a, b], \mathbb{R})$ denote the space of Lebesgue integrable real-valued functions on $[a, b]$.

Definition 1.4.27 Let $a,\ b \in \mathbb{R}$. A multi-valued mapping $Q : [a, b] \times \mathbb{R} \longrightarrow \mathscr{P}_{cp}(\mathbb{R})$ is called to be a $L^1_{\mathbb{R}}$-Carathéodory if, Q is a Carathéodory multi-valued mapping and, there exists a function $h \in L^1([a, b], \mathbb{R})$ such that

$$\|Q(t, x)\|_{\mathscr{P}} \leq h(t)\ \ a.e.\ t \in [a, b],$$

for all $x \in \mathbb{R}$. In this case, the function h is called the growth function of Q.

Let $a, b \in \mathbb{R}$. For any multi-valued mapping $Q : [a, b] \times \mathbb{R} \longrightarrow \mathscr{P}_{cp}(\mathbb{R})$, we denote

$$S^1_Q(x) = \{v \in L^1([a, b], \mathbb{R}),\ v(t) \in Q(t, x(t)),\ \ \text{for all } t \in [a, b]\},$$

for some $x \in C([a, b], \mathbb{R})$. The integral of the multi-valued function Q is defined as

$$\int_0^t Q(s, x(s))\, ds = \left\{ \int_0^t v(s)\, ds,\ v \in S^1_Q(x) \right\}.$$

We recall the following lemmas established by Lasota and Opital [70], which will play a crucial role below.

Lemma 1.4.28 *Let X be a finite-dimensional Banach space and let $a, b \in \mathbb{R}$. Assume that $Q : [a, b] \times X \longrightarrow \mathscr{P}_{cp}(X)$ is a L^1-Carathéodory multi-valued mapping. Then, $S^1_Q(x) \neq \emptyset$, for each $x \in X$.*

Lemma 1.4.29 *Let X be a finite-dimensional Banach space and let $a, b \in \mathbb{R}$. Assume that $Q : [a, b] \times X \longrightarrow \mathscr{P}_{cp}(X)$ is a Carathéodory multi-valued mapping with $S^1_Q(x) \neq \emptyset$, and let $\mathscr{L} : L^1([a, b], X) \longrightarrow C([a, b], X)$ be a linear continuous mapping. Then, the multi-valued mapping*

$$\mathscr{L} \circ S^1_Q : C([a, b], X) \longrightarrow \mathscr{P}_{cp,cv}(C([a, b], X))$$

has a closed graph.

Let X be a Banach space and (S, Σ, ν) be a finite measure space. A weakly ν-measurable function $f : S \longrightarrow X$ is said to be Pettis-integrable if, the composition $t \longrightarrow \langle x^*, f(t)\rangle$ is in $L^1(\nu)$, for every $x^* \in X^*$ and if, for every measurable set E there exists an element in X called the Pettis-integral of f over E and denoted by $\int_E f d\nu$ such that

$$\left\langle x^*, \int_E f d\nu \right\rangle = \int_E \langle x^*, f(t)\rangle d\nu(t), \quad \text{for all } x^* \in X^*.$$

For more details, we refer the reader to [71–73] and the original paper [74]. In the case where, ν is the Lebesgue measure in $[0, 1]$, the author B. J. Pettis showed that the space of Pettis-integrable functions $f : S \longrightarrow X$ endowed with norm

$$\|f\| = \sup \left\{ \int_E \langle x^*, f(t)\rangle d\nu(t), \|x^*\| \leq 1 \right\}$$

is not complete, see [74]. For any multi-valued function $Q : [0, \mathscr{T}] \times X \longrightarrow X$ and for each $x \in C([0, \mathscr{T}], X)$, we recall

$$\int_0^t Q(s, x(s))\, ds = \left\{ \int_0^t v(s)\, ds,\ v \text{ is Pettis-integrable},\ v(t) \in Q(t, x(t)) \right\}.$$

In the sequel, we will investigate some properties of the Pettis-integral. The reader may refer to [71, 74–79].

Lemma 1.4.30 *Let $a, b \in \mathbb{R}$. Let X be a Banach space and let $f : [a, b] \longrightarrow X$ be a Pettis-integrable function. Then,*

$$\int_a^b f(s)\, ds \in (b - a)\overline{co}\{f([a, b])\}.$$

Proposition 1.4.31 *Let X be a Banach space and let $a, b \in \mathbb{R}$. Let $f[a, b] \longrightarrow X$ and $h : [a, b] \longrightarrow \mathbb{R}$ be two mappings. If $f(\cdot)$ is Pettis-integrable and $h(\cdot)$ is a measurable and bounded real-valued function, then $h(\cdot)x(\cdot)$ is Pettis-integrable.*

Definition 1.4.32 Let X be a Banach space and let $\mathscr{T} \in \mathbb{R}$. The fractional Pettis-integral of a function $f : [0, \mathscr{T}] \longrightarrow X$ of order $\lambda \in \mathbb{R}_+$ is defined by

$$I^\lambda f(t) = \int_0^t \frac{(t-s)^{\lambda-1}}{\Gamma(\lambda)} f(s)\, ds.$$

Proposition 1.4.33 *Let X be a Banach space and let $\mathscr{T} \in \mathbb{R}$. If $f : [0, \mathscr{T}] \longrightarrow X$ is Riemann integrable, then $I^\lambda f$ exists on $[0, \mathscr{T}]$ and is a fractional Pettis-integral.*

Proof By using our assumptions, we infer that the operator $f(\cdot)$ is Pettis-integrable and so, it is also scalarly Riemann integrable, see [80]. Then, $f(\cdot)$ is weakly measurable and bounded on $[0, \mathscr{T}]$. Besides, it is easy to see that, the real function

$$s \mapsto \frac{(t-s)^{\lambda-1}}{\Gamma(\lambda)}$$

belongs to $L^1([0, t])$, for each $t \in [0, \mathscr{T}]$. By using Proposition 1.4.31, we can deduce that the partial

$$s \mapsto \frac{(t-s)^{\lambda-1}}{\Gamma(\lambda)} f(s)$$

is Pettis-integrable on $[0, t]$, for all $t \in [0, \mathscr{T}]$. Q.E.D.

According to Musial [81], we have the following lemma.

Lemma 1.4.34 *Let X be a Banach space and let $\mathscr{T} \in \mathbb{R}$. Let $f : [0, \mathscr{T}] \longrightarrow X$ be a function such that*
(i) There is a weakly convergent sequence of Pettis-integrable functions $(f_n)_n$ to f,
(ii) There exists $h \in L^1([0, \mathscr{T}])$ such that for each $\varphi \in X^*$ and each $n \in \mathbb{N}$, we have

$$|\langle \varphi, f_n \rangle| \leq h.$$

Then, f is Pettis-integrable and $\displaystyle\int_0^{\mathscr{T}} f_n(s)\, ds$ converges weakly to $\displaystyle\int_0^{\mathscr{T}} f(s)\, ds$.

Definition 1.4.35 Let (X, dist) be a metric space and let $Q : X \longrightarrow \mathscr{P}_p(X)$ be a multi-valued mapping. We say that Q is a multi-valued Lipschitzian mapping, if there exists $k \geq 0$ such that the inequality

$$\text{dist}_H(Qx, Qy) \leq k\, \text{dist}(x, y), \quad \text{for all } x,\ y \in X. \tag{1.4.1}$$

is satisfied, where $\text{dist}_H(\cdot, \cdot)$ is the Hausdorff distance in X. The minimal number k, for which equation (1.4.1) is satisfied is said to be the Lipschitz constant of Q. In particular, if $0 \leq k < 1$, we say that Q is a contraction multi-valued mapping.

Definition 1.4.36 A multi-valued mapping $Q : X \longrightarrow \mathscr{P}(X)$ is $\mathscr{D}$-Lipschitzian if, there exists a continuous and nondecreasing function $\Psi : \mathbb{R}^+ \longrightarrow \mathbb{R}^+$ with $\Psi(0) = 0$ such that

$$\operatorname{dist}_H(Q(u), Q(v)) \leq \Psi\left(\|u - v\|\right), \text{ for all } u, \ v \in X.$$

Specifically, if $\Psi(r) < r$, for $r > 0$, then Q is called to be a nonlinear $\mathscr{D}$-contraction multi-valued.

Example 1.4.6 Let $f : [0, 1] \longrightarrow [0, 1]$ be given by the formula

$$f(x) = \begin{cases} \frac{1}{2}x + \frac{1}{2} & \text{if} \quad 0 \leq x \leq \frac{1}{2} \\ -\frac{1}{2}x + 1 & \text{if} \quad \frac{1}{2} \leq x \leq 1. \end{cases}$$

Let us define the mapping $Q : [0, 1] \longrightarrow \mathscr{P}([0, 1])$ by putting

$$Q(x) = \{0\} \bigcup \{f(x)\}, \quad \text{for all } x \in [0, 1].$$

Notice that Q is a multi-valued contraction mapping and the set of fixed points of Q is $\mathscr{F}_Q = \{0, \frac{2}{3}\}$.

Remark 1.4.37 Let X be a Banach space and let $Q : X \longrightarrow \mathscr{P}_{bd}(X)$ be a multi-valued mapping. If Q is Lipschitzian with a Lipschitz constant k, then the representation of any bounded subset provided by Q is also bounded. Indeed, let S be a bounded subset of X. Then, there exists $r > 0$ such that $\|x\| \leq r$, for all $x \in S$. Since Q is Lipschitzian, then

$$\begin{aligned} \|Q(x)\|_{\mathscr{P}} &\leq \operatorname{dist}_H(Q(x), Q(\theta)) + \operatorname{dist}_H(Q(\theta), 0) \\ &\leq k\|x\| + \|Q(\theta)\|_{\mathscr{P}} \\ &\leq kr + \|Q(\theta)\|_{\mathscr{P}}. \end{aligned}$$

Since $x \in S$ is arbitrary, then $Q(S)$ is a bounded subset of X. For more detail concerning the multi-valued Lipschitzian mapping, see [82].

Remark 1.4.38 Let X be a Banach space and let $A, B \in \mathscr{P}_{cl,bd}(X)$ and $a \in A$. By using the definition of the Hausdorff distance, there exists $q > 0$ and there exists $b \in B$ such that

$$\operatorname{dist}(a, b) \leq \operatorname{dist}_H(A, B) + q.$$

However, there may not be a point $b \in B$ such that

$$\operatorname{dist}(a, b) \leq \operatorname{dist}_H(A, B).$$

Indeed, let ℓ^2 be the Hilbert space of all square-summable sequences of real numbers and choose

$$a = (-1, -\frac{1}{2}, \ldots, -\frac{1}{n}, \ldots).$$

For each $n = 1, 2, \ldots$, let e_n be a vector in ℓ^2 with zeros in all its coordinates except the nth coordinate, which is equal to one. Let

$$\begin{cases} A = (a, e_1, e_2, \ldots, e_n, \ldots) \\ B = (e_1, e_2, \ldots, e_n, \ldots). \end{cases}$$

It is easy to see that

$$\|a - e_n\| = \left(\|a\|^2 + 1 + \frac{2}{n} \right)^{\frac{1}{2}}$$

and

$$\text{dist}_H(A, B) = (\|a\|^2 + 1)^{\frac{1}{2}}.$$

Then, there is no e_n in B such that

$$\|a - e_n\| \leq \text{dist}_H(A, B).$$

Remark 1.4.39 If B is compact, then such a point b does exists, that is, there exists $b \in B$ such that

$$\text{dist}(a, b) \leq \text{dist}_H(A, B).$$

A set-valued version of Banach's fixed point theorem was proved in 1969 by H. Covitz and S. B. Jr. Nadler in the paper [83], utilizing the concept of Hausdorff metric. The set $\mathscr{F}_Q$ of fixed points of a multi-valued mapping Q is then described by the outcome that follows. Please see [22, 83–85] and the reference therein for additional information.

Theorem 1.4.40 *Let X be a Banach space and let $Q : X \longrightarrow \mathscr{P}_{cl}(X)$ be a multi-valued contraction mapping. Then, the set of fixed point $\mathscr{F}_Q$ of Q is a nonempty and closed in X.*

Proof Let $0 \leq k < 1$ be the Lipschitz constant for Q and let $x_0 \in X$ be fixed. Choose any $x_1 \in Q(x_0)$. By the definition of the Hausdorff distance, we can find a point $x_2 \in Q(x_1)$ such that

$$\text{dist}(x_2, x_1) \leq \text{dist}_H(Q(x_1), Q(x_0)) + k.$$

Similarly, we find a point $x_3 \in Q(x_2)$ such that

$$\text{dist}(x_3, x_2) \leq \text{dist}_H(Q(x_2), Q(x_1)) + k^2.$$

Continuing in this fashion, we produce a sequence $x_{n+1} \in Q(x_n)$ such that

$$\mathrm{dist}(x_{n+1}, x_n) \leq \mathrm{dist}_H(Q(x_n), Q(x_{n-1})) + k^n.$$

For fixed integer p, we have

$$\begin{aligned}
\mathrm{dist}(x_{p+1}, x_p) &\leq \mathrm{dist}_H(Q(x_p), Q(x_{p-1})) + k^p \\
&\leq k\ \mathrm{dist}(x_p, x_{p-1}) + k^p \\
&\leq k\left(\mathrm{dist}_H(Q(x_{p-1}), Q(x_{p-2})) + k^{p-1}\right) + k^p \\
&\leq k\left(\mathrm{dist}_H(Q(x_{p-1}), Q(x_{p-2}))\right) + 2k^p.
\end{aligned}$$

Proceeding by induction, we get

$$\mathrm{dist}(x_{p+1}, x_p) \leq k^p\,(\mathrm{dist}(x_0, x_1)) + pk^p.$$

This implies that,

$$\begin{aligned}
\mathrm{dist}(x_n, x_{n+p}) &\leq \sum_{i=n}^{n+p-1} \mathrm{dist}(x_i, x_{i+1}) \\
&\leq \sum_{i=n}^{n+p-1} pk^p + \sum_{i=n}^{n+p-1} k^i \mathrm{dist}(x_0, x_1).
\end{aligned}$$

This shows that $(x_n)_n$ is a Cauchy sequence and thus, it converges to a point $x^* \in X$, since X is complete. Therefore,

$$Qx_n \to Q(x^*).$$

Since $x_{n+1} \in Q(x_n)$, it follows that $x^* \in Q(x^*)$. Q.E.D.

Lemma 1.4.41 *Let* $(X, dist)$ *be a complete metric space and* $Q : X \longrightarrow \mathscr{P}_{cp}(X)$ *be a multi-valued contraction mapping with a constant* k*. Then, the set* $\mathscr{F}_Q$ *is compact.*

Proof Firstly, we prove that the set $\mathscr{F}_Q$ is closed. Let $(x_n)_n$ be a convergent sequence contained in the set $\mathscr{F}_Q$ to a point $x \in X$. Then,

$$\begin{aligned}
\mathrm{dist}(x, Q(x)) &\leq \mathrm{dist}(x, x_n) + \mathrm{dist}(x_n, Q(x)) \\
&\leq \mathrm{dist}(x, x_n) + \mathrm{dist}_H(Q(x_n), Q(x)) \\
&\leq (1+k)\mathrm{dist}(x, x_n).
\end{aligned}$$

Thus, $x \in Q(x)$. In other words, the set $\mathscr{F}_Q$ is closed and so, it is complete. Suppose that $\mathscr{F}_Q$ is not compact. Then, there exists $\delta > 0$ and there exists a sequence $(x_n)_n$ contained in the set $\mathscr{F}_Q$ such that

$$\mathrm{dist}(x_n, x_m) \geq \delta,$$

for any two different integers n and m. Let

$$\rho = \inf\left\{r > 0, \exists a \in X, \mathscr{B}_X(a, r) \text{ contains infinitely } x_n\right\}.$$

Since, for every $a \in X$, the ball $\mathscr{B}_X(a, \frac{\delta}{2})$ contains at most x_n, one has $\rho \geq \frac{\delta}{2}$. Fix $\varepsilon > 0$ with $\varepsilon < \rho\frac{1-k}{1+k}$ and choose $a \in X$ such that the set

$$J = \{n \in \mathbb{N},\ x_n \in \mathscr{B}_X(a, \rho + \varepsilon)\}$$

is infinite. For each $n \in J$, we have

$$\begin{aligned} \operatorname{dist}(x_n, Q(a)) &\leq \operatorname{dist}_H(Q(x_n), Q(a)) \\ &\leq k\operatorname{dist}(x_n, a) \\ &< k(\rho + \varepsilon). \end{aligned}$$

Moreover, we can choose a sequence $y_n \in Q(a)$ such that

$$\operatorname{dist}(x_n, y_n) < k(\rho + \varepsilon).$$

Since the set $Q(a)$ is compact, there exists $b \in Q(a)$ such that the set

$$J' = \{n \in J,\ \operatorname{dist}(y_n, b) < \varepsilon\}$$

is infinite. Then, for each $n \in J'$, we have

$$\begin{aligned} \operatorname{dist}(x_n, b) &< k(\rho + \varepsilon) \\ &\leq k\rho + \varepsilon(1 + \rho) \\ &< \rho, \end{aligned}$$

but this contradicts the definition of ρ, since $\mathscr{B}_X(b, k\rho + \varepsilon(1 + \rho))$ contains infinitely many x_n. Q.E.D.

Lemma 1.4.42 *Let X be a Banach space and let $Q_1, Q_2 : X \longrightarrow \mathscr{P}_{cl}(X)$ be multi-valued mappings. If Q_1 and Q_2 are two contractions with the same constant λ, then the following inequality is satisfied*

$$dist_H(\mathscr{F}_{Q_1}, \mathscr{F}_{Q_2}) \leq \frac{1}{1-\lambda}\sup_{x\in X}\left\{dist_H(Q_1(x), Q_2(x))\right\}.$$

Proof We may assume that $K = \sup\limits_{x\in X}\{\operatorname{dist}_H(Q_1(x), Q_2(x))\} < \infty$. Let $\varepsilon > 0$ and choose $p > 0$ such that

$$p\sum_{n\geq 1} n\lambda^n < 1.$$

Choose $\varepsilon_1 = \dfrac{p\varepsilon}{1-\lambda}$ and take $x_0 \in \mathscr{F}_{Q_1}$. Then, there exists $x_1 \in Q_2(x_0)$ such that

$$\mathrm{dist}(x_0, x_1) \le K + \varepsilon.$$

By induction, we can define a sequence $(x_n)_n$ such that $x_{n+1} \in Q_2(x_n)$ and

$$\begin{aligned}\mathrm{dist}(x_{n+1}, x_n) &\le \lambda \mathrm{dist}(x_n, x_{n-1}) + \lambda^n \varepsilon_1 \\ &\le \lambda \left(\lambda \mathrm{dist}(x_{n-1}, x_{n-2}) + \lambda^{n-1}\varepsilon_1\right) + \lambda^n \varepsilon_1 \\ &\le \lambda^2 \left(\lambda \mathrm{dist}(x_{n-2}, x_{n-3}) + \lambda^{n-2}\varepsilon_1\right) + 2\lambda^n \varepsilon_1.\end{aligned}$$

This implies that

$$\mathrm{dist}(x_{n+1}, x_n) \le \lambda^n \mathrm{dist}(x_1, x_0) + n\lambda^n \varepsilon_1,$$

and consequently, for all $n \in \mathbb{N}$, we have

$$\sum_{n\ge k} \mathrm{dist}(x_{n+1}, x_n) \le \frac{\lambda^k}{1-\lambda}\mathrm{dist}(x_1, x_0) + \sum_{n\ge k} n\lambda^n \varepsilon_1.$$

This show that $(x_n)_n$ is a Cauchy sequence with a limit x^*. The continuity of Q_2 guaranties that

$$Q_2(x_n) \to Q_2(x^*).$$

It follows $x_{n+1} \in Q_2(x_n)$, that $x^* \in Q_2(x^*)$ and so, $x^* \in \mathscr{F}_Q$. Furthermore, for all $n \in \mathbb{N}$, we have

$$\begin{aligned}\mathrm{dist}(x_n, x^*) &\le \sum_{n\ge 0} \mathrm{dist}(x_{n+1}, x_n) \\ &\le \frac{1}{1-\lambda}\mathrm{dist}(x_1, x_0) + \sum_{n\ge k} n\lambda^n \varepsilon_1 \\ &\le \frac{1}{1-\lambda}\left(\mathrm{dist}(x_1, x_0) + \varepsilon\right) \\ &\le \frac{1}{1-\lambda}\left(K + \varepsilon\right).\end{aligned}$$

This implies, by symmetry, that

$$\mathrm{dist}_H(Q_1(x), Q_2(x)) \le \frac{K}{1-\lambda}. \qquad \text{Q.E.D}$$

The equivalent theory of nonexpansive single-valued is far simpler and easier to understand than the fixed point theory of nonexpansive multi-valued mappings. The following generalization is obtained by T. C. Lim. For the proof, we refer to [86] and also [87–89].

Lemma 1.4.43 *Let S be a nonempty, closed, bounded, and convex subset of an uniformly convex Banach space X and let $Q : S \longrightarrow \mathscr{P}_{cp}(X)$ be a multi-valued mapping. If Q is nonexpansive, then Q has, at least, one fixed point.*

W. A. Kirk and S. Massa developed another significant theory for nonexpansive multi-valued mappings in [90].

Lemma 1.4.44 *Let S be a nonempty, closed, bounded, and convex subset of a Banach space X and let $Q : S \longrightarrow \mathscr{P}_{cp,cv}(X)$ be a nonexpansive multi-valued mapping. If the asymptotic center in S of each bounded sequence of X is nonempty and compact, then Q has a fixed point.*

Remark 1.4.45 In a nearly uniformly convex Banach space [91], the asymptotic center of a bounded sequence is not always compact, hence Lemma 1.4.44 not applicable in this context [92, 93].

The following result plays the role of a prototype in our discussion of this book. You can find a proof in [14].

Theorem 1.4.46 *Let S be a nonempty, closed, and convex subset of a Banach space X and let $Q : S \longrightarrow \mathscr{P}_{cl,cv}(S)$ be a compact multi-valued mapping. If Q is upper semi-continuous on S, then there exists $x \in S$ such that $x \in Q(x)$.*

In [94], L. Rybinski has proved the following result, which will be useful in the sequel.

Theorem 1.4.47 *Let S be a nonempty and closed subset of a Banach space X and let Y be a metric space. Assume that $Q : S \times Y \longrightarrow \mathscr{P}_{cl,cv}(S)$ is a multi-valued mapping satisfying the following conditions*
(i) For all $x_1, x_2 \in S$, there exists $q \in [0, 1)$ such that

$$dist_H(Q(x_1, y), Q(x_2, y)) \leq q\|x_1 - x_2\|,$$

whenever $y \in Y$,
(ii) The partial $y \mapsto Q(x, y)$ is lower semi-continuous on Y, for every $x \in S$.

Then, there exists a continuous function $f : S \times Y \longrightarrow S$ such that $f(x, y) \in Q(f(x, y), y)$ for each $(x, y) \in S \times Y$.

Dhage [82] has proved the following fixed point theorem involving three multi-valued mappings.

Theorem 1.4.48 *Let S be a nonempty, closed, convex, and bounded subset of a Banach algebra X and let $A, C : X \longrightarrow \mathscr{P}_{bd,cl,cv}(X)$ and $B : S \longrightarrow \mathscr{P}_{cp,cv}(X)$, where $\mathscr{P}_{bd,cl,cv}(X)$ denote the family of nonempty, bounded, closed, and convex subsets of X. be three multi-valued mappings such that*
(i) A and C are Lipschitzians with constants q_1 and q_2, respectively,
(ii) B is lower semi-continuous and compact,
(iii) $Ax \cdot By + Cx \in \mathscr{P}_{cl,cv}(X)$, for all $x, y \in S$,
(iv) $q_1 M + q_2 < 1$, where $M = \|B(S)\|_{\mathscr{P}} = \sup\{\|Bx\|_{\mathscr{P}}, x \in S\}$.

Then, the operator inclusion $x \in Ax \cdot Bx + Cx$ has a solution in S.

Proof Define the multi-valued mapping $Q : S \times S \longrightarrow \mathscr{P}_{cl,cv}(S)$ by putting

$$Q(x, y) = Ax \cdot By + Cx.$$

The use of definition of the Hausdorff distance as well as Remark 1.4.37 leads to

$$\begin{aligned}\operatorname{dist}_H\left(Q(x, y), Q(x', y)\right. &\leq M\operatorname{dist}_H(Ax, Ax') + \operatorname{dist}_H(Cx, Cx')\\ &\leq q_1 M\|x - x'\| + q_2\|x - x'\|.\end{aligned}$$

Since $q_1 M + q_2 < 1$, we can apply Theorem 1.4.40, in order to get $\mathscr{F}_{Q_y}$ is a nonempty and closed subset of S, for each $y \in S$, where $Q_y : X \longrightarrow \mathscr{P}(X)$ is defined by the formula

$$Q_y(x) = Ax \cdot By + Cx,$$

here, $\mathscr{P}(X)$ denote the family of nonempty subsets of X. Hence, an application of Theorem 1.4.47 yields that, there exists a continuous function f such that

$$f(x, y) \in Q(f(x, y), y) = A\left(f(x, y)\right) \cdot By + C\left(f(x, y)\right),$$

for each $(x, y) \in S \times S$. By using the continuity of the composition operator $f \circ \varphi = h$, where $\varphi : S \longrightarrow S \times S$ is defined by the formula $\varphi(x) = (x, x)$, we get

$$h(x) = f(x, x) \in A\left(f(x, x)\right) \cdot Bx + C\left(f(x, x)\right).$$

We claim that h is compact on S. To do this, it is sufficient to show that the multi-valued mapping $H : S \longrightarrow \mathscr{P}_{cl}(S)$ given by the formula

$$H(y) = \mathscr{F}_{Q_y}$$

is compact. Taking into account that A is Lipschitzian and using Remark 1.4.37, we infer that there exists a constant $\delta > 0$ such that $\|Ax\| \leq \delta$, for each $x \in S$. Let $\varepsilon > 0$. Since $B(S)$ is totally bounded, then there exists $Y = \{y_1, y_2, \ldots, y_n\} \subset S$ such that

$$B(S) \subseteq \bigcup_{i=1}^{n} \mathscr{B}_X(w_i, r)$$

where $r = \left(\dfrac{1 - (q_1 M + q_2)}{\delta}\right)\varepsilon$ and $w_i \in By_i$, for each $i = 1, 2, \ldots, n$. So, for any $y \in S$, there exists an element $y_k \in Y$ such that

$$\operatorname{dist}_H(By_k, By) < \left(\frac{1 - (q_1 M + q_2)}{\delta}\right)\varepsilon.$$

This implies that,

$$\begin{aligned}\operatorname{dist}_H(H(y_k), H(y)) &= \operatorname{dist}_H(\mathscr{F}_{Q_{y_k}}, \mathscr{F}_{Q_y})\\ &\leq \frac{1}{1-(q_1M+q_2)} \sup_{y\in Y} \operatorname{dist}_H(Ax\cdot By + Cx, Ax\cdot By_k + Cx)\\ &< \frac{\delta}{1-(q_1M+q_2)} \frac{1-(q_1M+q_2)}{\delta}\varepsilon.\end{aligned}$$

Thus, for each $u \in H(y)$, there exits an element $v_k \in H(y_k)$ such that

$$\|u - v_k\| < \varepsilon.$$

This means, in particular, that

$$h(S) \subset H(S) \subset \bigcup_{i=1}^{n} \mathscr{B}_X(v_i, \varepsilon).$$

This completes the proof. Q.E.D.

Definition 1.4.49 Let X be a Banach space and let $Q : \Omega \longrightarrow \mathscr{P}(X)$ be a multi-valued mapping. We say that

- (i) Q is weakly compact, if $Q(M)$ is a relatively weakly compact subset of X, for any bounded subset M of Ω.
- (ii) Q is sequentially weakly upper semi-compact in Ω (s.w.u.sco. for short) if, for any weakly convergent sequence $(x_n)_n$ of elements in Ω and, for any $y_n \in Q(x_n)$, the sequence $(y_n)_n$ has a weakly convergent subsequence. In particular, if Q is a single-valued operator, then we say that Q is sequentially weakly upper semi-compact if, for any weakly convergent sequence $(x_n)_n$, the sequence $(Q(x_n))_n$ has a weakly convergent subsequence.

Remark 1.4.50 Every weakly sequentially continuous single-valued mapping is sequentially weakly upper semi-compact. Moreover, if a multi-valued mapping Q is sequentially weakly upper semi-compact, then $Q(M)$ is relatively weakly compact, for any relatively weakly compact subset M of X.

Definition 1.4.51 Let X be a Banach space and let $Q : \Omega \longrightarrow \mathscr{P}(X)$ be a multi-valued mapping. We say that Q has a weakly closed graph in $\Omega \times X$, if for every weakly convergent sequence $(x_n)_n$ contained in the set Ω to a point $x \in \Omega$ and, for every sequence $y_n \in Qx_n$ with $y_n \to y$, we have $Qx \bigcap L(x, y) \neq \emptyset$, where

$$L(x, y) := \{\lambda y + (1-\lambda)x,\ \lambda \in [0, 1]\}. \tag{1.4.2}$$

Definition 1.4.52 Let X be a Banach space and let $Q : \Omega \longrightarrow \mathscr{P}(X)$ be a multi-valued mapping. We say that

(i) Q has w-weakly closed graph in $\Omega \times X$, if it has weakly closed graph in $\Omega \times X$ with respect to the weak topology,
(ii) Q has a weakly sequentially closed graph if, for every sequence $(x_n)_n$ contained in the set Ω with $x_n \rightharpoonup x$ in Ω and for every sequence $y_n \in Q(x_n)$ such that $y_n \rightharpoonup y$ in X, we have $y \in Q(x)$.
(iii) Q is hemi-weakly compact if each sequence $(x_n)_n$ has a weakly convergent subsequence, whenever there exists a sequence $(y_n)_n$, with $y_n \in Q(x_n)$ such that the sequence $(x_n - y_n)_n$ is weakly convergent.

Definition 1.4.53 ([95]) Let S be a nonempty subset of a Banach space X and let $Q : S \longrightarrow \mathscr{P}(Y)$ be a multi-valued mapping, where X and Y are two topological vector spaces. The graph $Gr(Q)$ is convex if the set S is convex and

$$\lambda Q(x) + (1-\lambda)Q(y) \subseteq Q(\lambda x + (1-\lambda)y),$$

for all $x, y \in S$ and $\lambda \in (0, 1)$.

The following result, which were established in [21], play the role of prototype in our discussion of this book.

Theorem 1.4.54 *Let S be a nonempty, closed, and convex subset of a Banach space X and let $Q_1, Q_2 : S \longrightarrow \mathscr{P}(X)$ be two multi-valued mappings satisfying the following conditions*
(i) Q_1 is weakly completely continuous.
(ii) Q_2 is hemi-weakly compact with weakly sequentially closed graph.
(iii) there exists a continuous and nondecreasing function $\phi : \mathbb{R}_+ \longrightarrow \mathbb{R}_+$ such that

$$\beta(Q(M) \leq \phi\left(\beta(M)\right),$$

for any bounded subset M of S,
(iv) there exists a bounded subset S' of S and there exists a sequence $(\lambda_n)_n$ contained in the set $[0, 1]$ with $\lambda_n \to 1$ such that

$$(Q_1 + \lambda_n Q_2)(x) \in \mathscr{P}_{cv}(S'),$$

for all $x \in S$.
Then, $Q_1 + Q_2$ has a fixed point in S.

In order to prove some fixed point theorems involving three multi-valued mappings in Banach algebras satisfying the sequential condition $(\mathscr{P})$, the authors in [96] have introduced the concept of multi-valued mappings of the form $\frac{I-C}{A}$, where A and C define two multi-valued mappings acting on Banach algebras.

Definition 1.4.55 Let X be a Banach algebra satisfying the sequential condition ($\mathscr{P}$) and let $A, C : X \longrightarrow \mathscr{P}(X)$ be two multi-valued mappings. The multi-valued mapping $\frac{I-C}{A}$ is well defined on $x \in X$ and we write

$$y \in \left(\frac{I-C}{A}\right)(x)$$

if $x \in yA(x) + C(x)$.

Now, we give the final result of this section. A proof can be found in [97].

Theorem 1.4.56 *Let S be a nonempty, closed, and convex subset of a Banach space X and let $Q : S \longrightarrow \mathscr{P}_{cv}(S)$ be a multi-valued mapping with a weakly sequentially closed graph. If $Q(S)$ is relatively weakly compact, then Q has a fixed point.*

Proof Let $\mathscr{K} = \overline{co}(Q(S))$. It follows from Theorem 1.3.34, that $\mathscr{K}$ is a weakly compact subset of X. This implies that the multi-valued mapping

$$Q : \mathscr{K} \longrightarrow \mathscr{P}_{cv}(\mathscr{K})$$

has a weakly sequentially closed graph and $Q(x)$ is weakly closed, for all $x \in \mathscr{K}$. Because X equipped with its weak topology is a Hausdorff locally convex space, then the mapping Q is upper semi-continuous on $\mathscr{K}$ with respect to the weak topologies. This leads to the expected conclusion when we use Theorem 1.4.22. Q.E.D.

1.5 Measure of Non-compactness and Measure of Weak Non-compactness

Assume that X is a given Banach space with norm $\|\cdot\|$. Functional analysis and operator theory both benefit greatly from the theory of measures of non-compactness, see for instance, [98–101]. In this section, we provide a summary of three key measures of non-compactness that are frequently encountered in applications.

Definition 1.5.1 A function $\beta : \mathscr{B}(X) \longrightarrow [0, +\infty[$ is said to be a measure of weak non-compactness (MNWC, for short), if it satisfies the following conditions

1. The family $\ker(\beta) = \{S \in \mathscr{B}(X),\ \beta(S) = 0\}$ is nonempty and $\ker(\beta) \subset \mathscr{W}(X)$, where $\mathscr{W}(X)$ denote the family of all relatively weakly compact subset of X.
2. For all $S_1, S_2 \in \mathscr{B}(X)$ with $S_1 \subset S_2$, we have $\beta(S_1) \leq \beta(S_2)$.
3. $\beta(\overline{co}(S)) = \beta(S)$.
4. $\beta(\lambda S_1 + (1-\lambda)S_2) \leq \lambda\beta(S_1) + (1-\lambda)\beta(S_2)$, for $\lambda \in [0, 1]$.
5. If $(S_n)_{n\geq 1}$ is a decreasing sequence of nonempty, and weakly closed subsets of X with S_1 is bounded and $\lim\limits_{n\to\infty} \beta(S_n) = 0$, then the set $S_\infty = \bigcap\limits_{n\geq 1} S_n$ is nonempty.

The family $\ker(\beta)$ is called to be the kernel of the measure β. Notice that the intersection set S_∞ belongs to $\ker(\beta)$, since $\beta(S_\infty) \leq \beta(S_n)$, for every n and

$$\lim_{n\to\infty} \beta(S_n) = 0.$$

Besides, it can be verified that $\beta(\overline{S}^w) = \beta(S)$. Further, if a measure of weak non-compactness β satisfies the condition

$$\beta(S) = 0 \Leftrightarrow S \in \mathscr{W}(X),$$

then β is called to be a regular measure of weak non-compactness. Moreover, if β satisfies the condition

$$\beta(S_1 + S_2) \leq \beta(S_1) + \beta(S_2),$$

it is called to be sub-additive, and if the equality

$$\beta(\lambda S) = |\lambda|\beta(S), \quad \text{for all } \lambda \in \mathbb{R}$$

is satisfied, then β is called to be homogeneous. Therefore, We say that β has the maximum property if

$$\beta(S_1 \bigcup S_2) = \max(\beta(S_1), \beta(S_2)).$$

An important example of an MNWC was defined by De Blasi [101] in the following way

$$\beta(S) = \inf\Big\{r > 0, \text{ there exists } K \in \mathscr{W}(X) \text{ such that } S \subseteq K + \overline{\mathscr{B}}_X(r)\Big\}.$$

Note that the De Blasi measure of weak non-compactness β is regular, homogeneous, sub-additive, and has the maximum property, see [101, 102]. Let Ω be a nonempty subset of a Banach space X and let $T : \Omega \longrightarrow X$ be a single-valued mapping. If $T(S)$ is bounded and the following inequality

$$\beta(T(S)) < \beta(S)$$

is satisfied, for any bounded subset S of Ω with $\beta(S) > 0$, then T is said to be condensing with respect to β. In order to recall the definition of convex-power condensing operator, we give some notations. Let Ω be a nonempty, closed, and convex subset of a Banach space X, $T : \Omega \longrightarrow \Omega$ and $x_0 \in \Omega$. For any subset S of Ω, let

$$\begin{cases} T^{(1,x_0)}(S) = T(S) \\ T^{(n,x_0)}(S) = T\left(\overline{co}\left\{T^{(n-1,x_0)}(S) \bigcup \{x_0\}\right\}\right). \end{cases}$$

Definition 1.5.2 An operator T is said to be convex-power condensing, if T is bounded and there exists a point $x_0 \in \Omega$ and a strictly positive integer n_0 such that the following estimate

$$\beta(T^{(n_0,x_0)}(S)) < \beta(S)$$

is satisfied, for any bounded subset S of Ω with $\beta(S) > 0$

Remark 1.5.3 If $\beta(T^{(n_0,x_0)}(S)) \geq \beta(S)$, then S is a relatively weakly compact subset of X. Further, every condensing mapping with respect to β is convex-power condensing, but the reverse implication may not hold.

Theorem 1.5.4 *Let S be a nonempty, closed, and convex subset of a Banach space X and let $T : S \longrightarrow S$ be a weakly sequentially continuous operator. If T is convex-power condensing with respect to β such that $T(S)$ is bounded, then T has, at least, one fixed point.*

Proof Since T is convex-power condensing, then there exists $x_0 \in S$ and a positive integer n_0 such that

$$\beta\left(T^{(n_0,x_0)}(V)\right) \leq \beta(V),$$

for any bounded subset V of X with $\beta(V) > 0$. Define the set

$$\Gamma = \{V \subset S,\ x_0 \in V,\ V \text{ is closed and convex and } T(V) \subset V\}.$$

Since $S \in \Gamma$, it follows that $\Gamma \neq \emptyset$. Choose $V_0 = \bigcap_{V \in \Gamma} V$. Then

$$\begin{cases} x_0 \in V_0 \subset S \\ V_0 \text{ is a closed and convex subset of } S \\ T(V_0) \subset V_0. \end{cases}$$

Since V_0 is a closed and convex subset of S, then

$$\overline{co}\{T(V_0) \bigcup \{x_0\}\} \subset V_0.$$

Consequently,

$$\begin{aligned} T^{(2,x_0)}(V_0) &= T\left(\overline{co}\{T(V_0) \bigcup \{x_0\}\}\right) \\ &\subset T(V_0) \\ &\subset \overline{co}\{T(V_0) \bigcup \{x_0\}\}. \end{aligned}$$

This yields that,

$$V_0 = \overline{co}\left\{T(V_0) \bigcup \{x_0\}\right\}.$$

By induction, one has

$$\overline{co}\left\{T^{(n_0,x_0)}(V_0) \bigcup \{x_0\}\right\} \subset \overline{co}\left\{T^{(n_0-1,x_0)}(V_0) \bigcup \{x_0\}\right\}.$$

This implies that

$$\begin{aligned} T\left(\overline{co}\left\{T^{(n_0,x_0)}(V_0)\bigcup\{x_0\}\right\}\right) &\subset T\left(\overline{co}\left\{T^{(n_0-1,x_0)}(V_0)\bigcup\{x_0\}\right\}\right) \\ &\subset T^{(n_0,x_0)} \\ &\subset \overline{co}\left\{T^{(n_0,x_0)}(V_0)\bigcup\{x_0\}\right\}. \end{aligned}$$

This means, in particular, that

$$\overline{co}\left\{T^{(n_0,x_0)}(V_0)\bigcup\{x_0\}\right\} \in \Gamma,$$

and so

$$V_0 \subset \overline{co}\left\{T^{(n_0,x_0)}(V_0)\bigcup\{x_0\}\right\}.$$

Using similar argument, we can deduce that

$$\overline{co}\left\{T^{(n_0,x_0)}(V_0)\bigcup\{x_0\}\right\} \subset V_0,$$

and then

$$\begin{aligned} V_0 &= \overline{co}\left\{T^{(n_0,x_0)}(V_0)\bigcup\{x_0\}\right\} \\ &= \overline{co}\left\{T(V_0)\bigcup\{x_0\}\right\}. \end{aligned}$$

Due to the fact that $T(S)$ is bounded, we infer that V_0 is a bounded, closed, and convex subset of S. If $\beta(V_0) > 0$, we get

$$\begin{aligned} \beta(V_0) &= \beta\left(\overline{co}\left\{T^{(n_0,x_0)}(V_0)\bigcup\{x_0\}\right\}\right) \\ &\leq \beta\left(\overline{co}\left\{T^{(n_0,x_0)}(V_0)\right\}\right) \\ &< \beta(V_0), \end{aligned}$$

which is a contradiction, and so $\overline{V_0}^w = V_0$ is a weakly compact of S. Thus, the operator $T : V_0 \longrightarrow V_0$ is weakly continuous and when we apply Theorem 1.2.42, we obtain T has, at least, one fixed point in V_0. Q.E.D.

Definition 1.5.5 Let X be a Banach space and let $T : X \longrightarrow X$ be a mapping. We say that T is ww-compact if, it is continuous and for any weakly convergent sequence $(u_n)_n$ of X, the sequence $(Tu_n)_n$ has a weakly convergent subsequence of X.

Remark 1.5.6 It is worth noting that ww-compact mappings are not necessarily weakly sequentially continuous. Moreover, the concepts of ww-compact and strongly continuous mappings arise naturally in the study of integral and partial differential equations. For more details, the reader may refer to [11, 14, 103, 104].

Proposition 1.5.7 *Let X be a Banach space and let $T : X \longrightarrow X$ be a $\mathscr{D}$-Lipschitzian mapping with $\mathscr{D}$-function ϕ. If T is ww-compact, then*

$$\beta(T(M)) \leq \phi(\beta(M)),$$

for any bounded subset M of X with $\beta(M) > 0$.

Proof Let M be a bounded subset of X with $0 < \beta(M) < r$. By the definition of β, we get there exists $0 \leq r_0 < r$ and there exists a relatively weakly compact subset K of X such that

$$M \subset K + \overline{\mathscr{B}}_X(r_0).$$

Since T is a $\mathscr{D}$-Lipschitzian mapping with $\mathscr{D}$-function ϕ, then

$$\begin{aligned} TM &\subset TK + \overline{\mathscr{B}}_X(\phi(r_0)) \\ &\subset \overline{TK}^w + \overline{\mathscr{B}}_X(\phi(r_0)). \end{aligned}$$

Taking into account that $\overline{TK}^w$ is a weakly compact subset of X, and using the subadditivity of the De Blasi measure, we arrive at the following

$$\begin{aligned} \beta(TM) &\leq \phi(r_0) \\ &\leq \phi(r). \end{aligned}$$

Using the continuity of ϕ, we get

$$\beta(TM) \leq \phi(\beta(M)). \qquad \text{Q.E.D}$$

Definition 1.5.8 Let S be a nonempty subset of a Banach space X and $T : S \longrightarrow X$ be a mapping. We say that T is
(i) countably $\mathscr{D}$-set-Lipschitzian if there exists a continuous and nondecreasing function $\psi : \mathbb{R}_+ \longrightarrow \mathbb{R}_+$ such that, for any countable bounded subset M of S with $T(M)$ being bounded, we have

$$\beta(T(M)) \leq \psi(\beta(M)).$$

When $\psi(r) = kr,\ k > 0$, T is called to be a countably k-set-Lipschitzian mapping.
(ii) countably β-nonexpansive, if T is bounded and, for any countable bounded subset M of S with $\beta(M) > 0$, we have

$$\beta(T(M)) \leq \beta(M).$$

(iii) countably condensing with respect to β, if for any countable bounded subset M of S with $\beta(M) > 0$, we have

$$\beta(T(M)) < \beta(M).$$

Lemma 1.5.9 *Let X be a Banach space and let $T : X \longrightarrow X$ be a weakly sequentially continuous mapping. If T is Lipschitzian with a Lipschitz constant $q \geq 0$, then*

$$\beta(T(V)) \leq q\beta(V),$$

for any bounded subset V of X.

Proof Let V be a bounded subset of X with $\beta(V) > 0$ and let $\varepsilon > 0$. It follows from the definition of β, that there exists a relatively weakly compact subset K of X such that

$$V \subset K + \left(\beta(V) + \frac{\varepsilon}{q}\right)\overline{\mathscr{B}}_X.$$

Then,

$$T(V) \subsetneq T(K) + (q\beta(V) + \varepsilon)\overline{\mathscr{B}}_X.$$

In other words, for $x \in V$ be fixed, there exists an element $y \in K$ and there exists an element $z \in \overline{\mathscr{B}}_X$ such that

$$x = y + \left(\beta(V) + \frac{\varepsilon}{q}\right)z.$$

Since T is Lipschitzian, then

$$\begin{aligned}\|Tx - Ty\| &\leq q\|x - y\| \\ &\leq q\beta(V) + \varepsilon.\end{aligned}$$

This means that

$$Tx \in T(K) + (q\beta(V) + \varepsilon)\overline{\mathscr{B}}_X.$$

If we consider the weak sequential continuity of T and deploy Theorem 1.3.44, we get $T(K)$ is weakly compact, and so

$$\beta(T(K)) \leq q\beta(V) + \varepsilon.$$ Q.E.D

In the sequel, we will use the following result.

Lemma 1.5.10 *Let X be a Banach algebra satisfying the sequential condition $(\mathscr{P})$. If $V \in \mathscr{B}(X)$ and $S \in \mathscr{W}(X)$, then*

$$\beta(V \cdot S) \leq \|S\|\beta(V).$$

Proof Let $\varepsilon > 0$. By using the definition of β, we get there exists a relatively weakly compact subset K of X such that

$$V \subset K + \left(\beta(V) + \frac{\varepsilon}{\|S\|}\right)\overline{\mathscr{B}}_X.$$

This implies that

$$V \cdot S \subset K \cdot S + \left(\beta(V) + \frac{\varepsilon}{\|S\|}\right) \|S\| \overline{\mathscr{B}}_X.$$

Using Lemma 1.3.62, we obtain

$$\beta(V \cdot S) \leq \|S\| \beta(V) + \varepsilon. \qquad \text{Q.E.D}$$

The following lemma is proved in [60].

Lemma 1.5.11 *Let V and S be two nonempty and bounded subsets of a Banach algebra X satisfying the sequential condition $(\mathscr{P})$. Then,*

$$\beta(V \cdot S) \leq \|S\| \beta(V) + \|V\| \beta(S) + \beta(V) \beta(S).$$

Proof Let V and S be two nonempty and bounded subsets of X with $r_1 > \beta(V) > 0$ and $r_2 > \beta(S) > 0$. By the definition of the De Blasi measure β, we can find two weakly compact subsets K_1 and K_2 of X such that

$$V \subseteq K_1 + \overline{\mathscr{B}}_X(r_1) \ \text{ and } \ S \subseteq K_2 + \overline{\mathscr{B}}_X(r_2).$$

Let $z \in V \cdot S$, then z can be represented in the form

$$z = x \cdot y,$$

for some $x \in V$ and $y \in S$. From the above discussion, we get there exist $k_1 \in K_1$, $k_2 \in K_2$, $b_1 \in \overline{\mathscr{B}}_X(r_1)$ and $b_2 \in \overline{\mathscr{B}}_X(r_2)$ such that

$$\begin{aligned} z &= (k_1 + b_1) \cdot (k_2 + b_2) \\ &= k_1 k_2 + k_1 b_2 + b_1 k_1 + b_1 b_2 \\ &= k_1 k_2 + x b_2 + b_1 y - b_1 b_2. \end{aligned}$$

This implies that

$$\begin{aligned} V \cdot S &\subseteq K_1 \cdot K_2 + V \cdot \overline{\mathscr{B}}_X(r_2) + \overline{\mathscr{B}}_X(r_1) \cdot S + \overline{\mathscr{B}}_X(r_1) \cdot \overline{\mathscr{B}}_X(r_2) \\ &\subseteq K_1 \cdot K_2 + \overline{\mathscr{B}}_X\left(\|V\| r_2 + \|S\| r_1 + r_1 r_2\right). \end{aligned}$$

Making use of the subadditivity of β, together with Lemma 1.3.62, we acquire that

$$\beta\left(V \cdot S\right) \leq \|V\| r_2 + \|S\| r_1 + r_1 r_2.$$

Letting $r_1 \to \beta(V)$ and $r_2 \to \beta(S)$, we get

$$\beta(V \cdot S) \leq \|S\|\beta(V) + \|V\|\beta(S) + \beta(V)\beta(S).$$

The desired result is deduced. Q.E.D.

Theorem 1.5.12 *Let S be a nonempty, closed, and convex subset of a Banach space X and let $T : S \longrightarrow S$ be a weakly sequentially continuous mapping such that $T(S)$ is a bounded subset of S. If T is condensing with respect to β, then T has a fixed point in S.*

Proof Fix an arbitrary $x_0 \in S$ and consider the sequence $(S_n)_n$, consisting of subsets such that

$$S_0 = \{x_0\} \text{ and } S_{n+1} = co\left(\{x_0\} \bigcup T(S_n)\right), \quad \text{for } n \in \mathbb{N}^*.$$

It is easy to see that $(S_n)_n$ is a increasing sequence of nonempty and convex subsets of X and the set $S_\infty = \bigcup\limits_{n \geq 0} S_n$ is convex, since $S_n \subset S_{n+1}$.
It follows from the inclusion $T(S_n) \subset S_\infty$, for all $n \in \mathbb{N}$, that

$$\begin{aligned} T(S_\infty) &\subset \bigcup_{n\geq 0} T(S_n) \\ &\subset S_\infty. \end{aligned}$$

This implies that

$$\begin{aligned} S_\infty &= \bigcup_{n\geq 0} co\left(\{x_0\} \bigcup T(S_n)\right) \\ &\subseteq co\left(\{x_0\} \bigcup T(S_\infty)\right). \end{aligned}$$

Therefore, S_∞ is a nonempty, bounded, and convex subset of X. If $\beta(S_\infty) > 0$, we obtain

$$\begin{aligned} \beta(S_\infty) &= \beta(T(S_\infty)) \\ &< \beta(S_\infty), \end{aligned}$$

which is a contradiction and so, the set S_∞ is relatively weakly compact and convex. An application of Theorem 1.3.44, shows that $T : \overline{S_\infty}^w \longrightarrow S_\infty$ is weakly continuous. Seeing that

$$T(\overline{S_\infty}^w) \subset \overline{T(S_\infty)}^w \subset \overline{S_\infty}^w,$$

we may apply Theorem 1.2.42, to infer that the operator T has a fixed point. Q.E.D.

Definition 1.5.13 Let X be a Banach space and let $T : X \longrightarrow X$ be a mapping. We say that T is k-set-weakly contractive if, T maps bounded sets into bounded sets, and there exists $k \in [0, 1)$ such that

$$\beta(T(S)) \leq k\beta(S),$$

for any bounded subset S in X with $\beta(S) > 0$.

Definition 1.5.14 Let S be a nonempty subset of a non-separable Banach space X and let $T : S \longrightarrow X$ be an operator. We say that
(i) T is countably k-β-contractive if, $T(S)$ is bounded and there exists $k \in [0, 1)$ such that

$$\beta(T(M)) \leq k\beta(M),$$

for any countably bounded subset M of S.
(ii) T is weakly countably 1-set-contractive if, $T(S)$ is bounded and

$$\beta(T(M)) \leq \beta(M),$$

for any countably bounded subset M of S.

Remark 1.5.15 Every k-β-contractive is countably k-β-contractive, but the converse may not be true.

The next result is the analog of Darbo's fixed point [100], see also [105, 106] for strongly contractive maps.

Theorem 1.5.16 *Let S be a nonempty, bounded, convex, and closed subset of a Banach space X, and let $T : S \longrightarrow S$ be a weakly sequentially continuous operator. If T is k-β-contractive, then T has a fixed point.*

Proof Consider the sequence $(S_n)_n$ consisting of subsets of the set S such that

$$\begin{cases} S_1 = S, \\ S_{n+1} = \overline{co}(T(S_n)), \quad \text{for } n = 1, 2, \ldots. \end{cases}$$

Let us pay attention to the fact that the operator T is contractive, then

$$\beta(S_2) \leq k\beta(S_1).$$

Proceeding by induction, we get

$$\beta(S_{n+1}) \leq k^n\beta(S_1).$$

Since $k < 1$, we have $\beta(S_n) = 0$. Therefore, the set

$$S_\infty = \bigcap_{n \geq 1} S_n$$

is a nonempty, weakly closed, and convex subset of S, since each S_n is weakly closed. Since $\beta(S_\infty) \leq \beta(S_n)$, it follows that S_∞ is a relatively weakly compact subset of S. An application of Theorem 1.3.44 yields that T has a fixed point in S_∞. Q.E.D.

The following lemma is needed later. A proof can be found in [107].

Lemma 1.5.17 *Let S be a nonempty, closed, and convex subset of a Banach space X and let $T : S \longrightarrow S$ be a countably condensing mapping with respect to β such that $T(S)$ is bounded. If T maps weakly compact sets into relatively weakly compact sets, then there exists a convex and relatively weakly compact subset K of S such that $T(K) \subset K$.*

Theorem 1.5.18 *Let S be a nonempty, closed, and convex subset of a non-separable Banach space X and let $T : S \longrightarrow S$ be a weakly sequentially continuous mapping such that $T(S)$ is bounded. If T is countably condensing with respect to β, then the set of fixed points $\mathscr{F}_T$ of T is nonempty.*

Proof By using Lemma 1.5.17, we can show that there exists a convex and relatively weakly compact subset K of S such that $T(K) \subset K$. Then, $\overline{K}^w$ is weakly compact and so, the restriction mapping $T_{|\overline{K}^w}$ is weakly continuous. The desired conclusion of the theorem follows by a direct application of Theorem 1.3.44. Q.E.D.

Definition 1.5.19 Let S be a nonempty and weakly closed subset of a Banach space X and let $T : S \longrightarrow E$ be a weakly sequentially continuous operator. We say that T is weakly semi-closed at θ if, for every sequence $(x_n)_n$ contained in the set S such that $x_n - T(x_n) \rightarrow \theta$, there exists $x \in S$ such that $T(x) = x$.

The following results was proved in [107], which is needed later.

Theorem 1.5.20 *Let S be a nonempty, closed, and convex subset of a Banach space X and let $T : S \longrightarrow S$ be a mapping such that*
(i) T is weakly sequentially continuous,
(ii) T is countably condensing with respect to β,
(iii) $T(S)$ is bounded.
Then, T has a fixed point in S.

Theorem 1.5.21 *Let S be a nonempty, closed, and convex subset of a Banach space X and let $T : S \longrightarrow S$ be a weakly sequentially continuous mapping such that*
(i) T is weakly countably 1-set-contractive,
(ii) T is weakly semi-closed at θ,
(iii) $T(S)$ is bounded.
Then, T has a fixed point in S.

Proof Let $z \in S$ be fixed and consider $(\lambda_n)_n$ as a sequence of $[0, 1]$ such that $\lambda_n \to 1$. Define the mapping $T_n : S \longrightarrow X$ by putting

$$T_n(x) = \lambda_n T(x) + (1 - \lambda_n)z.$$

Since S is convex, then $T_n(S) \subset S$. Let M be a countable subset of S with $\beta(M) > 0$. Making use of the subadditivity of the De Blasi measure, together with assumption (i) enables us to have

$$\begin{aligned}\beta(T_n(M)) &\leq \lambda_n \beta(T(M)) \\ &\leq \lambda_n \beta(M) \\ &< \beta(M).\end{aligned}$$

This means that T_n is a countably condensing mapping with respect to β. It follows from Theorem 1.5.20, that there exists a sequence $(x_n)_n$ contained in the set S such that

$$x_n = \lambda_n T(x_n) + (1 - \lambda_n)z.$$

In other words, we have

$$x_n - T(x_n) = (1 - \lambda_n)(z - T(x_n)).$$

Since $\lim_{n \to +\infty} \lambda_n = 1$, then there exists $x^* \in S$ such that $x^* = T(x^*)$. Q.E.D.

Definition 1.5.22 Let S be a nonempty subset of a Banach space X. We say that a mapping $T : S \longrightarrow X$ satisfies the condition $(\mathfrak{H}_1)$ if, for every weakly convergent sequence $(x_n)_n$ contained in the set S, the sequence $(T(x_n))_n$ has a strongly convergent subsequence in X.

Definition 1.5.23 Let S be a nonempty subset of a Banach space X. We say that a mapping $T : S \longrightarrow X$ satisfies the condition $(\mathfrak{H}_2)$ if, for every weakly convergent sequence $(x_n)_n$ contained in the set S, the sequence $(T(x_n))_n$ has a weakly convergent subsequence in X.

It should be noted that fundamental articles have already utilized the criteria $(\mathfrak{H}_1)$ and $(\mathfrak{H}_2)$, see, for instance, [108–111]. For some properties and results of operators meeting the requirements $(\mathfrak{H}_1)$ and $(\mathfrak{H}_2)$, we refer to [11, 112, 113]. An operator T satisfies $(\mathfrak{H}_1)$ or $(\mathfrak{H}_2)$ is not necessarily weakly continuous. Moreover, an operator T satisfies $(\mathfrak{H}_1)$ if, and only, if it maps relatively weakly compact sets into relatively compact ones, Latrach, Taoudi, and Zeghal [111] provides the proof.

Theorem 1.5.24 *Let S be a nonempty, closed, and convex subset of a Banach space X and let $T : S \longrightarrow S$ be a continuous mapping, which satisfies condition $(\mathfrak{H}_1)$. If $T(S)$ is relatively weakly compact, then there exists $x \in S$ such that $Tx = x$.*

Proof Let $\mathscr{K} = \overline{co}(T(S))$. Since S be a closed and convex subset of X such that $T(S) \subset S$, then $\mathscr{K} \subseteq S$. This implies that

$$T(\mathscr{K}) \subseteq T(S) \subseteq \mathscr{K}. \tag{1.5.1}$$

The use of Theorem 1.3.34 as well as the inclusion (1.5.1) leads to $\mathscr{K}$ is weakly compact and $T : \mathscr{K} \longrightarrow \mathscr{K}$ is well defined. Let $(x_n)_n$ be a sequence contained in the set $\mathscr{K}$. Then, $(x_n)_n$ has a weakly convergent subsequence, say $(x_{n_k})_k$. The condition $(\mathfrak{H}_1)$ guaranties that $(T(x_{n_k}))_k$ has a strongly convergent subsequence and therefore, the set $T(\mathscr{K})$ is relatively compact and when we apply Theorem 1.2.33, we reach the desired conclusion. Q.E.D.

Theorem 1.5.25 *Let S be a nonempty, closed, and convex subset of a Banach space X and let $T : S \longrightarrow S$ be a continuous mapping such that $T(S)$ is bounded. If T is condensing with respect to β, which satisfies $(\mathfrak{H}_1)$, then the set of fixed points $\mathscr{F}_T$ of T in S is nonempty and compact.*

Definition 1.5.26 Let S be a nonempty subset of a Banach space X. A multi-valued mapping $Q : S \longrightarrow \mathscr{P}(X)$ is called to be a

(i) $\mathscr{D}$-set-Lipschitzian with respect to β if, there exists a continuous nondecreasing function $\phi : \mathbb{R}_+ \longrightarrow \mathbb{R}_+$ with $\varphi(0) = 0$ such that the following inequality

$$\beta(Q(M)) \leq \phi(\beta(M))$$

is satisfied, for any bounded subset M of S such that $Q(M) \in \mathscr{P}_{bd}(X)$. In particular, if $\phi(r) < r$ for $r > 0$, then Q is called nonlinear $\mathscr{D}$-set-contraction multi-valued mapping.

(ii) β-nonexpansive if, Q is bounded and the following inequality

$$\beta(Q(M)) \leq \beta(M)$$

is satisfied, for any bounded subset M of S.

Remark 1.5.27 Let X be a Banach space and let $Q : X \longrightarrow \mathscr{P}(X)$ be a sequentially weakly upper semi-compact multi-valued mapping. If Q is $\mathscr{D}$-Lipschitzian, then Q is $\mathscr{D}$-set-Lipschitzian with respect to β, for more detail see [114]. In particular, if Q is weakly sequentially continuous single-valued mapping and $\mathscr{D}$-Lipschitzian, then Q is $\mathscr{D}$-set-Lipschitzian with respect to β, see [107, 115].

In the following, an axiomatic approach to the notion of a measure of weak non-compactness is presented.

Definition 1.5.28 Let X be a Banach space and let $\nu : \mathscr{B}(X) \longrightarrow \mathbb{R}_+$ be a function, where $\mathscr{B}(X)$ is the family of all nonempty bounded subset of X. We say that ν is a measure of non-compactness if ν satisfies the following conditions

1. The family $\ker(\nu)$ is nonempty and it is contained in the set of relatively compact sets of X,
2. $\nu(\overline{S}) = \nu(S)$, where $\overline{S}$ denote the closure of the set S,
3. $\nu(co(S)) = \nu(S)$,
4. ν is nondecreasing, and
5. If $\{S_n\}$ is a decreasing sequence of nonempty subsets in $\mathscr{P}_{cl,bd}(X)$ satisfying $\lim_{n\to+\infty} \nu(S_n) = 0$. Then, the set $S_\infty = \bigcap_{n\geq 1} S_n$ is nonempty and $\nu(S_\infty) = 0$.

Let us recall that the concept of a measure of non-compactness was initiated by fundamental papers, see [64, 116, 117] and the references therein. It turns out that measures of non-compactness create very useful tools in nonlinear analysis. Let X be a Banach space. The Kuratowski measure of non-compactness of a bounded subset S of X is a nonnegative real number $\alpha(S)$ defined by

$$\alpha(S) = \inf\left\{r > 0,\ S \subseteq \bigcup_{i=1}^{n} S_i,\ \ \operatorname{diam}(S_i) \leq r\ \ \forall i\right\}. \tag{1.5.2}$$

The Hausdorff measure of non-compactness of a bounded subset S of X is a nonnegative real number $\mu(S)$ defined by

$$\mu(S) = \inf\left\{r > 0,\ S \subseteq \bigcup_{i=1}^{n} \mathscr{B}_X(x_i, r),\ \ \text{for some } x_i \in X\right\}. \tag{1.5.3}$$

Now, we are equipped with the necessary details to define the measures of non-compactness of a bounded subset of a Banach space. Let X be a Banach space. A function $\nu : \mathscr{B}(X) \longrightarrow \mathbb{R}_+$ is called nondecreasing if, for any subsets $S, S' \in \mathscr{B}(X)$ with $S \subseteq S'$, we have

$$\nu(S) \leq \nu(S').$$

Note that Kuratowski and Hausdorff measures defined by (1.5.2) and (1.5.3) satisfy the conditions 1 through 5 of Definition 1.5.28. Some nice properties of Kuratowski and Hausdorff measures have been discussed by Alexander [118]. The Kuratowski measure and the Hausdorff measure are discussed in [98, 119–122] and the references therein. These measures of non-compactness are mutually equivalent in the sense that

$$\mu(S) \leq \alpha(S) \leq 2\mu(S), \tag{1.5.4}$$

for any nonempty and bounded subset S in X, see [123].

Lemma 1.5.29 *Let S_1 and S_2 be two nonempty and bounded subsets of a Banach space X. Then, the following assertions are satisfied*
(i) $\mu(S_1) = 0$ if, and only, if S_1 is totally bounded,
(ii) If $S_1 \subset S_2$ then, $\mu(S_1) \leq \mu(S_2)$.
(iii) $\mu(S_1 + S_2) \leq \mu(S_1) + \mu(S_2)$.

Lemma 1.5.30 *Let S_1 and S_2 be two nonempty and bounded subsets of a Banach algebra X. Then,*

$$\mu(S_1 \cdot S_2) \leq \mu(S_1)\|S_2\|_{\mathscr{P}} + \mu(S_2)\|S_1\|_{\mathscr{P}} + \mu(S_1)\mu(S_2).$$

Remark 1.5.31 Let X be a Banach space and let $T : X \longrightarrow X$ be a mapping. If T is $\mathscr{D}$-Lipschitzian with $\mathscr{D}$-function ϕ_T, then

$$\mu(T(M)) \leq \phi_T(\mu(M)),$$

for any bounded subset M of X with $\mu(M) > 0$. In fact, let M be a bounded subset of X with $r > \mu(M) > 0$. By using the definition of the Hausdorff measure of non-compactness μ, we obtain there exists a finite subset K of X such that

$$M \subset K + \overline{\mathscr{B}}_X(r).$$

Let $x \in M$ be fixed. Then, there exist $y \in K$ and $b \in \overline{\mathscr{B}}_X(r)$ such that

$$x = y + b.$$

Due to the fact that T is $\mathscr{D}$-Lipschitzian with a $\mathscr{D}$-function ϕ_T, we infer that

$$\begin{aligned} \|Tx - Ty\| &\leq \phi_T\left(\|x - y\|\right) \\ &\leq \phi_T\left(\|b\|\right) \\ &\leq \phi_T\left(r\right). \end{aligned}$$

This implies that

$$TM \subset TK + \overline{\mathscr{B}}_X(\phi_T(r)).$$

Using the continuity of ϕ_T, we deduce that

$$\mu(TM) \leq \phi_T(\mu(M)).$$

Remark 1.5.32 Let X be a Banach space and let $Q : X \longrightarrow \mathscr{P}_{cp}(X)$ be a multi-valued mapping. If Q is a contraction with a constant k, then

$$\mu(Q(S)) \leq k\mu(S),$$

for any bounded subset S of X. Going back to (1.5.4), we get

$$\alpha(Q(S)) \leq 2k\alpha(S).$$

Similarly, if Q is a single-valued contraction on X with a constant k, then

$$\alpha(Q(S)) \leq k\alpha(S)$$

for all $S \in \mathscr{P}_{bd}(X)$, see [67].

The following lemma is frequently used in the analytical fixed point theory of metric spaces. For more details, we can see [124–126].

Lemma 1.5.33 *If ψ is a $\mathscr{D}$-function with $\psi(r) < r$, for $r > 0$, then*

$$\lim_{n \to \infty} \psi^n(t) = 0,$$

for all $t \in [0, +\infty[$.

Proof Since ψ is nondecreasing, it follows that

$$\psi^2(t) \leq \psi(t), \quad \text{for all } \ t > 0.$$

Proceeding by induction, we get

$$\psi^{n+1}(t) \leq \psi^n(t),$$

for each $n \in \mathbb{N}$. This implies that $(\psi^n(t))_n$ is a sequence of positive real numbers, which is bounded below by 0, and so $\lim_{n \to \infty} \psi^n(t)$ exists. Suppose that $\lim_{n \to \infty} \psi^n(t) \neq 0$. By upper semi-continuity of ψ, we obtain

$$\begin{aligned} \lim_{n \to \infty} \psi^{n+1}(t) &= \lim_{n \to \infty} \psi(\psi^n(t)) \\ &\leq \psi\left(\lim_{n \to \infty} \psi^n(t)\right) \\ &< \lim_{n \to \infty} \psi^n(t) \end{aligned}$$

which is a contradiction, and consequently $\lim_{n \to \infty} \psi^n(t) = 0$. Q.E.D.

In the sequel, we will use the following results, which were established in [97].

Theorem 1.5.34 *Let S be a nonempty, closed, and convex subset of a Banach space X and let $Q : S \longrightarrow \mathscr{P}_{cv}(X)$ be a multi-valued mapping with weakly sequentially closed graph. If $Q(S)$ is relatively weakly compact, then Q has a fixed point.*

Before stating the generalization of Theorem 1.5.34, we recall a useful definition

Definition 1.5.35 Let S be a nonempty subset of a Banach space X. A multi-valued mapping $Q : S \longrightarrow X$ is called to be condensing with respect to β, if for any bounded subset M of S with $\beta(M) > 0$, we have $Q(M)$ is bounded and

$$\beta(Q(M)) < \beta(M).$$

Theorem 1.5.36 (see [97]) *Let S be a nonempty, closed and convex subset of a Banach space X and let $Q : S \longrightarrow \mathscr{P}_{cv}(S)$ be a multi-valued mapping with a weakly sequentially closed graph such that $Q(S)$ is bounded. If Q is condensing with respect to β, then Q has a fixed point.*

Using the concept of multi-valued mappings with w-weakly closed graphs, the authors in [107] have established the following.

Theorem 1.5.37 *Let S be a nonempty, closed, and convex subset of a Banach space X and let $Q : S \longrightarrow \mathscr{P}_{cl,cv}(S)$ be a multi-valued mapping such that*
(i) Q maps weakly compact sets into relatively weakly compact sets, and
(ii) Q has w-weakly closed graph.
(iii) Q is countably condensing with respect to β.
Then, Q has a fixed point.

By using the concept of $\mathscr{D}$-set-contraction multi-valued mappings in Banach spaces, Dhage [124] has proved the following result, which is very much useful in the study of multi-valued analysis. For more details see [67].

Theorem 1.5.38 *Let S be a nonempty, closed, convex, and bounded subset of a Banach space X and let $Q : S \longrightarrow \mathscr{P}_{cl,cv}(X)$ be a nonlinear $\mathscr{D}$-set-contraction multi-valued mapping with respect to the Hausdorff measure μ. If Q is closed, then Q has a fixed point.*

The next theorem extends Theorem 1.5.38 to the case of multi-valued mappings in the context of weak topology.

Theorem 1.5.39 *Let S be a nonempty, closed, and convex subset of a Banach space X and let $Q : S \longrightarrow \mathscr{P}_{cv}(X)$ be a multi-valued mapping such that*
(i) Q has a weakly sequentially closed graph,
(ii) $Q(S)$ is bounded,
(iii) Q is a nonlinear $\mathscr{D}$-set-contraction multi-valued mapping with respect to the De Blasi measure β with $\mathscr{D}$-function ϕ_Q.
Then, Q has a fixed point.

Proof Let $(S_n)_n$ be a sequence contained in the set $\mathscr{P}_{bd,cv}(S)$, consisting of subsets of the set S such that

$$\begin{cases} S_1 = S \\ S_{n+1} = \overline{co}(Q(S_n)), \end{cases}$$

where $\mathscr{P}_{bd,cv}(S)$ denote the family of nonempty, bounded and convex subsets of S. Obviously all sets of this sequence are nonempty, convex, and closed subsets of S. Taking into account assumption (iii), and using the definition of the sequence $(S_n)_n$, we obtain

$$\begin{aligned} \beta(S_{k+1}) = & \ \beta(Q(S_k)) \\ \leq & \ \phi_Q(\beta(S_k)). \end{aligned}$$

Proceeding by induction, for $n = 1, 2, \dots,$ we infer that

$$\beta(S_{n+1}) \leq \phi_Q^n(\beta(S)).$$

Making use of Lemma 1.5.33, together with the assumptions on Q enables us to have

$$\lim_{n\to\infty} \beta(S_n) = 0.$$

Thus, the set

$$S_\infty = \bigcap_{n\geq 1} S_n$$

is a convex and weakly compact subset of X. Therefore, the multi-valued mapping $Q : S_\infty \longrightarrow \mathscr{P}_{cv}(S_\infty)$ satisfies the assumptions of Theorem 1.5.34. Q.E.D.

Definition 1.5.40 Let S be a nonempty subset of a Banach space X. A multi-valued mapping $Q : S \longrightarrow \mathscr{P}(X)$ is weakly completely continuous if, it has a weakly sequentially closed graph and $Q(M)$ is relatively weakly compact, for any bounded subset M of S.

Lemma 1.5.41 *Let X be a Banach algebra and let H be a bounded and equicontinuous subset of $C([0, T], X)$. Then, the partial $t \longrightarrow \beta(H(t))$ is continuous on $[0, \eta]$, for $\eta > 0$ arbitrary and, we have*

$$\beta(H) = \sup_{t\in[0,\eta]} \beta(H(t)) = \beta(H([0, \eta])),$$

where β is the De Blasi measure, $H(t) = \{x(t),\ \ x \in H\}$ and $H([0, \eta]) = \bigcup_{t\in[0,\eta]} H(t)$.

Proof It is easy to check that

$$\sup_{t\in[0,\eta]} \beta(H(t)) \leq \beta(H([0, \eta])).$$

Let $\varepsilon > 0$ and $0 = s_0 < s_1 < \cdots < s_n = \eta$ be a partition of $[0, \eta]$ such that

$$\|u(t) - u(s)\| \leq \varepsilon$$

for every $t, s \in [s_i, s_{i+1}]$ and $u \in H$. Based on the definition of the measure β, it follows that, for any $i = 1, \dots, n-1$, there exists a relatively weakly compact K_i such that

$$H(s_i) \subset K_i + \overline{\mathscr{B}}_X\left(\varepsilon + \beta(H(s_i))\right).$$

Now, for any $s \in [0, \eta]$ and $u \in H$, we have

$$\begin{aligned} u(s) = u(s_i) + (u(s) - u(s_i)) &\in K_i + \overline{\mathscr{B}}_X(\varepsilon + \beta(H(s_i))) + \overline{\mathscr{B}}_X(\varepsilon) \\ &\subset K_i + \overline{\mathscr{B}}_X(2\varepsilon + \beta(H(s_i))) \\ &\subset K + \overline{\mathscr{B}}_X\left(2\varepsilon + \sup_{t\in[0,T]} \beta(H(t))\right), \end{aligned}$$

where $K = \bigcup_{i=1}^{n} K_i$. This implies that

$$\beta(H([0, \eta])) \leq 2\varepsilon + \sup_{s\in[0,\eta]} \beta(H(s)).$$

Since ε is arbitrary, we deduce that

$$\beta(H([0, \eta])) \leq \sup_{s\in[0,\eta]} \beta(H(s)).$$

This completes the proof. Q.E.D.

In [96], the authors combine Theorem 1.5.34 and Lemma 1.5.10, in order to obtain the following fixed point theorems in Banach algebras satisfying condition $(\mathscr{P})$.

Theorem 1.5.42 *Let S be a nonempty, closed, and convex subset of a Banach algebra X satisfying condition $(\mathscr{P})$ and let $A, B, C : S \longrightarrow \mathscr{P}(X)$ be three multi-valued mappings satisfying the following properties*
(i) A, B, and C have weakly sequentially closed graphs
(ii) A, B, and C are $\mathscr{D}$-set-Lipschitzians with $\mathscr{D}$-functions ϕ_A, ϕ_B and ϕ_C respectively,
(iii) A, B, and C are s.w.u.sco.,
(iv) $Ax \cdot Bx + Cx$ is a convex subset of S, for all $x \in S$,
(v) $\|A(S)\|\phi_B(r) + \|B(S)\|\phi_A(r) + \phi_A(r)\phi_B(r) + \phi_C(r) < r$, for all $r > 0$.
Then, the operator inclusion $x \in Ax \cdot Bx + Cx$ has, at least, one solution in S.

1.6 Generalized Banach Algebra

Let $(X, \|\cdot\|)$ be a generalized Banach space in the sense of Perov, that is, the vector-valued norm $\|\cdot\| : X \longrightarrow \mathbb{R}^n_+$ is given by

$$\|x\| = \begin{pmatrix} \|x\|_1 \\ \vdots \\ \|x\|_n \end{pmatrix}, \quad x \in X$$

with $\| \cdot \|_i$, $i = 1, \ldots, n$ define n norm on X. For more details, the reader may consult the monograph [127] and the references therein. Let $(Y_i, \tau_i)_{i \in I}$ be a family of topological spaces and let $f_i : X \longrightarrow Y_i$, $i \in I$ be a linear and continuous mapping. We define the smallest topology on X such that all the mappings f_i remain continuous, with respect to this topology. Its basic open sets are of the form

$$\bigcap_{i \in J} f_i^{-1}(u_i),$$

with J is a finite subset of I and $u_i \in \tau_i$, for each i. This topology is called the weak topology on X, generated by the $(f_i)_{i \in I}$ and we denote it by $\sigma(X, (f_i)_{i \in I})$. So that a sequence $(x_n)_n$ in X converges to x in $\sigma(X, (f_i)_{i \in I})$ if, the sequence $(f_i(x_n))_n$ converges to $f_i(x)$, for all $i \in I$. In a generalized Banach space, weakly open subset, weakly closed subset, and weak compactness, are similar to those for usual Banach space. If $x, y \in \mathbb{R}^n$, that is, $x = (x_1, \ldots, x_n)$, $y = (y_1, \ldots, y_n)$ by the inequality $x \leq y$ we means $x_i \leq y_i$, for all $i = 1, \ldots, n$.

Definition 1.6.1 A square matrix of real numbers M is said to be convergent to zero if its spectral radius $\rho(M)$ is strictly less than 1. In other words, $|\lambda| < 1$, for every $\lambda \in \mathbb{C}$, with $det(M - \lambda I) = 0$, where I denotes the unit matrix of $\mathcal{M}_{n \times n}(\mathbb{R}_+)$.

The following results is very much useful in the study of fixed point theory in generalized Banach algebra, the details appear in [128–130] and the reference therein

Lemma 1.6.2 *Let $M \in \mathbb{M}_{n,n}(\mathbb{C})$. The following assertions are equivalent*
(i) M is a matrix convergent to zero,
(ii) $M^k \to 0$ as $k \to \infty$,
(iii) the matrix $I - M$ is nonsingular and

$$(I - M)^{-1} = I + M + M^2 + \cdots + M^k + \cdots,$$

(iv) the matrix $I - M$ is nonsingular and $(I - M)^{-1}$ has nonnegative elements.

Definition 1.6.3 A generalized normed algebra X is an algebra endowed with a norm satisfying the following property

$$\|x \cdot y\| \leq \|x\| \|y\|, \quad \text{for all } x, \ y \in X,$$

where

$$\|x.y\| = \begin{pmatrix} \|x.y\|_1 \\ \vdots \\ \|x.y\|_n \end{pmatrix} \quad \text{and} \quad \|x\| \|y\| = \begin{pmatrix} \|x\|_1 \|y\|_1 \\ \vdots \\ \|x\|_n \|y\|_n \end{pmatrix}.$$

A complete generalized normed algebra is called to be a generalized Banach algebra.

Because the product of two weakly sequentially continuous single-valued mappings in generalized Banach algebra is not necessarily weakly sequentially continuous, Jeribi, Kaddachi, and Laouar [131] introduced a class of generalized Banach algebras satisfying a sequential condition known as $(\mathcal{G}\mathcal{P})$.

Definition 1.6.4 We will say that a generalized Banach algebra X satisfies the sequential condition $(\mathcal{G}\mathcal{P})$ if $(\mathcal{G}\mathcal{P})\begin{cases} \text{for any sequences } (x_n)_n \text{ and } (y_n)_n \text{ of } X \text{ such that} \\ x_n \rightharpoonup x \text{ and } y_n \rightharpoonup y, \text{ then } x_n \cdot y_n \rightharpoonup x \cdot y. \end{cases}$

Lemma 1.6.5 *If K and K' are two weakly compact subsets of a generalized Banach algebra X satisfying the condition $(\mathcal{G}\mathcal{P})$, then $K \cdot K'$ is weakly compact.*

Proof We will show that $K \cdot K'$ is weakly sequentially compact. To prove the above-announced fact, let $(z_n)_n$ be a sequence contained in the set $K \cdot K'$. So, there exist two sequences $(x_n)_n$ and $(x'_n)_n$ of K and K', respectively, such that

$$z_n = x_n \cdot x'_n.$$

Since K and K' are weakly compact, then there exists a renamed subsequence of $(x_n)_n$ in the set K such that $x_n \rightharpoonup x$, for some $x \in K$, and there exists a renamed subsequence of $(x'_n)_n$ in the set K' such that $x'_n \rightharpoonup x'$, for some $x \in K'$. The sequential condition $(\mathcal{G}\mathcal{P})$ guarantees that

$$z_n \rightharpoonup z = x \cdot x'.$$

This implies that $K \cdot K'$ is weakly sequentially compact. Hence, an application of Theorem 1.3.29 yields that $K \cdot K'$ is weakly compact. Q.E.D.

Definition 1.6.6 Let X be a generalized Banach space. A mapping $T : X \longrightarrow X$ is said to be contractive if, there exists a convergent matrix $M \in \mathcal{M}_{n\times n}(\mathbb{R}_+)$ to zero such that

$$\|Tx - Ty\| \leq M\|x - y\|, \quad \text{for all } x, \ y \text{ in } X.$$

Precup [132] obtained the following fixed point theorem.

Theorem 1.6.7 (Perov's fixed point theorem) *Let $(X, dist)$ be a complete generalized metric space and let $T : X \longrightarrow X$ be a contractive operator with a matrix M. Then, T has an unique fixed point x^*, and for each $x_0 \in X$, we have*

$$dist(T^k(x_0), x^*) \leq M^k(I - M)^{-1}dist(x^*, T(x^*)), \ \textit{for all } k \in \mathbb{N}.$$

Proof Let $x_0 \in X$ be fixed and define the sequence $x_{n+1} = T(x_n)$. It is easy to verify that $x_n = T^n x_0$ and

$$\text{dist}(x_n, x_{n+1}) \leq M^n \text{dist}(x_0, x_1).$$

Using the induction, we infer that

$$\text{dist}(x_n, x_{n+p}) \leq M^n(I - M)^{-1}\text{dist}(x_0, x_1), \quad \text{for all } p \in \mathbb{N}.$$

Thus, $(x_n)_n$ is a Cauchy sequence. Due to the fact that X is complete, we get the sequence $(x_n)_n$ has a limit, say x^*. After simple calculations, we can see that $x^* = Tx^*$. From the above discussion, we infer that

$$\text{dist}(T^k(x_0), x^*) \leq M^k(I - M)^{-1}\text{dist}(x^*, T(x^*)), \text{ for all } k \in \mathbb{N}.$$

Now, it remains to prove the uniqueness of x^*. Assume that there exists another point $y^* \in X$ such that $y^* = Ty^*$. Then,

$$\begin{aligned} \text{dist}(x^*, y^*) &\leq \text{dist}(Tx^*, Ty^*) \\ &\leq M^k \text{ dist}(x^*, y^*). \end{aligned}$$

Since $M^k \to 0$ as $k \to \infty$, it follows that $x^* = y^*$. Q.E.D.

As a consequence of Perov's theorem, Petre and Petruşel [133] have established the following.

Theorem 1.6.8 *Let X be a generalized Banach space X and let $T : S \longrightarrow S$ be a contractive operator with a matrix M. Then, the operator $I - T$ is a homeomorphism.*

Proof Let $y \in X$ be fixed. Since T is contractive, then the map which assigns to each $x \in X$, the value

$$T_1(x) = T(x) + y$$

is contractive. It follows from Theorem 1.6.7, that T_1 has an unique fixed point $x^* \in X$. This means that $I - T$ is bijective. The continuity of $(I - T)^{-1}$ follows in a similar way to the case of the usual Banach space. Q.E.D.

Similar to Theorem 1.6.7, we have the following result concerning the fixed point theory in generalized Banach space. We omit their proofs. For more detail, the reader may refer to [128, 134, 135].

Theorem 1.6.9 *Let S be a nonempty, compact, and convex subset of a generalized Banach space X and let $T : S \longrightarrow S$ be a continuous operator. Then, T has, at least, one fixed point in S.*

Definition 1.6.10 Let X be a generalized Banach space. The generalized De Blasi measure of weak non-compactness of X is the map $\beta_G : \mathscr{B}(X) \longrightarrow \mathbb{R}^n_+$ defined in the following way

$$\beta_G(S) = \inf \left\{ r \in \mathbb{R}^n_+, \text{ there exists } K \in \mathscr{W}(X) \text{ such that } S \subseteq K + \overline{\mathscr{B}}_X(r) \right\},$$

for all $S \in \mathscr{B}(X)$.

Lemma 1.6.11 *Let S_1 and S_2 be two bounded subsets of a generalized Banach space X. Then, the generalized De Blasi measure of weak non-compactness β_G satisfies the following properties*
(i) $S_1 \subseteq S_2$ *implies* $\beta_G(S_1) \leq \beta_G(S_2)$,
(ii) $\beta_G(\overline{S_1}^w) = \beta_G(S_1)$,
(iii) $\beta_G(S_1) = 0$, *if, and only, if* $\overline{S_1}^w \in \mathscr{W}(X)$,
(iv) $\beta_G(co(S_1)) = \beta_G(S_1)$,
(v) $\beta_G(S_1 + S_2) \leq \beta_G(S_1) + \beta_G(S_2)$,
(vi) $\beta_G(\{a\} + S_1) \leq \beta_G(S_1)$, *for all* $a \in X$,
(vii) $\beta_G(S_1 \bigcup S_2) = \max\{\beta_G(S_1), \beta_G(S_2)\}$,
(viii) $\beta_G(\lambda S_1) = |\lambda|\beta_G(S_1)$, *for all* $\lambda \in \mathbb{R}$.

Proof From the definition of β_G, there exists a subset $K_1 \in \mathscr{W}(X)$ and there exists $r_1 \in \mathbb{R}^n_+ \setminus \{0_{\mathbb{R}^n}\}$ such that

$$S_1 \subseteq K_1 + \overline{\mathscr{B}}_X\left(r_1 + \beta_G(S_1)\right).$$

It follows from Theorem 1.3.34, that $\overline{co}(K_1)$ is weakly compact. Due to the fact that $\overline{\mathscr{B}}_X\left(r_1 + \beta_G(S_1)\right)$ is weakly closed, we infer that

$$\overline{S_1}^w \subseteq \overline{co}(K_1) + \overline{\mathscr{B}}_X\left(r_1 + \beta_G(S_1)\right).$$

Thus, $\beta_G(\overline{S_1}^w) \leq r_1 + \beta_G(S_1)$ and consequently,

$$\beta_G(\overline{S_1}^w) \leq \beta_G(S_1).$$

The reverse inequality is obvious.

(iii) It is easy to see that, there exists a $K_1 \in \mathscr{W}(X)$ and $r_1 \in \mathbb{R}^n_+ \setminus \{0_{\mathbb{R}^n}\}$ such that

$$\overline{S_1}^w \subseteq \overline{K_1}^w + \overline{\mathscr{B}}_X(r_1). \tag{1.6.1}$$

Consider $(x_n)_n$ as an arbitrary sequence contained in the set $\overline{S_1}^w$ and take two sequences $y_n \in \overline{K_1}^w$ and $z_n \in \overline{\mathscr{B}}_X(r_1)$ such that

$$x_n = y_n + z_n.$$

Going back to the inclusion (1.6.1), we get the sequence $(x_n)_n$ has a weakly convergent subsequence. In other words, we have $\overline{S_1}^w$ is weakly compact. Q.E.D.

Remark 1.6.12 For $n = 1$, we recover the classical De Blasi measure of weak non-compactness.

Definition 1.6.13 Let S be a nonempty subset of a generalized Banach space X and let $M \in \mathscr{M}_{n\times n}(\mathbb{R}_+)$ be a convergent matrix to zero. A mapping $T : S \longrightarrow X$ is said to be
(i) M-set-contraction with respect to β_G if, T is bounded and the following inequality

$$\beta_G(T(V)) \leq M\beta_G(V)$$

is satisfied, for any bounded subset V in S with $\beta_G(V) > 0_{\mathbb{R}^n}$.
(ii) condensing with respect to β_G if, T is bounded and the following inequality

$$\beta_G(T(V)) < \beta_G(V)$$

is satisfied, for any bounded subset V in S with $\beta_G(V) > 0_{\mathbb{R}^n}$.

Remark 1.6.14 Let S be a nonempty subset of a generalized Banach space X and let $T : S \longrightarrow X$ be a mapping. If T is M-set-contraction, then T is condensing with respect to β_G. In fact, let V be a bounded subset of S with $\beta_G(V) > 0_{\mathbb{R}^n}$. We claim that

$$M\beta_G(V) < \beta_G(V).$$

If not the case, we obtain $(I - M)\beta_G(V) \leq 0_{\mathbb{R}^n}$. Due to the fact that $(I - M)^{-1}$ has nonnegative elements, we get $\beta_G(V) = 0_{\mathbb{R}^n}$, which is a contradiction and the claim is proved.

References

1. Agarwal, R.P., Meehan, M., O'Regan, D.: Fixed Point Theory and Applications. Cambridge Tracts in Mathematics, 141. Cambridge University Press, Cambridge (2001)
2. Assad, N.A., Kirk, W.A.: Fixed point theorems for setvalued mappings of contractive type. Pacific J. Math. **43**, 533–562 (1972)
3. Aubin, J.P., Frankowska, H.: Set Valued Analysis. Birkhuser, Boston (1990)
4. Boyd, D.W., Wong, J.S.W.: On nonlinear contractions. Proc. Amer. Math. Soc. **20**, 458–464 (1969)
5. Brezis, H.: Analyse fonctionnelle. Théorie et Applications. Masson, Paris (1983)
6. Browder, F.E.: Nonlinear mappings of nonexpansive and accretive type in Banach spaces. Bull. Am. Math. Soc. **73**, 875–882 (1967)
7. Brooks, R.M., Schmitt, K.: The contraction mapping principle and some applications. Electronic J. Diff. Equ. Monograph 09 (2009)
8. Dunford, N., Schwartz, J.T.: Linear Operators. I. General Theory. With the assistance of W. G. Bade and R. G. Bartle. Pure and Applied Mathematics, vol. 7. Interscience Publishers, Inc., New York, London (1958)
9. Edelstein, M.: Fixed point theorems in uniformly convex Banach spaces. Proc. Amer. Math. Soc. **44**, 369–374 (1947)
10. Garcia-Falset, J., Morales, C.H.: Existence theorems for m-accretive operators in Banach spaces. J. Math. Anal. Appl. **309**, 453–461 (2005)
11. Jeribi, A., Krichen, B.: Nonlinear Functional Analysis in Banach Spaces and Banach Algebras Fixed Point Theory under Weak Topology for Nonlinear Operators and Block Operator

Matrices with Applications. Monographs and Research Notes in Mathematics. CRC Press, Boca Raton, FL (2016)
12. Kakutani, S.: A generalization of Brouwer's fixed point theorem. Duke Math. J. **8**, 457–459 (1941)
13. Laxmikantham, V., Leela, S.: Differential and Integral Inequalities. Theory and Application. Academic Press, New York-London (1969)
14. Zeidler, E.: Nonlinear Functional Analysis and its Applications. I. Fixed point theorems. Translated from the German by Peter R. Wadsack. Springer, New York (1986)
15. Smart, D.R.: Fixed Point Theorems. Cambridge University Press, Cambridge (1980)
16. Edelstein, M., O'Brien, R.C.: Nonexpansive mappings, asymptotic regularity and successive approximations. J. Lond. Math. Soc. **17**, 547–554 (1978)
17. Pata, V.: Fixed Point Theorems and Applications. La Matematica per il 3+2. Springer (2019)
18. Xiang, T., Yuan, R.: A class of expansive-type Krasnosel'skii fixed point theorems. Nonlinear Anal. **71**, 3229–3239 (2009)
19. Dotson, W.G., Mann, W.R.: The Schauder fixed point theorem for nonexpansive mappings. Amer. Math. Monthly **84**, 363–364 (1977)
20. Agarwal, R.P., Hussain, N., Taoudi, M.A.: Fixed point theorems in ordered Banach spaces and applications to nonlinear integral equations. Abstract Appl. Anal. **2012**, 1–15 (2012)
21. Ben Amar, A., O'Regan, D.: Topological Fixed Point Theory for single-valued and multi-valued Mappings and Applications. Springer, Cham (2016). x+194 pp. ISBN: 978-3-319-31947-6, 978-3-319-31948-3
22. Djebali, S., Górniewicz, L., Ouahab, A.: Solution sets for differential equations and inclusions, De Gruyter series in nonlinear analysis and applications, 18. Walter de Gruyter, Berlin (2013)
23. Browder, F.E., Petryshyn, W.V.: The solution by iteration of nonlinear functional equations in Banach spaces. Bull. Amer. Math. Soc. **72**, 571–575 (1966)
24. Goebel, K., Kirk, W.A.: Topics in Metric Fixed Point Theory. Cambridge Studies in Advanced Mathematics, vol. 28. Cambridge University Press, Cambridge (1990)
25. Edwards, R.E.: Functional Analysis. Inc, New York (1995)
26. Aksoy, A.G., Khamsi, M.A.: Nonstandard Methods in Fixed Point Theory. Springer, New York, Berlin (1990)
27. Chidume, C.: Geometric Properties of Banach Spaces and Nonlinear Iterations. Lecture Notes in Mathematics 1965, Springer (2009)
28. Browder, F.E., Gupta, G.P.: Topological degree and nonlinear mappings of analytic type in Banach spaces. J. Math. Anal. Appl. **26**, 390–402 (1969)
29. Göhde, D.: Zum Prinzip der Kontraktiven Abbildung. Math. Nachr. **30**, 251–258 (1965)
30. Granas, A., Guenther, R.B., Lee, J.W.: Some general existence principles in the Carathéodory theory of nonlinear differential systems. J. Math. Pures Appl. **70**, 153–196 (1991)
31. Agarwal, R.P., Gorniewicz, L., O'Regan, D.: Aronszajn type results for Volterra equations and inclusions. Top. Meth. Nonlin. Anal. **23**, 149–159 (2004)
32. Browder, F.E.: nonexpansive nonlinear operators in a Banach space. Proc. Nat. Acad. Sci. USA **54**, 1041–1044 (1965)
33. Goebel, K.: Concise Course on Fixed Point Theorems. Yokohama Publishers (2002)
34. Dhage, B.C.: On a fixed point theorem in Banach algebras with applications. Appl. Math. Lett. **18**, 273–280 (2005)
35. Dhage, B.C.: A fixed point theorem in banach algebras involving three operators with applications. KYUNGPOOK Math. J. **44**, 145–155 (2004)
36. Tychonoff, A.: Ein Fixpunktsatz. (German) Math. Ann. **111**(1), 767–776 (1935)
37. Singbal, B.V.: Generalized form of Schauder-Tychnoff fixed point principle. In: Bonsall, F.F. (ed.) Lectures on Some fixed point Theorems of Functional Analysis. Mimeographed Notes. TFIR, Bombay (1962)
38. Cobzas, S.: Fixed point theorems in locally convex spaces the Schauder mapping method. Fixed Point Theory Appl, Art. ID 57950, 13 pp (2006)
39. Liu, Y., Li, Z.: Schaefer type theorem and periodic solutions of evolution equations. J. Math. Anal. Appl. **316**, 237–255 (2006)

40. Liu, Y., Li, Z.: Krasnoselskii type fixed point theorems and applications. Proc. Amer. Math. Soc. **136**, 1213–1230 (2008)
41. Kato, T.: Nonlinear semigroups and evolution equation. J. Math. Soc. Japan **16**, 508–520 (1967)
42. Diestel, J.: A survey of results related to the Dunford-Pettis property. Contemp. Math. **2**, 15–60 (1984)
43. Megginson, R.E.: An Introduction to Banach Space Theory, Graduate texts in Mathematics. Springer (1998)
44. Jeribi, A.: Spectral Theory and Applications of Linear Operators and Block Operator Matrices. Springer, Cham (2015). xvi+599 pp. ISBN: 978-3-319-17565-2, 978-3-319-17566-9
45. Jeribi, A.: Denseness, bases and frames in Banach spaces and applications. (English) Zbl 06849137 Berlin : De Gruyter (ISBN 978-3-11-048488-5/hbk , 978-3-11-049386-3/ebook). xv, 406 p. (2018)
46. Jeribi, A.: Perturbation Theory for Linear Operators. Denseness and Bases with Applications. Springer, Singapore (2021). ISBN 978-981-16-2527-5
47. Vrabie, I.O.I., Cuza, A.I.I., Mayer, O.: *C_0-Semigroups and Applications.* Elseiver, New-York (2003)
48. Ball, J.M.: Weak continuity properties of mapping and semi-groups. Proc. Royal Soc. Edinburgh Sect. A **72**, 275–280 (1979)
49. Banaś, J.: Demicontinuity and weak sequential continuity of operators in the Lebesgue space. In: Proceeedings of the 1st Polish Symposuim on Nonlinear Analysis, pp. 124–129 (1997)
50. Arino, O., Gautier, S., Penot, J.-P.: A fixed point theorem for sequentially continuous mappings with application to ordinary differential equations. Funkcial. Ekvac. **27**, 273–279 (1984)
51. Fan, K.: A generalization of Tychonoff's fixed point theorem. Math. Ann. **142**, 305–310 (1961)
52. Ben Amar, A., Jeribi, A., Mnif, M.: On a generalization of the Schauder and Krasnosel'skii fixed points theorems on Dunford-Pettis spaces and applications. Math. Methods Appli. Sci. **28**(14), 1737–1756 (2005)
53. Ben Amar, A., Jeribi, A., Mnif, M.: Some fixed point theorems and application to biological model. Numer. Funct. Anal. Optim. **29**, 1–23 (2008)
54. Dobrokov, I.: On representation of linear operators on $C_0(T, X)$. Czechoslovak Math. J. **21**, 13–30 (1971)
55. Bochner, S., Taylor, A.E.: Linear functional on certain spaces of abstractly-valued functions. Ann. Math. **39**, 913–944 (1938)
56. Bogdanowicz, W.M.: Representations of linear continuous functional on the space $C(X, Y)$ of continuous functions from compact X into locally convex. Proc. Japan Acad. **10**, 1122–1127 (1967)
57. Grothendieck, A.: Sur les applications linéaires faiblement compactes d'espaces du type $C(K)$. Canad. J. Math. **5**, 129–173 (1953)
58. Diestel, J.: Sequences and Series in Banach Spaces. Springer, Berlin Heidelberg New York (1984)
59. Ryan, R.A.: Dunford-Pettis properties. Bull. Acad. Polon. Sci. **27**, 373–379 (1979)
60. Banaś, J., Taoudi, M.A.: Fixed points and solutions of operator equations for the weak topology in Banach algebras. Taiwanese J. Math. **18**(3), 871–893 (2014)
61. Banas, J., Olszowy, L.: On the equivalence of some concepts in the theory of Banach algebras. Ann. Funct. Anal. **10**, 277–283 (2019)
62. Ben Amar, A., Chouayekh, S., Jeribi, A.: Fixed point theory in a new class of Banach algebras and application. Afr. Mat. **24**, 705–724 (2013)
63. Ghiocel, M., Petrusel, A., Petrusel, G.: Topics in Nonlinear Analysis and Application to Mathematical Economics. Cluj-Napoca (2007)
64. Dhage, B.C.: Some generalizations of mulit-valued version of Schauder's fixed point theorem with applications. [Some generalizations of multi-valued version of Schauder's fixed point theorem with applications]. Cubo **12**, 139–151 (2010)

65. Dhage, B.C.: Multi-valued operators and fixed point theorems in Banach algebras. Comput. Math. Appl. **48**, 1461–1476 (2004)
66. Górniewicz, L.: Topological fixed point theory of multi-valued mappings. Mathematics and Its Applications, 495. Kluwer Academic Publishers, Dordrecht (1999)
67. Deimling, K.: Nonlinear Functional Analysis. Springer, Berlin (1985). xiv+450 pp. ISBN: 3-540-13928-1
68. Appel, J., De Pascale, E., Nguyên, H.T., Zabreiko, P.P.: Multi-valued superpositions. Dissertationes Math. (Rozprawy Mat.) (1995)
69. Bohnenblust, H.F., Karlin, S.: On a theorem of Ville. In: Kuhn, W.H., Tucker, A.W. (eds.) Contribution to the Theory of Games, pp. 155–160. Princeton, Princeton University Press (1950)
70. Lasota, A., Opial, Z.: An application of the Kakutani-Ky Fan theorem in the theory of ordinary differential equations. Bull. Acad. Polon. Sci. Sér. Sci. Math. Astronom. Phys. **13**, 781–786 (1965)
71. Diestel, J., Uhl, J.J.: Vector Measures. Mathematical Surveys, vol. 15. American Mathematical Society, Providence, R.I. (1977)
72. Huff, R.E., Phillips, R.S.: Functional Analysis and Semi-groups. Colloquium Publications, vol. 31, American Mathematical Society, Providence, RI (1957)
73. Lorentz, G.G., Wertheim, D.G.: Representation of linear functionals on Köthe spaces. Canad. J. Math **5**, 568–575 (1953)
74. Pettis, B.J.: On integration in vector spaces. Trans. Amer. Math. Soc. **44**, 277–304 (1938)
75. Agarwal, R.P., Lupulescu, V., O'Regan, D., ur Rahman, G.: Weak solutions for fractional differential equations in nonreflexive Banach spaces via Riemann-Pettis-integrals. Math. Nachr. **289**, 395–409 (2016)
76. Cichon, M., Kubiaczyk, I., Sikorska, A.: The Henstock-Kurzweil-Pettis integrals and existence theorems for the Cauchy problem. Czechoslovak. Math. J. **54**, 279–289 (2004)
77. Geitz, R.F.: Pettis integration. Proc. Amer. Math. Soc. **82**, 81–86 (1981)
78. Mitchell, A.R., Smith, C.K.L., An existence theorem for weak solutions of differential equations in Banach spaces. Nonlinear Equations in Abstract Spaces (Proceedings of an International Symposium, The University of Texas, Arlington, Tex, 1977), pp. 387–404. Academic, New York (1978)
79. O'Regan, D.: Operator equations in Banach spaces relative to the weak topology. Arch. Math. **71**, 123–136 (1998)
80. Gordon, R.: Riemann integration in Banach spaces. Rocky Mountain J. Math **21**, 923–949 (1991)
81. Musial, K.: Pettis integral. Handbook of Measure Theory, pp. 531–586. North-Holland, Amsterdam (2002)
82. Dhage, B.C.: On solvability of operator inclusions $x \in AxBx + Cx$ in Banach algebras and differential inclusions. Commun. Appl. Anal. **14**, 567–596 (2010)
83. Covitz, H., Nadler, S.B.: Multi-valued contraction mappings in generalized metric spaces. Israel J. Math. **8**, 5–11 (1970)
84. Lim, T.C.: On fixed point stability for set-valued contraction mappings with applications to generalized differential equations. J. Math. Anal. Appl. **110**, 436–441 (1985)
85. Markin, J.T.: Continuous dependence of fixed point set. Proc. Amer. Math. Soc. **38**, 545–547 (1973)
86. Cheong, L.T.: A fixed point theorem for multi-valued nonexpansive mappings in a uniformly convex banach space. Bull. Amer. Math. Soc. **80**, 1123–1126 (1974)
87. Goebel, K.: On a Fixed Point Theorem for multi-valued nonexpansive Mappings. Ann. Univ. Mariae Curie-Sklodowska Sect. A **29**, 69–72 (1975)
88. Lim, T.C.: A fixed point theorem for multi-valued nonexpansive mappings in a uniformly convex Banach space. Bull. Amer. Math. Soc. **80**, 1123–1126 (1974)
89. Lim, T.C.: Remarks on some fixed point theorems. Proc. Amer. Math. Soc. **60**, 179–182 (1976)

90. Kirk, W.A., Massa, S.: Remarks on asymptotic and Chebyshev centers. Houston J. Math. **16**, 357–364 (1990)
91. Huff, R.E.: Banach spaces which are nearly uniformly convex. Rocky Mountain J. Math. **10**, 743–749 (1980)
92. Kuczumow, T., Prus, S.: Compact asymptotic centers and fixed points of multi-valued nonexpansive mappings. Houston J. Math. **16**, 465–468 (1990)
93. Sullivan, F.: A generalization of uniformly rotund Banach spaces. Can. J. Math. **31**, 628–636 (1979)
94. Rybinski, L.: An application of the continuous selection theorem to the study of the fixed points of multi-valued mappings. J. Math. Anal. Appl. **153**, 391–396 (1990)
95. Ursescu, C.: Multifunctions with convex closed graph. Czechoslovak Math. J. **25**(100), 438–441 (1975)
96. Ben Amar, A., Boumiza, M., O'Regan, D.: Hybrid fixed point theorems for multi-valued mappings in Banach algebras under a weak topology setting. J. Fixed Point Theory Appl. **18**, 327–350 (2016)
97. Ben Amar, A., Sikorska-Nowak, A.: On some fixed point theorems for 1-set weakly contractive multi-valued mappings with weakly sequentially closed graph. Adv. Pure Math. **1**, 163–169 (2011)
98. Banaś, J., Mursaleen, M.: Sequence Spaces and Measures of Non-compactness with Applications to Differential and Integral Equations. Springer, New Delhi (2014)
99. Banaś, J., Rivero, J.: On measures of weak non-compactness. Ann. Mat. Pura Appl. (4) **151**, 213–224 (1988)
100. Banaś, J., Goebel, K.: Measures of non-compactness in Banach spaces. Lecture Notes in Pure and Applied Mathematics, vol. 60. Marcel Dekker, New York and Basel (1980)
101. De Blasi, F.S.: On a property of the unit sphere in a Banach space. Bull. Math. Soc. Sci. Math. R. S. Roumanie (N.S.) **21**(69), 259–262 (1977)
102. Appell, J., De Pascale, E.: Some parameters associated with the Hausdorff measure of non-compactness in spaces of measurable functions. (Italien) Boll. Un. Mat. Ital. B (6) **3**, 497–515 (1984)
103. Djebali, S., Sahnoun, Z.: Nonlinear alternatives of Schauder and Krasnosel'skii types with applications to Hammerstein integral equations in L^1-spaces. J. Differ. Equ. **249**, 2061–2075 (2010)
104. Latrach, K., Taoudi, M.A.: Existence results for a generalized nonlinear Hammerstein equation on L^1-spaces. Nonlinear Anal. **66**, 2325–2333 (2007)
105. O'Regan, D.: New fixed point results for 1-set contractive set valued maps. Comput. Math. Appl. **35**, 27–34 (1998)
106. O'Regan, D.: fixed point theory for weakly sequentially continuous mapping. Math. Comput. Modelling. **27**, 1–14 (1998)
107. Ben Amar, A., Derbel, S., O'Regan, D., Xiang, T.: Fixed point theory for countably weakly condensing maps and multimaps in non-separable Banach spaces. J. Fixed Point Theory Appl. **21**, 8–25 (2019)
108. Garcia-Falset, J.: Existence of fixed points and measures of weak non-compactness. Nonlinear Anal. **71**, 2625–2633 (2009)
109. Garcia-Falset, J., Guran, L., Llorens-Fuster, E.: Fixed points for multi-valued contractions with respect to a w-distance. Sci. Math. Jpn. **71**, 83–91 (2010)
110. Jeribi, A., Krichen, B., Mefteh, B.: Fixed point theory in WC-Banach algebras. Turkish J. Math. **40**, 283–291 (2016)
111. Latrach, K., Taoudi, M.A., Zeghal, A.: Some fixed point theorems of the Schauder and the Krasnosel'skii type and application to nonlinear transport equations. J. Diff. Equ. **221**, 256–271 (2006)
112. Ben Amar, A., Garcia-Falset, J.: Fixed point theorems for 1-set weakly contractive and pseudo-contractive operators on an unbounded domain. Port. Math. **68**, 125–147 (2011)
113. Chlebowicz, A., Taoudi, M.A.: Measures of weak non-compactness and fixed points. Advances in Nonlinear Analysis via the Concept of Measure of Non-compactness, pp. 247–296, Springer, Singapore (2017)

114. Banaś, J., Lecko, M.: Fixed points of the product of operators in Banach algebras. PanAmer. Math. J. **12**(2), 101–109 (2002)
115. Kim, In-Sook.: Index formulas for countably k-set contractive operators. Nonlinear Anal. **69**, 4182–4189 (2008)
116. Aubin, J.P., Cellina, A.: Differential Inclusions. Springer (1984)
117. Goldenštein, L.S., Gohberg, I.T., Markus, A.S.: Investigations of some properties of bounded linear operators with their q-norms. Ucen. Zap. Kishinevsk. Univ. **29**, 29–36 (1957)
118. Alexander, J.W.: On transformations with invariant points. Trans. Amer. Math. Soc. **23**, 89–95 (1922)
119. Banaś, J., Olszowy, L.: On a class of measures of non-compactness in Banach algebras and their application to nonlinear integral equations. Z. Anal. Anwend. **28**, 475–498 (2009)
120. Hu, S., Papageorgiou, N.S.: Handbook of Multi-valued Analysis. Vol. I: Theory. Mathematics and its Applications, vol. 419, pp. 7923–4682. Kluwer Academic Publishers, Dordrecht (1997)
121. Malkowsky, E., Rakočević, V., Živković, S.: Matrix transformations between the sequence space $W_0^p(\Lambda)$, $V_0^p(\Lambda)$, $C_0^p(\Lambda)(1<p<\infty)$ and certain BK spaces. Appl. Math. Comput. **147**(2), 377–396 (2004)
122. Malkowsky, E.: Compact matrix operators between some BK-spaces. In: Mursaleen, M. (ed.) Modern Methods of Analysis and Its Applications, pp. 86–120. Anamaya Publ., New Delhi (2010)
123. Akhmerov, P.P., Kamenskii, M.I., Potapov, A.S., Rodkina, A.E., Sadovskii, B.N.: Measures of non-compactness and Condensing Operators. Birkhäuser, Basel, Boston, Berlin (1992)
124. Dhage, B.C.: Asymptotic stability of nonlinear functional integral equations via measures of non-compactness. Comm. Appl. Nonlinear Anal. **15**, 89–101 (2008)
125. Dhage, B.C.: Global attractivity results for nonlinear integral equations via a Krasnoselskii type fixed point theorem. Nonlinear Anal. **70**, 2485–2493 (2009)
126. Dhage, B.C.: Hybrid fixed point theory in partially ordered normed linear spaces and applications to fractional integral equations. Differ. Equ. Appl. **5**, 155–184 (2013)
127. Graef, J.R., Henderson, J., Ouahab, A.: Topological Methods for Differential Equations and Inclusions. Monographs and Research Notes in Mathematics, CRC Press (2019)
128. Nieto, J.J., Ouahab, A., Rodríguez-López, R.: Fixed point theorems in generalized Banach algebras and applications. Fixed Point Theory **19**, 707–732 (2018)
129. Rus, I.A.: Generalized Contractions and Applications. Cluj University Press, Cluj-Napoca (2001)
130. Varga, R.S.: Matrix Iterative Analysis. Second revised and expanded edition, Springer Series in Computational Mathematics, 27. Springer, Berlin (2000)
131. Jeribi, A., Kaddachi, N., Laouar, Z.: Generalized form of fixed point theorems in generalized Banach algebra relative to the weak topology with an application. Filomat **36**(18), 6253–6268 (2022)
132. Precup, R.: Methods in Nonlinear Integral Equations. Kluwer Academic Publishers, Dordrecht (2002), xiv+218 pp. ISBN: 1-4020-0844-9
133. Petre, I.R., Petrusel, A.: Krasnoselskii's theorem in generalized Banach spaces and application. Elec. J. Qual. Theo. Diff. Equ. **85**, 1–20 (2012)
134. Granas, A., Dugundji, J.: Fixed Point Theory. Springer, New York (2003)
135. Viorel, A.: Contributions to the study of nonlinear evolution equations. Ph.D. Thesis, Babeş-Bolyai University Cluj-Napoca, Department of Mathematics, (2011)

Chapter 2
Fixed Point Theory in Suitable Banach Algebras

In this chapter, we shall establish a survey of the fundamental fixed point theory of both for single-valued mapping and multi-valued mappings. Precisely, we study some new variants of fixed point results of multi-valued mapping in many spaces such as Banach space, Banach algebra, Banach algebra satisfying the condition $(\mathscr{P})$, generalized Banach algebra, and RWC-Banach algebra, by using an analog of Dhage's fixed point theorems and, by using an analog the well-known Perov's fixed point theorem. Several Perov's results were developed and extended by many authors in different directions, see for example [1–3]. Their arguments were essentially based on the vector-valued metric properties. Further progress was made in [4–6], where the authors established Krasnoselskii type of fixed point theorems of both single-valued mapping and multi-valued mappings. Perov's fixed point theorem is then employed in this chapter, to prove the existence results of the operator equation

$$x = Ax \cdot Bx + Cx, \tag{2.0.1}$$

where A, B, and C are nonlinear single-valued mapping. The objective of the last section of this chapter is to characterize the notion of relatively weakly compact Banach algebra, in short RWC-Banach algebras, which was introduced in [7] and generalize the condition $(\mathscr{P})$, may not be satisfied for a large class of Banach algebras. For more details concerning the material of this chapter, we recommend [8–20] and the references therein.

2.1 Existence Results for $\mathscr{D}$-Lipschitzian Mappings.

We state some new fixed point theorems involving three operators in Banach algebra. Our approach here is based on Boyd's fixed point theorem, see Theorem 1.2.26, combined with Schauder's fixed point theorem, see Theorem 1.2.33. You can find

A. Jeribi and N. Kaddachi, *Topological Fixed-Point Theory in Suitable Banach Algebras with Applications*, Infosys Science Foundation Series,
https://doi.org/10.1007/978-981-96-5542-7_2

the findings of this section in [19]. The concept of $\mathcal{D}$-Lipschitzian mappings was introduced in the first key finding, which is as follows.

Theorem 2.1.1 *Let S be a nonempty, bounded, closed, and convex subset of a Banach algebra X and let A, $C : X \longrightarrow X$ and $B : S \longrightarrow X$ be three operators such that*

(i) A and C are $\mathcal{D}$-Lipschitzians with $\mathcal{D}$-functions ϕ_A and ϕ_C, respectively,
(ii) B is completely continuous, and
(iii) $x = Ax \cdot By + Cx,\ y \in S \Rightarrow x \in S$.

Then, the operator equation (2.0.1) *has a solution in S provided that*

$$M\phi_A(r) + \phi_C(r) < r, \text{ for } r > 0,$$

where $M = \|B(S)\|$.

Proof Let $y \in S$ be fixed and define the mapping $\varphi_y : X \longrightarrow X$ by putting

$$\varphi_y(x) = Ax \cdot By + Cx.$$

Notice that φ_y is a nonlinear contraction operator, since for all $x_1, x_2 \in X$, we have

$$\begin{aligned} \|\varphi_y(x_1) - \varphi_y(x_2)\| &\leq \|Ax_1 \cdot By - Ax_2 \cdot By\| + \|Cx_1 - Cx_2\| \\ &\leq \|Ax_1 - Ax_2\| \|By\| + \|Cx_1 - Cx_2\| \\ &\leq \big(M\phi_A + \phi_C\big)(\|x_1 - x_2\|). \end{aligned}$$

An application of Theorem 1.2.26 yields that the equation $\varphi_y(x) = x$ has an unique solution $z \in X$. In other words, we have

$$z = Az \cdot By + Cz. \tag{2.1.1}$$

By assumption (iii), we get $z \in S$. Define the mapping N on S by $N(y) = z$, where z is the unique solution of the equation (2.1.1). We claim that N is continuous. Indeed, let $(y_n)_n$ be any sequence of S converging to a point $y \in X$. Since the set S is closed, then $y \in S$. Using assumption (i) one sees that

$$\begin{aligned} \|N(y_n) - N(y)\| &= \|A(z_n) \cdot B(y_n) + C(z_n) - Az \cdot By - Cz\| \\ &\leq \|A(z_n) \cdot B(y_n) - Az \cdot By\| + \|C(z_n) - Cz\| \\ &\leq (M\phi_A + \phi_C)(\|z_n - z\|) + \|ANy\| \|B(y_n) - By\|. \end{aligned}$$

This proves that N is continuous and the claim is proved. Our next task is to demonstrate that N is compact on S. In fact, for all $m \in S$, we have

$$
\begin{aligned}
\|Am\| &\leq \|Aa\| + \phi_A(\|m-a\|) \\
&< \|Aa\| + \frac{\|m-a\|}{M} \\
&\leq c,
\end{aligned}
$$

where $c = \|Aa\| + \frac{\text{diam}(S)}{M}$, for some $a \in S$. Let $(z_n)_n$ be any sequence contained in the set $N(S)$, then there exists a sequence $(y_n)_n$ of S such that $z_n = N(y_n)$. Since B is a compact operator, there exists a subsequence, say $(y_{n_k})_k$, of the sequence $(y_n)_n$ such that $(B(y_{n_k}))_k$ is convergent. Besides, for all $k_1, k_2 \in \mathbb{N}$, we have

$$
\|N(y_{n_{k_1}}) - N(y_{n_{k_2}})\| \leq (M\phi_A + \phi_C)(\|N(y_{n_{k_1}}) - N(y_{n_{k_2}})\|) + c\|B(y_{n_{k_1}}) - B(y_{n_{k_2}})\|.
$$

Due to the fact that $\phi = M\phi_A + \phi_C$ is continuous and nondecreasing, we obtain

$$
\begin{aligned}
\limsup_{k_1, k_2 \to +\infty} \phi\left(\|N(y_{n_{k_1}}) - N(y_{n_{k_2}})\|\right) &= \inf_{k\in\mathbb{N}} \sup_{k_1, k_2 \geq k} \phi\left(\|N(y_{n_{k_1}}) - N(y_{n_{k_2}})\|\right) \\
&= \phi\left(\inf_{k\in\mathbb{N}} \sup_{k_1, k_2 \geq k} \|N(y_{n_{k_1}}) - N(y_{n_{k_2}})\|\right) \\
&= \phi\left(\limsup_{k_1, k_2 \to +\infty} \|N(y_{n_{k_1}}) - N(y_{n_{k_2}})\|\right).
\end{aligned}
$$

This implies that

$$
\limsup_{k_1, k_2 \to +\infty} \|N(y_{n_{k_1}}) - N(y_{n_{k_2}})\| \leq \phi\left(\limsup_{k_1, k_2 \to +\infty} \|N(y_{n_{k_1}}) - N(y_{n_{k_2}})\|\right).
$$

Since $\phi(r) < r$, for all $r > 0$, we conclude that

$$
\limsup_{k_1, k_2 \to +\infty} \phi\left(\|N(y_{n_{k_1}}) - N(y_{n_{k_2}})\|\right) = 0.
$$

This means, in particular, that $(N(y_{n_k}))_k$ is a Cauchy sequence, and then $(N(y_{n_k}))_k$ is convergent, since X is a complete space. As a result, $N(S)$ is a sequentially relatively compact subset in S. Q.E.D.

Remark 2.1.2 By using Remark 1.2.25, we can deduce that Theorem 2.1.1 generalize and improve Theorem 1.2.34.

Theorem 2.1.3 *Let S be a nonempty, bounded, convex, and closed subset of a Banach algebra X and let $A, B, C : S \longrightarrow S$ be three mappings such that*

(i) A and C are $\mathscr{D}$-Lipschitzians with $\mathscr{D}$-functions ϕ_A and ϕ_C, respectively,
(ii) $\left(\frac{I}{A}\right)^{-1}$ exists on $B(S)$,
(iii) B is continuous and totally bounded,
(iv) $x = Ax \cdot By + Cx,\ y \in S \Rightarrow x \in S$.

Then, the operator equation (2.0.1) *has a solution in S provided that*

$$M\phi_A(r) + \phi_C(r) < r, \quad for\, r > 0,$$

where $M = \|B(S)\|$.

Proof Define the mapping $N : S \longrightarrow S$ by putting

$$Nx = \left(\frac{I - C}{A}\right)^{-1} Bx.$$

We first claim that the operator inverse $\left(\frac{I - C}{A}\right)^{-1}$ is continuous. Let $(y_n)_n$ be a sequence contained in the set $B(S)$, which converges to a point y. Let us denote

$$\begin{cases} x_n = A(x_n) \cdot y_n + C(x_n) \\ x = Ax \cdot y + Cx. \end{cases}$$

This implies that

$$\begin{aligned} \|x_n - x\| &= \|A(x_n) \cdot y_n + C(x_n) - Ax \cdot y - Cx\| \\ &\leq \|A(x_n) \cdot y_n - Ax \cdot y\| + \|C(x_n) - Cx\| \\ &\leq \|A(x_n) \cdot y_n - Ax \cdot y_n\| + \|Ax \cdot y_n - Ax \cdot y\| + \|C(x_n) - Cx\| \\ &\leq (M\phi_A + \phi_C)(\|x_n - x\|) + \|Ax\|\|y_n - y\|. \end{aligned}$$

This means, in particular, that

$$(I - (M\phi_A + \phi_C))\,(\|x_n - x\|) \leq \|Ax\|\|y_n - y\|.$$

Then, $\lim_{n\to\infty} \|x_n - x\| = 0$, since $M\phi_A + \phi_C$ is continuous and the claim is proved. The desired result follows from Theorem 1.2.33. Q.E.D.

Remark 2.1.4 Note that the operator inverse of $\frac{I}{A}$ exists if the latter is well defined and one-to-one on X. Further, $\frac{I}{A}$ is well defined if A is a regular mapping, i.e., A map into the family of all invertible elements of X.

2.2 Fixed Point Results for Hemi-Weakly Compact Maps

By using the notions of countably condensing, that is, the condition (0.0.4) holds only for countable bounded sets, and hemi-weakly compact multi-valued mappings, see Definition 1.4.52, we prove in this section some hybrid fixed point theorems for multi-valued mappings of the form $A + B$ and $A \cdot B + C$ acting on a nonempty, closed, and convex subset of a Banach space. The obtained results can be found in [15]. Assume that S is a nonempty, convex, and closed subset of a Banach space X. To any multi-valued mapping $Q : S \longrightarrow \mathscr{P}(S)$, consider the decreasing sequence $\left(S_n^Q\right)_n$, consisting of subsets of the set $\mathscr{P}_{cl,cv}(S)$ such that

$$S_0^Q = S \quad \text{and} \quad S_{n+1}^Q = \overline{co}\left\{Q\left(S_n^Q\right)\right\}. \tag{2.2.1}$$

2.2.1 *Krasnoselskii Fixed Point Theorem Under Weak Topology*

Based on the concept of countably β-nonexpansive and the well-known Sadovskii's fixed point theorem, we establish some new fixed point theorems for multi-valued mapping . Our first theorem is an extension of the results obtained in [20] in the sense of weak topologies setting.

Theorem 2.2.1 *Let S be a nonempty, convex, and closed subset of a Banach space X and let $Q : S \longrightarrow \mathscr{P}(S)$ be a multi-valued mapping. Suppose that there exists a bounded subset Ω of S_n^Q, where the set S_n^Q as is given in* (2.2.1)*, such that*

(i) $Qx \in \mathscr{P}_{cl,cv}(\Omega)$, for all $x \in \Omega$,
(ii) Q is hemi-weakly compact with a weakly sequentially closed graph on Ω,
(iii) Q is countably $\mathscr{D}$-set-Lipschitzian on Ω with $\mathscr{D}$-function φ.

Then, Q has a fixed point in Ω provided that $\varphi(r) \leq r$, for $0 < r \leq \|\Omega\|$.

Proof Let $n_0 \in \mathbb{N}$ be a fixed number such that $\Omega = S_{n_0}^Q$. Proceeding by induction, we obtain $\Omega = \overline{co}\{Q^{n_0}(S)\}$. It is easy to verify that, the sequence $(Q^n(S))_n$ is decreasing, since $S \neq \emptyset$ and $Q(S) \subset S$. Consider $\gamma \in (0, 1)$ and take $y \in Q^{n_0}(S)$. Define the multi-valued mapping $Q_\gamma : S \longrightarrow \mathscr{P}(S)$ by putting

$$Q_\gamma(x) = \gamma Q(x) + (1 - \gamma)y.$$

Based on the convexity of S, we infer that $Q_\gamma(S) \subset S$ and so, $\left(S_n^{Q_\gamma}\right)_n$ is a decreasing sequence of nonempty subsets of S. After simple calculations, we can see that

$$\begin{aligned} Q_\gamma\left(S_{n_0}^{Q_\gamma}\right) &\subset \overline{co}\left\{Q_\gamma\left(S_{n_0}^{Q_\gamma}\right)\right\} \\ &\subset S_{n_0}^{Q_\gamma}. \end{aligned}$$

This proves that $Q_\gamma : S_{n_0}^{Q_\gamma} \longrightarrow \mathscr{P}\left(S_{n_0}^{Q_\gamma}\right)$ is well defined. We show that $Q_\gamma x \in \mathscr{P}_{cv}(X)$, for all $x \in S_{n_0}^{Q_\gamma}$. To do this, we claim first that

$$S_k^{Q_\gamma} \subset S_k^Q, \text{ for all } k \in \{0, 1, \ldots, n_0\}. \tag{2.2.2}$$

The property is valid for $k = 0$. Assume that $S_j^{Q_\gamma} \subset S_j^Q$, for all $j \in \{1, \ldots, n_0 - 1\}$. Then,

$$\begin{aligned} S_{j+1}^{Q_\gamma} &= \overline{co}\left\{Q_\gamma\left(S_j^{Q_\gamma}\right)\right\} \\ &= \overline{co}\left\{\gamma Q\left(S_j^{Q_\gamma}\right) + (1-\gamma)y\right\} \\ &\subset \overline{co}\left\{\gamma Q\left(S_j^{Q}\right) + (1-\gamma)y\right\} \\ &\subset \overline{co}\left\{\gamma \overline{co}\left(Q\left(S_j^{Q}\right)\right) + (1-\gamma)y\right\} \\ &= \overline{co}\left\{\gamma S_{j+1}^{Q} + (1-\gamma)y\right\}. \end{aligned}$$

Since $y \in Q^{j+1}(S) \subset S_{j+1}^Q$ and S_{j+1}^Q is a closed and convex subset of S, we get

$$\overline{co}\left\{\gamma S_{j+1}^{Q} + (1-\gamma)y\right\} \subset \overline{co}\left\{\gamma S_{j+1}^{Q} + (1-\gamma)S_{j+1}^{Q}\right\} \subset S_{j+1}^{Q}.$$

This show the inclusion (2.2.2) and the claim is proved, and so $S_{n_0}^{Q_\gamma}$ is bounded, since $S_{n_0}^{Q_\gamma} \subset \Omega$. Let $x \in S_{n_0}^{Q_\gamma}$ be fixed. Consider $t \in (0, 1)$ and take $y_1, y_2 \in Q_\gamma(x)$. Then, there exist two elements $u_1, u_2 \in Q(x)$ such that

$$\begin{cases} y_1 = \gamma u_1 + (1-\gamma)y, \\ y_2 = \gamma u_2 + (1-\gamma)y. \end{cases}$$

This implies that

$$ty_1 + (1-t)y_2 = \gamma(tu_1 + (1-t)u_2) + (1-\gamma)y.$$

Since $Q(x) \in \mathscr{P}_{cv}(S)$, we deduce that

$$ty_1 + (1-t)y_2 \in Q_\gamma(x).$$

Now, we demonstrate that $Q_\gamma(x) \in \mathscr{P}_{cl}(S)$, for all $x \in S_{n_0}^{Q_\gamma}$. For this purpose, let $x \in S_{n_0}^{Q_\gamma}$ be fixed and consider $(z_n)_n$ as a convergent sequence contained in the set $Q_\gamma(x)$ to an element $z \in S_{n_0}^{Q_\gamma}$. Then, there exists a sequence $w_n \in Q(x)$ such that

$$z_n = \gamma w_n + (1-\gamma)y,$$

for all $n \in \mathbb{N}$. This implies that

$$w_n \to \frac{1}{\gamma}(z - (1-\gamma)y).$$

In other words, we have

$$\frac{1}{\gamma}(z - (1-\gamma)y) \in Q(x).$$

Thus, $z \in Q_\gamma(x)$. This achieves the proof of our claim. Let D be a countably subset contained in the set $S_{n_0}^{Q_\gamma}$ with $\beta(D) > 0$. Then, $Q_\gamma(D)$ is bounded and

$$\begin{aligned}\beta(Q_\gamma(D)) &\leq \beta(Q(D)) + \beta(\{(1-\gamma)y\}) \\ &\leq \gamma\varphi(\beta(D)).\end{aligned} \tag{2.2.3}$$

This means that Q_γ is countably condensing with respect to β. Our next task is to show that Q_γ maps weakly compact sets into relatively weakly compact sets. Let M be a weakly compact subset of $S_{n_0}^{Q_\gamma}$. Consider $(x_n)_n$ as an arbitrary sequence of M and take a sequence $y_n \in Q_\gamma(x_n)$. The use of inequality (2.2.3) allows us to obtain

$$\begin{aligned}\beta(\{y_n,\ n \in \mathbb{N}\}) &\leq \beta\left(Q_\gamma(\{x_n,\ n \in \mathbb{N}\})\right) \\ &\leq \gamma\varphi\left(\beta(\{x_n,\ n \in \mathbb{N}\})\right).\end{aligned}$$

It follows from Theorem 1.3.29, that $Q_\gamma(M)$ is relatively weakly compact. Since Q_γ has a weakly sequentially closed graph, we may invoke Theorem 1.5.37, in order to conclude that, there exists an element $u_\gamma \in S_{n_0}^{Q_\gamma}$ verifying $u_\gamma \in Q_\gamma(u_\gamma)$. Let us assume that $(\gamma_m)_m$ is a sequence contained in the set $]0, 1[$, which is convergent to 1. Going back to the inclusion (2.2.2), we obtain there exists a sequence $(u_m)_m$ contained in the set Ω with $u_m \in Q_{\gamma_m}(u_m)$. This leads to the conclusion that there exists a sequence $w_m \in Q(u_m)$ such that

$$u_m = \gamma_m w_m + (1-\gamma_m)y.$$

Besides, the set $\{Q(u_m),\ m \in \mathbb{N}\}$ is bounded, since

$$\begin{aligned}Q(\{u_m,\ m \in \mathbb{N}\}) &\subset Q(\Omega) \\ &\subset S_{n_0+1}^{Q} \\ &\subset \Omega.\end{aligned}$$

After simple calculations, we can see that

$$u_m - w_m = (\gamma_m - 1)w_m + (1-\gamma_m)y \rightharpoonup \theta.$$

Due to the fact that Q is hemi-weakly compact, we conclude that $(u_m)_m$ has a weakly convergent subsequence to some u. Since Q has a weakly sequentially closed graph, we reach the result that $u \in Q(u)$. Q.E.D.

Corollary 2.2.2 *Let S be a nonempty, convex, and closed subset of a Banach space X and let $Q : S \longrightarrow \mathscr{P}(S)$ be a multi-valued mapping with a weakly sequentially closed graph such that*

(i) there exists a bounded subset Ω contained in the set S_n^Q, $n \in \mathbb{N}$,
(ii) $Qx \in \mathscr{P}_{cl,cv}(\Omega)$, for all $x \in \Omega$,

(iii) Q is hemi-weakly compact
(iv) Q is countably nonexpansive on Ω with respect to β.

Then, the operator inclusion $x \in Qx$ has a solution in Ω.

Corollary 2.2.3 *Let S be a nonempty, convex, and closed subset of a Banach space X and let $Q : S \to \mathscr{P}(S)$ be a multi-valued mapping with a weakly sequentially closed graph such that*

(i) there exists a bounded subset Ω contained in the set S_n^Q, $n \in \mathbb{N}$,
(ii) $Qx \in \mathscr{P}_{cl,cv}(\Omega)$, for all $x \in \Omega$,
(iii) Q is countably $\mathscr{D}$-set-contraction on Ω with $\mathscr{D}$-function φ.

Then, the operator inclusion $x \in Qx$ has solution in Ω.

Proof Let $\Omega = S_{n_0}^Q$, for some $n_0 \in \mathbb{N}$. In order to achieve the proof, we will apply Theorem 2.2.1. Then, we only have to demonstrate that $Q : \Omega \longrightarrow \mathscr{P}(\Omega)$ is hemi-weakly compact. Consider $(x_n)_n$ as a sequence contained in the set Ω and take a sequence $y_n \in Q(x_n)$ such that $(x_n - y_n)_n$ possesses a weakly convergent subsequence. Taking into account that

$$x_n \in \{x_n - y_n,\ n \in \mathbb{N}\} + \{Q(x_n),\ n \in \mathbb{N}\},$$

we obtain

$$\begin{aligned}\beta(\{x_n,\ n \in \mathbb{N}\}) &\leq \beta(\{x_n - y_n,\ n \in \mathbb{N}\}) + \beta(\{Q(x_n),\ n \in \mathbb{N}\})\\ &\leq \beta(\{Q(x_n),\ n \in \mathbb{N}\}).\end{aligned}$$

If $\beta(\{x_n,\ n \in \mathbb{N}\}) > 0$, we get a contradiction and so, the sequence $(x_n)_n$ has a weakly convergent subsequence by a direct application of Theorem 1.3.29. Q.E.D.

Remark 2.2.4 Corollary 2.2.3 is an extension of Theorem 1.5.39 and proves that the assumption $\mathscr{D}$-set-contraction can be replaced by Q is countably $\mathscr{D}$-set-contraction. Similarly, the assumption $Q(S)$ is bounded can be replaced by, there exists $n_0 \in \mathbb{N}$ such that $S_{n_0}^Q$ is bounded.

The following theorem regarding the concepts of countably condensing multi-valued mapping with respect to β and is a generalization of Theorem 1.5.37, because it does not require the property $Q(S)$ is bounded

Corollary 2.2.5 *Let S be a nonempty, convex, and closed subset of a Banach space X and let $Q : S \longrightarrow \mathscr{P}(S)$ be a multi-valued mapping with weakly sequentially closed graph such that*

(i) there exists a bounded subset Ω contained in the set S_n^Q, $n \in \mathbb{N}$,
(ii) $Qx \in \mathscr{P}_{cl,cv}(\Omega)$, for all $x \in \Omega$,
(iii) Q is countably condensing on Ω with respect to β,
(iv) Q maps weakly compact sets into relatively weakly compact sets.

Then, the inclusion $x \in Qx$ has solution in Ω.

Proof Let $\Omega = S^{Q}_{n_0}$, for some $n_0 \in \mathbb{N}$. By using our assumptions, we infer that $\beta(Q(M)) = 0$, for any subset $M \in \mathscr{W}(\Omega)$. This means, in particular, that

$$\beta(Q(M)) \leq \beta(Q(\overline{M}^{w})) = 0. \tag{2.2.4}$$

Using assumption (iii) as well as inequality (2.2.4), we deduce that Q is countably $\mathscr{D}$-set-Lipschitzian with $\mathscr{D}$-function φ, where $\varphi(r) = r$, for all $0 \leq r \leq \|\Omega\|$. An argument similar to that in the proof of Corollary 2.2.3 leads to the hemi-weakly compactness of Q, and when we apply Theorem 2.2.1, we reach the desired conclusion. Q.E.D.

In the following, we prove some new variants of Krasnoselskii type of fixed point of two multi-valued mappings.

Theorem 2.2.6 *Let S be a nonempty, convex, and closed subset of a Banach space X and let A, $B : S \longrightarrow \mathscr{P}_{cl,cv}(E)$ be two multi-valued mappings with weakly sequentially closed graphs such that*

(i) A is hemi-weakly compact,
(ii) A is countably $\mathscr{D}$-set-nonlinear contraction with $\mathscr{D}$-function Φ,
(iii) B is weakly compact,
(iv) $(A + B)(S)$ is a bounded subset of S.

Then, the operator inclusion $u \in Au + Bu$ has a solution in S.

Proof Define the multi-valued mapping $Q : S_1^{A+B} \longrightarrow \mathscr{P}_{cl,cv}\left(S_1^{A+B}\right)$ by putting

$$Q(u) = Au + Bu,$$

where $S_1^{A+B} = \overline{co}\,((A + B)(S))$ is the second term of the sequence described in (2.2.1), associate to the sum of A and B. Let M be a countably bounded subset contained in the set S_1^{A+B}. Taking into account that $B(M)$ is relatively weakly compact and using the subadditivity of the De Blasi measure β, we get

$$\begin{aligned}\beta(Q(M)) &\leq \beta\,(A(M) + B(M))\\ &\leq \beta(A(M)) + \beta(B(M))\\ &\leq \beta(A(M)) + \beta(\overline{B(M)}^{w})\\ &\leq \Phi(\beta(M)).\end{aligned}$$

This means, in particular, that Q is countably $\mathscr{D}$-set-Lipschitzian multi-valued mapping on S_1^{A+B} with $\mathscr{D}$-function Φ. Let $(u_n)_n$ be a sequence contained in the set S_1^{A+B} and take a sequence $v_n \in Q(u_n)$ such that

$$w_n = u_n - v_n \rightharpoonup w,$$

for some $w \in S_1^{A+B}$. Since $v_n \in A(u_n) + B(u_n)$, there exists a sequence $a_n \in A(u_n)$ and there exists a sequence $b_n \in B(u_n)$ such that

$$v_n = a_n + b_n.$$

Then, $u_n - a_n = w_n + b_n$. Due to the fact that $b_n \in B\left(S_1^{A+B}\right)$ and Theorem 1.3.29, we obtain the sequence $(b_n)_n$ possesses a weakly convergent subsequence. This proves that $(u_n)_n$ possesses a weakly convergent subsequence, since A is hemi-weak compact. The desired result follows from Theorem 2.2.1. Q.E.D.

As an immediate consequence of Theorem 2.2.6, we obtain the following result.

Corollary 2.2.7 *Let S be a nonempty, convex, and closed subset of a Banach algebra X and let A, $B : S \longrightarrow \mathscr{P}_{cl,cv}(E)$ be two multi-valued mappings with weakly sequentially closed graphs such that*

(i) A is hemi-weakly compact and countably β-nonexpansive,
(ii) B is weakly compact,
(iii) $(A + B)(S)$ is a bounded subset of S.

Then, the operator inclusion $u \in Au + Bu$ has, at least, one solution in S.

Remark 2.2.8 Let X be a Banach space. If there exists a subset $S_0 \in \mathscr{B}(X)$ and there exists a converging sequence $(\lambda_n)_n$ contained in the set $(0, 1)$ to 1 such that

$$(A + \lambda_n B)x \in S_0, \text{ for all } x \in S_0.$$

Then, the set S_1^{A+B} is bounded. Using this implication, we obtain Theorem 2.2.6 becomes as an extension of Theorem 1.4.54.

2.2.2 *Hybrid Fixed Point Results Relative to the Weak Topology*

We establish some new fixed point results on nonempty, convex, and closed subset of a Banach algebra satisfying the condition $(\mathscr{P})$, for the sum and the product of three multi-valued mappings having weakly sequentially closed graphs. Precisely, we study the existence of fixed point by looking at the existence of solutions of the operator inclusion

$$x \in Ax \cdot Bx + Cx \tag{2.2.5}$$

with multi-valued entries by using the concepts of countably nonexpansive property and countably $\mathscr{D}$-set-Lipschitzian property with respect to β.

Theorem 2.2.9 *Let S be a nonempty, convex, and closed subset of a Banach algebra X and let A, B, $C : S \longrightarrow \mathscr{P}(X)$ be three multi-valued mappings with weakly sequentially closed graphs such that*

(i) there exists a bounded subset Ω contained in the set $\left\{S_n^{A\cdot B+C},\ n \in \mathbb{N}\right\}$, where $S_1^{A\cdot B+C} = \overline{co}\left((A \cdot B + C)(S)\right)$ is the second term of the sequence described in (2.2.1) *associate to $A \cdot B + C$,*

(ii) $A(u) \cdot B(u) + C(u) \in \mathscr{P}_{cl,cv}(\Omega)$, for all $u \in \Omega$,
(iii) A, B, and C are countably $\mathscr{D}$-set-Lipschitzians with $\mathscr{D}$-functions Φ_A, Φ_B and Φ_C, respectively,
(iv) For any $0 < r \leq \|\Omega\|$, the following estimate is satisfied

$$\|B(S)\|\Phi_A(r) + \|A(S)\|\Phi_B(r) + \Phi_A(r)\Phi_B(r) + \Phi_C(r) < r.$$

Then, the operator inclusion (2.2.5) *has a solution in S.*

Proof Let $\Omega = S_{n_0}^{A\cdot B+C}$, for some $n_0 \in \mathbb{N}$ and define the multi-valued mapping $Q : \Omega \longrightarrow \mathscr{P}_{cl,cv}(\Omega)$ by putting

$$Qu = Au \cdot Bu + Cu.$$

Let M be a countably subset of Ω. Using Lemma 1.5.11, we obtain

$$\beta(AM \cdot BM) \leq \|BM\|\ \Phi_A(\beta(M)) + \|AM\|\ \Phi_B(\beta(M)) + \Phi_A(\beta(M))\ \Phi_B(\beta(M)).$$

This implies that

$$\begin{aligned}\beta(QM) &\leq \beta(AM \cdot BM) + \beta(CM)\\ &\leq \|BM\|\ \Phi_A(\beta(M)) + \|AM\|\ \Phi_B(\beta(M)) + \Phi_A(\beta(M))\ \Phi_B(\beta(M)) + \Phi_C(\beta(M)).\end{aligned}$$

This means that Q is countably $\mathscr{D}$-set-Lipschitzian on Ω with $\mathscr{D}$-function Ψ, where

$$\Psi(r) = \|B(\Omega)\|\ \Phi_A(r) + \|A(\Omega)\|\ \Phi_B(r) + \Phi_A(r)\ \Phi_B(r) + \Phi_C(r).$$

We demonstrate that Q has a weakly sequentially closed graph. In fact, consider $(u_n)_n$ as a weakly convergent sequence of Ω to a point $u \in \Omega$ and choose a sequence $v_n \in Q(u_n)$ such that $v_n \rightharpoonup v \in \Omega$. By the definition of Q, we get there exist three sequence $\sigma_n \in A(u_n)$, $\gamma_n \in B(u_n)$ and $\rho_n \in C(u_n)$ such that

$$v_n = \sigma_n \cdot \gamma_n + \rho_n.$$

From the above discussion, we have

$$\beta\left(\{\sigma_n,\ n \in \mathbb{N}\}\right) \leq \Phi_A\left(\beta\left(\{u_n,\ n \in \mathbb{N}\}\right)\right) = 0.$$

By using Theorem 1.3.29, we can assume that $\{\sigma_n,\ n \in \mathbb{N}\}$ and $\{\gamma_n,\ n \in \mathbb{N}\}$ converge weakly to some points $\sigma, \gamma \in X$, respectively. Due to the fact that A and B have a weakly sequentially closed graph, we get $\sigma \in Au$ and $\gamma \in Bu$ and so,

$$\rho_n \rightharpoonup v - \sigma \cdot \gamma.$$

The desired conclusion follows by a direct application of Corollary 2.2.3. Q.E.D.

Remark 2.2.10 Theorem 2.4.10 is an extension of Theorem 1.5.42, and proves that the hypothesis $\mathcal{D}$-set-Lipschitzian can be made more flexible by assuming that Q is countably $\mathcal{D}$-set-Lipschitzian. Moreover, it shows that the hypothesis that A, B, and C are sequentially weakly upper semi-compact in Theorem 1.5.42, in the statement of that theorem is not needed.

Remark 2.2.11 Every sequentially weakly upper semi-compact and Lipschitzian multi-valued mapping with a Lipschitz constant k is $\mathcal{D}$-set-Lipschitzian with $\mathcal{D}$-function ϕ, where $\phi(r) = k$ and so, it is countably $\mathcal{D}$-set-Lipschitzian. Besides, condition $(\mathcal{P})$ guaranties that the multi-valued mappings $A, B, C : X \longrightarrow \mathcal{P}_{cl,cv}(X)$. Thus, Theorem 2.2.9 becomes as an extension of Theorem 1.5.42, by taking $n_0 = 0$.

Corollary 2.2.12 *Let S be a nonempty, convex, and closed subset of a Banach algebra X and let $A, B, C : S \longrightarrow \mathcal{P}(X)$ be three multi-valued mappings with weakly sequentially closed graphs such that*

(i) there exists a bounded subset Ω contained in the set $\left\{S_n^{A\cdot B+C},\ n \in \mathbb{N}\right\}$,
(ii) $A(u) \cdot B(u) + C(u) \in \mathcal{P}_{cl,cv}(\Omega)$, for all $u \in \Omega$,
(iii) A and C are countably $\mathcal{D}$-set-Lipschitzians with $\mathcal{D}$-functions Φ_A and Φ_C, respectively,
(iv) B is weakly compact

Then, the operator inclusion (2.2.5) has a solution in S provided that

$$\|B(S)\|\Phi_A(r) + \Phi_C(r) < r,$$

for all $0 < r \leq \|\Omega\|$.

Corollary 2.2.13 *Let S be a nonempty, convex, and closed subset of a Banach algebra X and let $A, B, C : S \longrightarrow \mathcal{P}(X)$ be three multi-valued mappings with weakly sequentially closed graphs such that*

(i) there exists a bounded subset Ω contained in the set $\left\{S_n^{A\cdot B+C},\ n \in \mathbb{N}\right\}$,
(ii) $A(u) \cdot B(u) + C(u) \in \mathcal{P}_{cl,cv}(\Omega)$, for all $u \in \Omega$,
(iii) A is countably nonexpansive with respect to β on Ω,
(iv) B is weakly compact on Ω,
(v) C is countably set-contraction with constant q on Ω.

Then, the operator inclusion (2.2.5) *has, at least, one solution in* Ω *provided that*

$$\|B(\Omega)\| + q < 1.$$

Note that the operator inverse $\frac{I-C}{A}$ was used in many papers to study the existence of some fixed point theorems for single-valued mappings in suitable Banach algebras, see for example [9, 10, 12]. In many of these papers, the operator $\frac{I-C}{A}$ play an important role in their arguments. We are now able to establish the following fixed point theorem.

Theorem 2.2.14 *Let* S *be a nonempty, convex and closed subset of a Banach algebra* X *and let* $A, C : X \longrightarrow \mathscr{P}_{cl,cv}(X)$ *and* $B : S \longrightarrow \mathscr{P}_{cl,cv}(X)$ *be three multi-valued mappings with weakly sequentially closed graphs such that*

(i) *A is weakly compact,*
(ii) *$A(X)$, $B(S)$ and $C(X)$ are bounded*
(iii) *B is countably $\mathscr{D}$-set-contraction with $\mathscr{D}$-function Φ,*
(iv) *C is $\mathscr{D}$-set-contraction with $\mathscr{D}$-function Ψ,*
(v) $B(S) \subset \left(\frac{I-C}{A}\right)(S)$,
(vi) $\left(\frac{I-C}{A}\right)^{-1} B(u) \in \mathscr{P}_{cl,cv}(S)$, *for all* $u \in S$.

Then, the operator inclusion (2.2.5) *has, at least, one solution in* S *provided that*

$$\|A(S)\|\ \Phi(r) + \Psi(r) < r, \quad \text{for } r > 0.$$

Proof Define the multi-valued mapping $Q : S_1^{A\cdot\gamma+C} \longrightarrow \mathscr{P}_{cl,cv}\left(S_1^{A\cdot\gamma+C}\right)$ by putting

$$Qx = Ax \cdot \gamma + Cx,$$

for some $\gamma \in B(S)$, where the set

$$S_1^{A\cdot\gamma+C} = \overline{co}((A \cdot \gamma + C)(E))$$

is the second term of the sequence described in (2.2.1) associate to $A \cdot \gamma + C$. By the definition of Q, there exist two sequences $\sigma_n \in A(u_n)$ and $\rho_n \in C(u_n)$ such that

$$v_n = \sigma_n \cdot \gamma + \rho_n.$$

Making use of assumption (i), together with Theorem 1.3.29 enables us to have the sequence $(\sigma_n)_n$ is weakly convergent to a point $\sigma \in Au$. This implies that

$$v - \sigma \cdot \gamma \in Cu.$$

Thus, $v \in Qu$. Invoking Corollary 2.2.3, we get there exists $u \in S_1^{A\cdot\gamma+C}$ such that

$$u \in Au \cdot \gamma + Cu.$$

In other words, we have $\gamma \in \left(\frac{I-C}{A}\right)(u)$. This implies that

$$\left(\frac{I-C}{A}\right)(u) \bigcap B(S) \neq \emptyset.$$

This prove that the operator inverse $\left(\frac{I-C}{A}\right)^{-1}$ is well defined on $B(S)$. From the above discussion, we can define a nonempty, closed, and convex subset

$$\mathfrak{R} := \overline{co}\left(\left(\frac{I-C}{A}\right)^{-1} B(S)\right) \subset S, \tag{2.2.6}$$

which is invariant by the multi-valued mapping

$$x \longrightarrow \left(\frac{I-C}{A}\right)^{-1} Bx.$$

Now, define the multi-valued mapping $F : \mathfrak{R} \longrightarrow \mathscr{P}_{cl,cv}(\mathfrak{R})$ by putting

$$Fu = \left(\frac{I-C}{A}\right)^{-1} B(u).$$

Making use of assumption (v), together with the following inclusion

$$F(S) \subset AF(S) \cdot B(S) + CF(S), \tag{2.2.7}$$

enables us to have $\mathfrak{R}$ is bounded in S. Our next task is to demonstrate that the multi-valued mapping F is countably condensing with respect to β. Indeed, let D be a countably subset of $\mathfrak{R}$. Taking into account that $A(\mathfrak{R})$ is relatively weakly compact, and using Lemma 1.5.11, we obtain

$$\begin{aligned} \beta(FD) &\leq \|A(\mathfrak{R})\| \, \beta(B(D)) + \beta(C(FD)) \\ &\leq \|A(S)\| \, \Phi\left(\beta(D)\right) + \Psi\left(\beta(FD)\right), \end{aligned} \tag{2.2.8}$$

which is a contradiction and so, we can prove that $F(D)$ is relatively weakly compact. So, we may assume that $\Phi(\beta(D)) > 0$. The use of inequality (2.2.8) allows us to obtain that

$$\beta(FD) - \Psi\left(\beta(FD)\right) \leq \|A(S)\| \, \Phi\left(\beta(D)\right).$$

This implies that $\Phi(\beta(F(D))) < \Phi(\beta(D))$. This means, in particular, that Q is countably condensing with respect to β, since the function Φ is nondecreasing. By using a contradiction argument, we can show that $F(M)$ is relatively weakly compact, for any weakly compact subset M of X. Let us consider a sequence $(u_n)_n$ contained in the set $\mathfrak{R}$, which converges weakly to a point u and take a sequence $v_n \in Fu_n$, which converges weakly to a point v. Then, there exists a sequence $(w_n)_n$ such that

$$w_n \in \left(\frac{I-C}{A}\right)(v_n) \bigcap B(u_n), \text{ for all } n \in \mathbb{N}. \tag{2.2.9}$$

Using the fact that $(u_n)_n$ is weakly convergent, we get

$$\begin{aligned} \beta\left(\{w_n,\ n \in \mathbb{N}\}\right) &\leq \beta\left(B\left(\{u_n,\ n \in \mathbb{N}\}\right)\right) \\ &\leq \Phi\left(\beta\left(\{u_n,\ n \in \mathbb{N}\}\right)\right) = 0. \end{aligned}$$

Consequently, the sequence $(w_n)_n$ possesses a weakly convergent renamed subsequence with limit $w \in B(u)$, in view Theorem 1.3.29.

Going back to the inclusion (2.2.9), we get

$$v_n \in w_n \cdot A(v_n) + C(v_n),\ n \in \mathbb{N}.$$

The condition $(\mathscr{P})$ guaranties that

$$v \in w \cdot A(v) + C(v).$$

This imply that

$$w \in \left(\frac{I-C}{A}\right)(v).$$

In other words, we have $v \in Fu$. Hence, by applying Theorem 1.5.37, we deduce that F has a fixed point. Q.E.D.

Theorem 2.2.15 *Let S be a nonempty, convex, and closed subset of a Banach algebra X and let $A, C : X \longrightarrow \mathscr{P}_{cl,cv}(X)$ and $B : S \longrightarrow \mathscr{P}(X)$ be three multi-valued mappings with weakly sequentially closed graph such that*

(i) A is $\mathscr{D}$-set-Lipschitzian with $\mathscr{D}$-function Φ,
(ii) $A(X)$, $B(S)$, and $C(X)$ are bounded,
(iii) B is countably q-set-contraction,
(iv) C is γ-set-contraction,
(v) for each $u \in S$, $\left(\frac{I-C}{A}\right)^{-1} B(u) \in \mathscr{P}_{cl,cv}(S)$.

Then, the operator inclusion $x \in Ax \cdot Bx + Cx$ has a solution in S provided that

$$\|B(S)\|\Phi(r) < (1 - 2q\|A(S)\| - \gamma)r, \ \textit{for}\ r > 0.$$

Proof Let $v \in B(S)$ be fixed and define the mapping $Q : X \longrightarrow \mathscr{P}_{cl,cv}(X)$ by putting

$$Qx = Ax \cdot v + Cx.$$

In the similar way as in the proof of Theorem 2.2.14, we can show that Q is countably $\mathscr{D}$-set-contraction with $\mathscr{D}$-function φ given by

$$\varphi(r) = \|v\|\Phi(r) + \gamma r,$$

and has a weakly sequentially closed graph. Applying Theorem 1.5.37, we can deduce that there exists an element $u \in X$ such that

$$u \in Au \cdot v + Cu,$$

or equivalently

$$v \in \left(\frac{I - C}{A}\right)(u).$$

This implies that

$$\left(\frac{I - C}{A}\right)(u) \bigcap B(S) \neq \emptyset.$$

Then, the operator inverse $\left(\frac{I-C}{A}\right)^{-1}$ exists on $B(S)$. Define the multi-valued mapping $F : \mathfrak{R} \longrightarrow \mathscr{P}_{cl,cv}(\mathfrak{R})$ by putting

$$F(u) = \left(\frac{I - C}{A}\right)^{-1} B(u),$$

where the set $\mathfrak{R}$ as is described in (2.2.6). In the next step of our proof, we prove that F is countably $\frac{1}{2}$-set-contraction. To this end, let us fix an arbitrary countably subset D contained in the set $\mathfrak{R}$ with $\beta(D) > 0$. Going back to the inclusion (2.2.7), we get $\mathfrak{R}$ is bounded. Using Lemma 1.5.11, we obtain

$$\begin{aligned}
\beta(F(D)) &\leq \beta(AF(D) \cdot B(D)) + \beta(CF(D)) \\
&\leq \|B(D)\|\beta(AF(D)) + \|AF(D)\|\beta(B(D)) + \beta(CF(D)) \\
&\leq \|B(S)\|\Phi(\beta(FD)) + q\|AF(D)\|\beta(D) + \gamma\beta(F(D)) \\
&\leq (1 - 2q\|A(S)\| - \gamma)\beta(FD) + q\|AF(D)\|\beta(D) + \gamma\beta(F(D)) \\
&\leq (1 - 2q\|A(S)\|)\beta(FD) + q\|AF(D)\|\beta(D).
\end{aligned}$$

This implies that

$$2\beta(F(D)) \leq \beta(D).$$

This means, in particular, that F is countably $\frac{1}{2}$-set-contraction. The result follows from Corollary 2.2.3. Q.E.D.

As a consequence of Theorem 2.2.15, we have the following result.

Corollary 2.2.16 *Let S be a nonempty, convex and closed subset of a Banach algebra X and let $A, C : X \longrightarrow \mathscr{P}_{cl,cv}(X)$ and $B : S \longrightarrow \mathscr{P}(X)$ be three multi-valued mappings with weakly sequentially closed graphs such that*

(i) A is countably nonexpansive with respect to β,
(ii) $A(X)$, $B(S)$ and $C(X)$ are bounded
(iii) B is countably q_1-set-contraction,
(iv) C is q_2-set-contraction,
(v) $\left(\frac{I-C}{A}\right)^{-1} B(u) \in \mathscr{P}_{cl,cv}(S)$, for all $u \in S$,
(vi) there exists $\delta > 0$ such that the following estimate is satisfied

$$2\delta\|A(S)\| + \|B(S)\| + q_2 < 1.$$

Then, the operator inclusion (2.2.5) has, at least, one solution in S.

2.3 Existence Results in Generalized Banach Algebra Under Weak Topology

This section is devoted to studying of new version of the fixed point result involving three nonlinear operators in generalized Banach algebra by using an analog of the well-known Perov's fixed point theorems. First of our main results is the following.

Theorem 2.3.1 *Let S be a nonempty, convex, and closed subset of a generalized Banach space X and let $T : S \longrightarrow S$ be a weakly sequentially continuous mapping such that $T(S)$ is relatively weakly compact. Then, T has a fixed point.*

Proof Let $M = \overline{co}(T(S))$. Because $T(S)$ is relatively weakly compact, then M is a weakly compact, convex subset of X, in view of Theorem 1.3.34. On the other hand, we can verify that

$$\begin{aligned} T(M) &\subset T(S) \\ &\subset \overline{co}(T(S)). \end{aligned}$$

It follows from Theorem 1.3.44, that T is weakly continuous on M. Consider $X = (X, \sigma(X, (f_i)_{i\in I}))$ the space endowed with the weak topology and note that $T : M \longrightarrow M$ is continuous with M is compact. The use of Theorem 2.4.3 achieves the proof. Q.E.D.

By using the concept of generalized De Blasi measure of weak non-compactness in vector-valued Banach spaces, we obtain the following fixed point theorem.

Theorem 2.3.2 *Let S be a nonempty, closed, and convex subset of a generalized Banach space X and let $T : S \longrightarrow S$ be a weakly sequentially continuous mapping such that $T(S)$ is bounded. If T is condensing with respect to β_G, then T has a fixed point in S.*

Proof Let x_0 be fixed in S and define the set

$$\Sigma = \{K \subset S, \quad x_0 \in K, \quad K \text{ is closed, convex and bounded with } T(K) \subseteq K\}.$$

Notice that Σ is nonempty, since $\overline{co}\left(T(S)\bigcup\{x_0\}\right)$ is a nonempty, closed, convex, and bounded subset of S, and

$$T\left(\overline{co}\left(T(S)\bigcup\{x_0\}\right)\right) \subset T(S) \subset \overline{co}\left(T(S)\bigcup\{x_0\}\right).$$

This shows that $\overline{co}(T(S)\bigcup\{x_0\}) \in \Sigma$. Put $M = \bigcap\limits_{K\in\Sigma} K$. Then,

$$\begin{cases} x_0 \in M \subseteq S, \\ M \text{ is a nonempty, bounded, closed, and convex subset of } S, \\ T(M) \subseteq M. \end{cases}$$

This means, in particular, that $M \in \Sigma$. We prove that M is weakly compact. Denoting by $M_0 = \overline{co}\left(T(M)\bigcup\{x_0\}\right)$. From the above discussion, we get $M_0 \subseteq M$ and

$$T(M_0) \subseteq T(M) \subseteq M_0.$$

Therefore, $M_0 \in \Sigma$ and so, $M \subseteq M_0$. Hence, $M = M_0$. Since M is weakly closed, it suffices to show that M is relatively weakly compact. If not the case, we have $\beta_G(M) > 0_{\mathbb{R}^n}$. This implies that

$$\begin{aligned} \beta_G(M) &= \beta_G(\overline{co}(T(M)\bigcup\{x_0\})) \\ &< \beta_G(M), \end{aligned}$$

which is a contradiction and so, M is relatively weakly compact. Besides, T is weakly continuous on M, since M is weakly compact. Consider $(X, \sigma(X, (f_i)_{i\in I}))$ the space endowed with the weak topology and when we apply Theorem 1.3.44, we reach that T has, at least, one fixed point in M. Q.E.D.

Theorem 2.3.3 *Let S be a nonempty, convex and closed subset of a generalized Banach algebra X satisfying the sequential condition $(\mathcal{G}\mathcal{P})$ and let $A, C : X \longrightarrow X$ and $B : S \longrightarrow X$ be three weakly sequentially continuous operators such that*

(i) A and C are contractive with matrix M_A and M_C, respectively,
(ii) A is regular and $B(S)$ is bounded,

(iii) If $x = Ax \cdot By + Cx$, *for* $y \in S$, *then* $x \in S$,

(iv) $\left(\frac{I-C}{A}\right)^{-1}$ *is weakly compact.*

Then, the operator equation (2.0.1) *has, at least, one solution in S provided that*

$$\|B(S)\|M_A + M_C$$

is a matrix converging to zero.

Proof Let $y \in S$ be fixed and define the mapping $\varphi_y : X \longrightarrow X$ by putting

$$\varphi_y(x) = Ax \cdot By + Cx.$$

Based on assumption (i), we have

$$\begin{aligned}\|\varphi_y(x_1) - \varphi_y(x_2)\| &\leq \|Ax_1By - Ax_2By\| + \|Cx_1 - Cx_2\| \\ &\leq \|Ax_1 - Ax_2\|\|By\| + \|Cx_1 - Cx_2\| \\ &\leq M_A\|x_1 - x_2\|\|By\| + M_C\|x_1 - x_2\| \\ &\leq (\|B(S)\|M_A + M_C)\|x_1 - x_2\|,\end{aligned}$$

whenever $x_1, x_2 \in X$. Then, φ_y is contractive with a matrix $\|B(S)\|M_A + M_C$. In view of Theorem 1.6.7, we obtain that there exists an unique point $x_y \in X$ such that

$$x_y = Ax_y \cdot By + Cx_y.$$

Define the mapping $T : S \longrightarrow S$ by putting

$$Ty = \left(\frac{I-C}{A}\right)^{-1} By.$$

Let $(x_n)_n$ be a weakly convergent sequence contained in the set S to a point $x \in S$ and take $y_n = T(x_n)$. Then

$$y_n = A(y_n) \cdot B(x_n) + C(y_n).$$

Based on assumption (iii), we obtain $y_n \in S$. It is easy to see that the sequence $(y_n)_n$ possesses a weakly convergent subsequence, say $(y_{n_k})_k$, to a point $y \in S$. The condition $(\mathscr{G}\mathscr{P})$ guaranties that

$$y_{n_k} = Ay_{n_k} \cdot Bx_{n_k} + Cy_{n_k} \rightharpoonup Ay \cdot Bx + Cy.$$

This means that $y = Ay \cdot Bx + Cy$. Now, we are going to show that

$$T(x_n) \rightharpoonup Tx.$$

If not the case, there exists a weakly neighborhood V^w of Tx and there exists a subsequence $(x_{n_j})_j$ with $Tx_{n_j} \notin V^w$. Arguing as above, there exists a weakly convergent subsequence $(Tx_{n_{j_k}})_{k\in\mathbb{N}}$ to some y, but this contradicts the above discussion and so, T is a weakly sequentially continuous operator. The desired conclusion of the theorem follows by a direct application of Theorem 1.3.44. Q.E.D.

As an immediate consequence of Theorem 2.3.3, we derive the following.

Corollary 2.3.4 *Let S be a nonempty, convex and closed subset of a generalized Banach algebra X satisfying the sequential condition $(\mathcal{GP})$ and let $A, C : X \longrightarrow X$ and $B : S \longrightarrow X$ be three weakly sequentially continuous operators such that*

(i) A and C are contractive with matrix M_A and M_C respectively,
(ii) A is regular,
(iii) $A(S)$, $B(S)$ and $C(S)$ are relatively weakly compact,
(iv) If $x = Ax\cdot By + Cx$, for $y \in S$ then, $x \in S$.

Then, the operator equation (2.0.1) *has, at least, one solution in S provided that*

$$\|B(S)\|M_A + M_C$$

is a matrix converging to zero.

Now our attention is directed toward the case when, only B is weakly compact, in order to obtain the following result.

Theorem 2.3.5 *Let S be a nonempty, convex and closed subset of a generalized Banach algebra X satisfying condition $(\mathcal{GP})$ and let $A, C : X \longrightarrow X$ and $B : S \longrightarrow X$ be three weakly sequentially continuous operators such that*

(i) A and C are contractive with matrix M_A and M_C, respectively,
(ii) A is regular,
(iii) B is weakly compact,
(iv) If $x = Ax \cdot By + Cx$, for $y \in S$, then $x \in S$.

Then, the operator equation (2.0.1) *has, at least, one solution in S provided that*

$$\|B(S)\|M_A + M_C$$

is a matrix converging to zero.

Before the proof of Theorem 2.3.5, we shall formulate two lemmas.

Lemma 2.3.6 *Let R and K be two nonempty subset of a generalized Banach algebra X with $S \in \mathcal{B}(X)$ and $K \in \mathcal{W}(X)$. Then,*

$$\beta_G(R\cdot K) \leq \|K\|\beta_G(R),$$

where β_G is the generalized De Blasi measure of weak non-compactness.

Proof By using the definition of β_G, there exists a subset $K' \in \mathscr{W}(X)$ and there exists a vector $r = (r_1, \dots, r_n)$, $r_i > 0$ such that the following inclusion is satisfied

$$R \cdot K \subseteq K' \cdot K + \overline{\mathscr{B}}_X \left(\beta_G(R) + r\|K^{-1}\|\right) \cdot K,$$

where

$$r\|K^{-1}\| := \begin{pmatrix} \dfrac{r_1}{\|K\|_1} \\ \vdots \\ \dfrac{r_n}{\|K\|_n} \end{pmatrix}.$$

After simple calculations, we can prove that

$$R \cdot K \subseteq K' \cdot K + \overline{\mathscr{B}}_X \left(\|K\|\beta_G(R) + r\right).$$

Making use of the subadditivity of the generalized De Blasi measure β_G, together with Lemma 1.6.5 enables us to have

$$\begin{aligned} \beta_G(R \cdot K) &\leq \beta_G\left(K \cdot K'\right) + \beta_G\left(\overline{\mathscr{B}}_X \left(\|K\|\beta_G(R) + r\right)\right) \\ &\leq \|K\|\beta_G(S) + r. \end{aligned}$$

Since r is arbitrary, we get $\beta_G(R \cdot K) \leq \|K\|\beta_G(R)$. Q.E.D.

Lemma 2.3.7 *Let X be a generalized Banach space and let $T : X \longrightarrow X$ be a weakly sequentially continuous mapping. If T is contractive with a matrix M. Then,*

$$\beta_G\left(T(D)\right) \leq M\beta_G\left(D\right),$$

for any bounded subset D of X.

Proof Let D be a bounded subset of X and let $\varepsilon = (\varepsilon_1, \dots, \varepsilon_n)$, $\varepsilon_i \in \mathbb{R}_+^*$ be such that $\varepsilon > \beta_G(D) > 0$, for $i = 1, \dots, n$. Then, there exists $r = (r_1, \dots, r_n)$ with $0 < r < \varepsilon$ and there exists a subset $K \in \mathscr{W}(X)$ such that

$$D \subseteq K + \overline{\mathscr{B}}_X(r).$$

Put $x \in D$ and $y = Tx$. Then,

$$\begin{aligned}\|y - Tk\|_i &= \|Tx - Tk\|_i \\ &\leq \sum_{j=1}^{n} a_{ij} \|x - k\|_j \\ &\leq \sum_{j=1}^{n} a_{ij} \|b\|_j \\ &\leq \sum_{j=1}^{n} a_{ij} r_j,\end{aligned}$$

for some $k \in K$ and some $b \in \overline{\mathscr{B}}_X(r)$. This implies that

$$\|y - Tk\| \leq \begin{pmatrix} \sum_{j=1}^{n} a_{1j} r_j \\ \sum_{j=1}^{n} a_{2j} r_j \\ \vdots \\ \sum_{j=1}^{n} a_{nj} r_j \end{pmatrix} = Mr.$$

Thus, $y \in TK + \overline{\mathscr{B}}_X(Mr)$ and so,

$$TD \subset TK + \overline{\mathscr{B}}_X(Mr).$$

Taking into account the weak compactness of $\overline{TK}^w$, and using the subadditivity of the generalized De Blasi measure of weak non-compactness β_G, it is shown that

$$\beta_G(TD) \leq Mr.$$

This completes the proof. Q.E.D.

Proof of Theorem 2.3.5 Arguing as in the proof of Theorem 2.3.3, we can show that the operator

$$T := \left(\frac{I - C}{A}\right)^{-1} B \tag{2.3.1}$$

is well defined on S. Now, we are going to show that the set $T(S)$ is relatively weakly compact. Based on the subadditivity of β_G and using the following equality

$$T = ATB + CT, \tag{2.3.2}$$

we arrive at the following

$$\beta_G(T(S)) \leq \beta_G\left(AT(S) \cdot B(S)\right) + \beta_G\left(CT(S)\right).$$

Further, applying Lemmas 2.3.6 and 2.3.7 and keeping in mind assumptions concerning the operators A, B and C, we get

$$\beta_G(T(S)) \leq \left(\|B(S)\|M_A + M_C\right)\beta_G(T(S)).$$

But this contradicts the fact that the matrix

$$\|B(S)\|M_A + M_C$$

is converging to zero. Q.E.D.

Corollary 2.3.8 *Let S be a nonempty, bounded, convex, and closed subset of a generalized Banach algebra X satisfying condition $(\mathscr{G}\mathscr{P})$ and let $A, C : X \longrightarrow X$ and $B : S \longrightarrow X$ be three weakly sequentially continuous operators such that*

(i) A is contractive with a matrix M,
(ii) A is regular,
(iii) B and C are weakly compact,
(iv) $\left(\frac{I-C}{A}\right)^{-1}$ is well defined on $B(S)$,
(v) If $x = Ax \cdot By + Cx$, for $y \in S$, then $x \in S$.

Then, the operator equation (2.0.1) *has, at least, one solution in S provided that*

$$\|B(S)\|M$$

is a matrix converging to zero.

Now, we can modify some hypotheses of Theorem 2.3.5, in order to obtain the following result.

Theorem 2.3.9 *Let S be a nonempty, bounded, convex, and closed subset of a generalized Banach algebra X satisfying condition $(\mathscr{G}\mathscr{P})$ and let A, B, $C : S \longrightarrow X$ be three operators such that*

(i) A and B are weakly sequentially continuous and weakly compact,
(ii) A is regular,
(iii) If $x = Ax \cdot By + Cx$, for $y \in S$ then, $x \in S$,

(iv) If $(I - C)(x_n) \rightharpoonup z$, *for some* $z \in X$, *then the sequence* $(x_n)_n$ *possesses a weakly convergent subsequence,*

(*v*) $\left(\dfrac{I-C}{A}\right)^{-1}$ *is well defined on* $B(S)$.

Then, the operator equation (2.0.1) *has, at least, one solution in* S.

Proof Let $(x_n)_n$ be any sequence contained in the set S and take $y_n = T(x_n)$, where T is described in equality (2.3.1). Then,

$$y_n = A(y_n) \cdot B(x_n) + C(y_n).$$

Making use the last part of assumption (i), together with equality (2.3.2) enables us to have the sequence $(A(y_n))_n$ possesses a renamed weakly convergent subsequence to some $x \in S$ and $(B(x_n))_n$ possesses a renamed weakly convergent subsequence to some $x' \in S$. The condition $(\mathcal{GP})$ guaranties that

$$A(y_n) \cdot B(x_n) = (I - C)(y_n) \rightharpoonup x \cdot x'.$$

Based on assumption (iv), we obtain the sequence $(y_n)_n$ possesses a weakly convergent subsequence and so, the set $T(S)$ is relatively weakly compact, in view of Theorem 1.3.29. The desired result follows from Theorem 2.3.3. Q.E.D.

Remark 2.3.10 Let S be a nonempty, bounded, convex, and closed subset of a generalized Banach algebra X satisfying condition $(\mathcal{GP})$ and let $C : S \longrightarrow X$ be a weakly sequentially continuous mapping. If C is a contractive mapping with a matrix M, then C verify the assumption (iv) of Theorem 2.3.9. Indeed, consider $(x_n)_n$ as a sequence contained in the set S such that

$$(I - C)(x_n) \rightharpoonup y, \quad \text{for some } y \in X.$$

Then,

$$\beta_G(\{x_n,\ n \in \mathbb{N}\}) \leq \beta_G(\{(I - C)(x_n),\ n \in \mathbb{N}\}) + \beta_G(\{C(x_n),\ n \in \mathbb{N}\}).$$

Taking into account that $((I - C)(x_n))_n$ is relatively weakly compact, and using Lemma 2.3.7, we deduce that

$$\beta_G(\{x_n,\ n \in \mathbb{N}\}) \leq M\beta_G(\{x_n,\ n \in \mathbb{N}\}).$$

Because M is a matrix converging to zero, it follows that $(x_n)_n$ possesses a weakly convergent subsequence.

As an immediate consequence of Theorem 2.3.9, we obtain the following result.

Corollary 2.3.11 *Let S be a nonempty, bounded, convex, and closed subset of a generalized Banach algebra X satisfying condition $(\mathcal{G}\mathcal{P})$ and let $A, B, C : S \longrightarrow X$ be three operators such that*

(i) A is regular,
(ii) A and B are weakly sequentially continuous,
(iii) B is weakly compact,
(iv) If $\left(\frac{I-C}{A}\right) x_n \rightharpoonup z$, for some $z \in X$, then $(x_n)_n$ possesses a weakly convergent subsequence,
(v) $\left(\frac{I-C}{A}\right)^{-1} B(S) \subseteq S$.

Then, the operator equation (2.0.1) *has, at least, one solution in S.*

Theorem 2.3.12 *Let S be a nonempty, bounded, convex, and closed subset of a generalized Banach algebra X satisfying condition $(\mathcal{G}\mathcal{P})$ and let $A, C : X \longrightarrow X$ and $B : S \longrightarrow X$ be three operators such that*

(i) A is regular,
(ii) B and $I - \frac{I-C}{A}$ are contractive with matrix M_1 and M_2, respectively,
(iii) B and $\frac{I-C}{A}$ are weakly sequentially continuous,
(iv) $\left(\frac{I-C}{A}\right)^{-1}$ exists on $B(S)$, and
(v) if $x = Ax \cdot By + Cx$, for $y \in S$, then $x \in S$.

Then, the operator equation (2.0.1) *has, at least, one solution in S provided that*

$$(I - M_2)^{-1} M_1$$

is a matrix converging to zero.

Proof Define the mapping $T : S \longrightarrow S$ by putting

$$Tx = \left(\frac{I-C}{A}\right)^{-1} Bx.$$

Consider $(x_n)_n$ as a weakly convergent sequence to an element $x \in S$. Going back to the following equality

$$T = B + \left(I - \frac{I-C}{A}\right) T,$$

and using the subadditivity of the generalized De Blasi measure, we obtain

$$\begin{aligned}
\beta_G(\{T(x_n),\ n \in \mathbb{N}\}) &\leq \beta_G(\{B(x_n),\ n \in \mathbb{N}\}) + \beta_G\left(\left(I - \frac{I-C}{A}\right)(\{T(x_n),\ n \in \mathbb{N}\})\right) \\
&\leq M_1 \beta_G(\{x_n,\ n \in \mathbb{N}\}) + M_2 \beta_G(\{T(x_n),\ n \in \mathbb{N}\}) \\
&\leq M_2 \beta_G(\{T(x_n),\ n \in \mathbb{N}\}).
\end{aligned}$$

This implies that, the sequence $(T(x_n))_n$ possesses a weakly subsequence, say $(T(x_{n_k}))_k$, to a point $y \in S$. Keeping in mind assumptions concerning the operators B and $\frac{I-C}{A}$, we obtain $y = Tx$. In the next step of our proof, we prove that T is condensing with respect to β_G. Then,

$$\begin{aligned}\beta_G(T(D)) &\leq \beta_G(B(D)) + \beta_G\left(\left(I - \frac{I-C}{A}\right)T(D)\right)\\ &\leq M_1\beta_G(V) + M_2\beta_G(T(D)),\end{aligned}$$

for any subset D of S with $\beta(D) > 0$. This means that

$$\beta_G(T(D)) \leq (I - M_2)^{-1}M_1\beta_G(D),$$

and when we apply Theorem 2.3.2, we reach the desired conclusion. Q.E.D.

By combining Theorem 2.3.12, together with Theorem 1.6.8, we obtain a new version of the Krasnoselskii type fixed point theorem.

Corollary 2.3.13 *Let S be a nonempty, bounded, convex, and closed subset of a generalized Banach algebra X satisfying condition $(\mathcal{G}\mathcal{P})$ and let $C : X \longrightarrow X$ and $B : S \longrightarrow X$ be two weakly sequentially continuous mapping such that*

(i) B and C are contractive with matrix M_1 and M_2, respectively,
(ii) $(I - C)^{-1}$ exists on $B(S)$,
(iii) If $x = By + Cx$, for $y \in S$, then $x \in S$.

Then, the operator equation (2.0.1) *has, at least, one solution in S provided that*

$$(I - M_2)^{-1}M_1$$

is a matrix converging to zero.

2.4 Fixed Point Theorems in RWC-Banach Algebra

In the paper [21], J. Banás and L. Olszowy have solved an open problem raised by Jeribi and Krichen [22], by establishing that every Banach algebra satisfying the sequential condition $(\mathcal{P})$ is a WC-Banach algebra, see Definition 1.3.47. Precisely, they demonstrated the equivalence between the two concepts. For this reason, they have introduced the class of relatively weakly compact Banach algebras, see Definition 1.3.61. So, we find firstly, a sequential characterization of RWC-Banach algebras and we indicate some principals Banach algebras satisfying this property. Secondly, we precise the importance of this development to give a new extension in fixed point theory for the product of two nonlinear operators. The results of this section can be found in [14].

2.4.1 *Sequential Characterization of RWC-Banach Algebras*

Because the product of two relatively weakly compact in a Banach algebra satisfying the condition $(\mathscr{P})$ is not necessarily a relatively weakly compact subset, we will introduce the following definition.

Definition 2.4.1 We will say that the Banach algebra X satisfies condition $(\overline{\mathscr{P}})$ if

$$(\overline{\mathscr{P}})\begin{cases}\text{For any weakly convergent sequences } (x_n)_n \text{ and } (y_n)_n \text{ of } X,\\ \text{then } (x_n \cdot y_n)_n \text{ has a weakly convergent subsequence.}\end{cases}$$

Notice that the class of Banach algebras satisfying $(\overline{\mathscr{P}})$ comprises those, which are satisfying $(\mathscr{P})$, but the converse may not be true. In fact, let us take the Banach algebra ℓ^2 endowed with its usual norm and with the multiplication defined in the following way

$$x \cdot y = \left(\sum_{n=1}^{\infty} x_n y_n, 0, 0, \ldots\right), \quad \text{for all } x,\ y \in \ell^2. \tag{2.4.1}$$

It is clear that the two sequences $(x_n = e_n)_n$ and $(y_n = e_n)_n$, where $(e_n)_n$ is the sequence of canonical vectors in ℓ^2, are weakly convergent to the null element of the space ℓ^2. In contrary $(x_n \cdot y_n)_n$ is weakly convergent to e_1, since $x_n \cdot y_n = e_1$.

A characterization of the RWC-Banach algebras was established as follows.

Theorem 2.4.2 *Let X be a Banach algebra. Then, the following assertions are equivalent*

(i) X is RWC-Banach algebra,
(ii) X satisfies the sequential condition $(\overline{\mathscr{P}})$.

Proof Let $(x_n)_n$ and $(y_n)_n$ be two weakly convergent sequences in a RWC-Banach algebra X to some point $x \in X$ and $y \in X$, respectively. Due to the fact that X is a RWC-Banach algebra and $\{x_n,\ n \in \mathbb{N}\}$ and $\{y_n,\ n \in \mathbb{N}\}$ are two relatively weakly compact subset of X, we get

$$\beta(\{x_n \cdot y_n,\ n \in \mathbb{N}\}) \leq \beta(\{x_n,\ n \in \mathbb{N}\} \cdot \{y_n,\ n \in \mathbb{N}\}) = 0.$$

It follows from Theorem 1.3.29, that $(x_n \cdot y_n)_n$ possesses a weakly convergent subsequence. This show that X satisfying the condition $(\overline{\mathscr{P}})$. Let M and N be two relatively weakly compact subsets of a Banach algebra X, which satisfies the condition $(\overline{\mathscr{P}})$. Let $(z_n)_n$ be a sequence of elements contained in the set $M \cdot N$. Then, there exist two sequences $x_n \in M$ and $y_n \in N$ such that

$$z_n = x_n \cdot y_n, \quad \text{for all } n \in \mathbb{N}.$$

By using Theorem 1.3.29, there exist two renamed weakly convergent subsequences of $(x_n)_n$ and $(y_n)_n$. The condition $(\overline{\mathscr{P}})$ guaranties that the sequence $(z_n = x_n \cdot y_n)_n$ possesses a weakly convergent subsequence. Using again Theorem 1.3.29, we infer that $M \cdot N$ is relatively weakly compact. Q.E.D.

Let us mentioned that Diestel [23] was asked if the Dunford–Pettis property can be lifted from a Banach space X to $C(K, X)$ or $L^1(J, X)$, where $L^1(J, X)$ denote the space of all Bochner integrable functions on closed interval J. In the paper [24], M. Talagrand constructed counter-examples for these questions so the answer is negative in general.

In the following result, the Banach algebra $C(K, X)$ is considered with the sup-norm and with the operation of the multiplication of elements defined pointwise

$$(f \cdot g)(t) = f(t)g(t), \quad \text{for all } t \in [0, 1]. \tag{2.4.2}$$

Theorem 2.4.3 *Let K be a compact Hausdorff space and let X be a Banach algebra. Then, the following assertions are equivalent*

(i) X is a RWC-Banach algebra.
(ii) $C(K, X)$ is a RWC-Banach algebra.

Proof Suppose that X is RWC-Banach algebra. In view of Theorem 2.4.2, it is sufficient to verify that $C(K, X)$ satisfies the condition $(\overline{\mathscr{P}})$. To this end, let $(u_n)_n$ and $(v_n)_n$ be two weakly convergent sequences contained in the set $C(K, X)$ to some elements u and v, respectively. Define a sequence $(\zeta_n)_n$ of $C(K, X)$ by

$$\zeta_n(t) = u_n(t) \cdot v_n(t), \quad \text{for all } t \in K, \ n \in \mathbb{N}.$$

The use of Remark 1.3.7 as well as Theorem 1.3.50 leads to

$$u_n(t) \rightharpoonup u(t) \text{ and } v_n(t) \rightharpoonup v(t), \quad \text{for each } t \in K.$$

Then, the sets $\{u_n(t),\ n \in \mathbb{N}\}$ and $\{v_n(t),\ n \in \mathbb{N}\}$ are relatively weakly compact in X. Since X is RWC-Banach algebra, it follows that the sequence product $\{u_n(t),\ n \in \mathbb{N}\} \cdot \{v_n(t),\ n \in \mathbb{N}\}$ is also relatively weakly compact. This together, with the following inclusion

$$\{u_n(t) \cdot v_n(t),\ n \in \mathbb{N}\} \subset \{u_n(t),\ n \in \mathbb{N}\} \cdot \{v_n(t),\ n \in \mathbb{N}\}$$

implies the relatively weak compactness of the sequence $(\zeta_n(t))_n$. One sees, as a direct application of Theorem 1.3.29, that the sequence $(\zeta_n(t))_n$ has a weakly convergent subsequence, say $\left(\zeta_{n_k}(t)\right)_k$, to a function $\zeta(t) \in X$. Next, we claim that $\zeta \in C(K, X)$. Let $t,\ t_0 \in K$. Without loss of generality, we assume that $\zeta(t) \neq \zeta(t_0)$. Then, on account of Theorem 1.3.18, there exists a function $x^* \in X^*$, with $\|x^*\| = 1$ such that

$$\begin{aligned}\|\zeta(t)-\zeta(t_0)\| &= x^*(\zeta(t)-\zeta(t_0))\\ &\leq x^*(\zeta(t)-\zeta_{n_k}(t))+x^*(\zeta_{n_k}(t)-\zeta_{n_k}(t_0))+x^*(\zeta_{n_k}(t_0)-\zeta(t_0))\\ &\leq x^*(\zeta(t)-\zeta_{n_k}(t))+\|\zeta_{n_k}(t)-\zeta_{n_k}(t_0)\|+x^*(\zeta_{n_k}(t_0)-\zeta(t_0)).\end{aligned}$$

If we consider the continuity of $\zeta_{n_k}: K \longrightarrow X$, for all $k \in \mathbb{N}$, together with

$$\zeta_{n_k}(s) \rightharpoonup \zeta(s), \quad \text{for all } s \in K,$$

we can prove that $\zeta \in C(K, X)$, which achieves the claim. Combining this fact with Theorem 1.3.50, we obtain

$$\zeta_{n_k} = x_{n_k} \cdot y_{n_k} \rightharpoonup \zeta.$$

Then, $C(K, X)$ satisfies the condition $(\overline{\mathscr{P}})$. Conversely, suppose that $C(K, X)$ is a RWC-Banach algebra. We will show that X is relatively weakly compact. If we suppose the contrary, by Theorem 2.4.2, we can find two sequences $(\xi_n)_n$ and $(\zeta_n)_n$ of X, which converge to some points ξ and ζ in the weak topology of X such that the sequence $(\xi_n \cdot \zeta_n)_n$ does not have any weakly convergent subsequence. Consider the sequences $(u_n)_n$ and $(v_n)_n$ of $C(K, X)$ given by

$$u_n(t) = \xi_n \quad \text{and} \quad v_n(t) = \zeta_n, \quad \text{for all } t \in K.$$

It is clear that $(u_n)_n$ and $(v_n)_n$ are bounded, and when we apply Theorem 1.3.50, we reach that $(u_n)_n$ and $(v_n)_n$ converge weakly in $C(K, X)$ to the functions ξ and ζ. Let us denote $z_n = u_n \cdot v_n$, for each $n = 0, 1, \ldots$. Then, on account of Theorem 2.4.2, we can extract a subsequence, say $\left(z_{n_k}\right)_k$, which converges weakly to an element $z \in C(K, X)$. By proceeding essentially as above, we can prove

$$z_{n_k}(t) \rightharpoonup z(t), \quad \text{for all } t \in K.$$

In other words, the subsequence $\left(\xi_{n_k} \cdot \zeta_{n_k}\right)_k$ is weakly convergent in X, which is a contradiction, and the claim is proved. Q.E.D.

Remark 2.4.4 Observe that if X is a WC-Banach algebra, then the space $C(K, X)$ is also a WC-Banach algebra. This is an immediate consequence of Theorems 1.3.50 and 1.3.59. For more detail, see for example [22, 25]. In the similar way as in the proof of Theorem 2.4.3, we can show that the reciprocal involvement. Indeed, if we suppose the contrary, then X does not satisfy the sequential condition $(\mathscr{P})$ in view of Theorem 1.3.59. Based on Proposition 1.3.58, we can show that there are two weakly convergent sequences $(\xi_n)_n$ and $(\zeta_n)_n$ of X to origin θ such that $(\xi_n \cdot \zeta_n)_n$ does not converge to θ in the weak topology of X. Defining now the two sequences $(u_n)_n$ and $(v_n)_n$ contained in the space $C(K, X)$ by $u_n(t) = \xi_n$ and $v_n(t) = \zeta_n$, for each $t \in K$. It is easily seen that $(u_n \cdot v_n)_n$ does not converge to the origin θ although $(u_n)_n$ and $(v_n)_n$ weakly converge to origin θ in $C(K, X)$. It follows from Proposition 1.3.58, that $C(K, X)$ does not satisfy $(\mathscr{P})$, or equivalently $C(K, X)$ is not a WC-Banach algebra in view of Theorem 1.3.59, which is absurd.

Let X be a Banach space and let $J = [0, 1]$. Recall that a subset $\Omega \subset L^1(J, X)$, the Banach space of all Bochner integrable functions $f : J \to X$ endowed with norm

$$\|f\|_1 := \int_J \|f(t)\|_X \, dt < \infty,$$

and the convolution product is uniformly integrable, if for each $\varepsilon > 0$, there exists $\delta > 0$ such that if $\nu(E) < \delta$, then

$$\int_E \|f\| d\nu \leq \varepsilon,$$

for all $f \in \Omega$, where ν is the Lebesgue measure on J, see [26]. Next, we prove that $L^1(J, X)$ is a RWC-Banach algebra, whenever X is reflexive.

Theorem 2.4.5 *Let X be a reflexive Banach algebra and let $J = [0, 1]$. Then, the space $L^1(J, X)$ is a RWC-Banach algebra.*

Proof Let us define the convolution product "$*$" on $L^1(J, X)$ by

$$(f * g)(t) = \int_0^t f(t - s)g(s) \, ds \text{ for all } t \in J. \tag{2.4.3}$$

In view of Theorem 2.4.2, it is sufficient to show that the Banach algebra $L^1(J, X)$ endowed with the norm $\| \cdot \|_1$ and with the product (2.4.3) satisfies the sequential condition $(\overline{\mathcal{P}})$. To see this, let us consider $(f_n)_n$ and $(g_n)_n$ as two weakly convergent sequences in $L^1(J, X)$. We start by showing that the set $\{f_n * g_n,\ n \in \mathbb{N}\}$ is uniformly integrable. Let $\varepsilon > 0$. Arguing as above, we obtain $\{f_n,\ n \in \mathbb{N}\}$ and $\{g_n,\ n \in \mathbb{N}\}$ are two relatively weakly compact subsets of $L^1(J, X)$. Let M be the bound of the set $\{g_n,\ n \in \mathbb{N}\}$. Recall that a relatively weakly compact subsets of $L^1(J, X)$ are characterized in Proposition 1.3.40, which tells us that the set $\{f_n,\ n \in \mathbb{N}\}$ is uniformly integrable. Choose $\delta > 0$ such that

$$\int_E \|f_n\| \nu(E) \leq \frac{\varepsilon}{M},$$

whenever $\nu(E) < \delta$, here ν is the Lebesgue measure on J. Let $E \subset J$ be such that $\nu(E) < \delta$. By using Fubini's theorem, we get

$$\begin{aligned}
\int_E \left\| \int_0^t f_n(t - s)g_n(s) \, ds \right\| dt &\leq \int_E \int_0^t \|f_n(t - s)\| \, \|g_n(s)\| \, ds dt \\
&\leq \int_0^1 \|g_n(s)\| \, ds \int_E \|f_n(u)\| \, du \\
&\leq M \int_E \|f_n(u)\| \, du.
\end{aligned}$$

This shows that

$$\int_E \|f_n * g_n\| \leq \varepsilon,$$

for all $n \in \mathbb{N}$, which means that the set $\{f_n * g_n,\ n \in \mathbb{N}\}$ is uniformly integrable. Hence, an application of Proposition 1.3.40 yields that $\{f_n * g_n,\ n \in \mathbb{N}\}$ is a relatively weakly compact subset of X and then, the sequence $(f_n * g_n)_n$ has a weakly convergent subsequence in view of Theorem 1.3.29. Q.E.D.

In what follows, we present some useful examples to illustrate the previous results and motivate the concept of RWC-Banach algebras.

Example 2.4.1 Every reflexive Banach algebra is an RWC-Banach algebra. This follows from the result any bounded subset in reflexive Banach space is relatively weakly compact, by using Theorem 1.3.24. But, it is not necessarily weakly compact, since ℓ^2 with the multiplication as is described in (2.4.1) is not a WC-Banach algebra, see Remark 1.3.61.

Example 2.4.2 The space $C([0, 1], X)$ is RWC-Banach algebra, whenever X is reflexive. This follows from Theorem 2.4.3. But, it is not necessarily a WC-Banach algebra. In fact, the space $C([0, 1], \ell^2)$ endowed with the norm

$$\begin{aligned}\|u\|_\infty &= \|(u_0, u_1, \ldots)\|_\infty \\ &= \sup_{t\in[0,1]} \left(\sum_{n=0}^{\infty} |u_n(t)|^2\right)^{\frac{1}{2}}\end{aligned}$$

and with a multiplication defined by

$$(u \cdot v)(t) = \left(\sum_{n=1}^{\infty} u_n(t)v_n(t), 0, 0, \ldots\right), \tag{2.4.4}$$

where $u = (u_n)_n$, $v = (v_n)_n$ and $t \in [0, 1]$. After simple calculations, we can verify that

$$\begin{aligned}\|(u \cdot v)(t)\| &= \left\|\left(\sum_{n=1}^{\infty} u_n(t)v_n(t), 0, 0, \ldots\right)\right\| \\ &= \left|\sum_{n=1}^{\infty} u_n(t)v_n(t)\right| \\ &\leq \left(\sum_{n=0}^{\infty} |u_n(t)|^2\right)^{\frac{1}{2}} \left(\sum_{n=0}^{\infty} |v_n(t)|^2\right)^{\frac{1}{2}} \\ &= \|u(t)\|\ \|v(t)\|.\end{aligned}$$

This implies that

$$\|u \cdot v\|_\infty \leq \|u\|_\infty \, \|v\|_\infty.$$

By using Remark 2.4.4, we can deduce that the Banach algebra $C([0, 1], \ell^2)$ with the multiplication (2.4.4) is not a WC-Banach algebra.

2.4.2 Hybrid Fixed Point Results in RWC-Banach Algebra

The following result gives sufficient conditions for the product of two operators, which are acting on an RWC-Banach algebra to have a fixed point. For our considerations, we need to prove the following lemmas.

Lemma 2.4.6 *Let X be a RWC-Banach algebra. If $K \in \mathscr{W}(X)$ and $M \in \mathscr{B}(X)$, then*

$$\beta(M \cdot K) \leq \|K\|\beta(M),$$

where β is the De Blasi measure of weak non-compactness.

Proof Let $\varepsilon > 0$ be fixed. By the definition of the measure β, there exists a subset $K' \in \mathscr{W}(X)$ verifying the following inclusion

$$M \subset K' + \left(\beta(M) + \frac{\varepsilon}{\|K\|}\right)\overline{\mathscr{B}}_X.$$

Then,

$$M \cdot K \subset K' \cdot K + (\beta(M)\|K\| + \varepsilon)\overline{\mathscr{B}}_X.$$

Since X is an RWC-Banach algebra, it follows that

$$\beta(M \cdot K) \leq (\beta(M)\|K\| + \varepsilon)\overline{\mathscr{B}}_X.$$

This completes the proof. Q.E.D.

Remark 2.4.7 Lemma 2.4.6 becomes as an extension of Lemma 1.5.10 to the RWC-Banach algebras.

In the following result, we extend Lemma 1.5.11 to the RWC-Banach algebra.

Lemma 2.4.8 *Let M and N be two bounded subsets of a RWC-Banach algebra X. Then,*

$$\beta(M \cdot N) \leq \|M\|\beta(N) + \|N\|\beta(M) + \beta(M)\beta(N).$$

Proof Let $t > \beta(N) > 0$. By using the definition of the measure β, we infer that there exists a subset $W \in \mathscr{W}(X)$ of X such that

$$N \subset W + \overline{\mathscr{B}}_X(t).$$

Define, a set W_N of W by

$$W_N := \bigcup_{x \in N} \left\{ y_x \in W, \ x = y_x + z_x, \ \text{ for some } z_x \in \overline{\mathscr{B}}_X(t) \right\}.$$

It is easy to show that W_N is a nonempty subset of W and satisfies

$$N \subset W_N + \overline{\mathscr{B}}_X(t).$$

Then,

$$M \cdot N \subset M \cdot W_N + t \, \|M\| \, \overline{\mathscr{B}}_X.$$

Making use of Lemma 2.4.6, together with the subadditivity of β, we infer that

$$\beta(M \cdot N) \leq \|W_N\| \beta(M) + t \, \|M\|.$$

This implies that

$$\beta(M \cdot N) \leq \|N\| \beta(M) + (\|M\| + \beta(M)) \ t.$$

Letting $t \to \beta(N)$, we get

$$\beta(M \cdot N) \leq \|M\| \beta(N) + \|N\| \beta(M) + \beta(M) \beta(N).$$

This completes the proof. Q.E.D.

In what follows, we are in position to study the existence of fixed points for the nonlinear operator equation $x = Ax \cdot Bx$ in RWC-Banach algebras.

Theorem 2.4.9 *Let S be a nonempty, convex, and closed subset of an RWC-Banach algebra X and let $A, B : \Omega \longrightarrow X$ be two strongly continuous operators such that*

(i) A and B are countably $\mathscr{D}$-set-Lipschitzians with $\mathscr{D}$-functions Φ and Ψ, respectively,
(ii) $A(S)$ and $B(S)$ are bounded,
(iii) $(A \cdot B)(x) \in S$, for each $x \in S$,
(iv) For any $r > 0$, the following estimate is satisfied

$$\|B(S)\| \Phi(r) + \|A(S)\| \Psi(r) + \Phi(r) \Psi(r) < r.$$

Then, the operator equation $x = Ax \cdot Bx$ has a solution in S.

Proof Define the multi-valued mapping $Q : S \longrightarrow S$ by putting

$$Qx = Ax \cdot Bx.$$

Notice that Q is countably $\mathscr{D}$-set-contraction. In fact, let $\mathscr{M} \subset S$ be a countably bounded subset of S. By using Lemma 2.4.8, we get

$$\begin{aligned}\beta(A\mathscr{M} \cdot B\mathscr{M}) &\leq \|B\mathscr{M}\|\ \beta(A\mathscr{M}) + \|A\mathscr{M}\|\ \beta(B\mathscr{M}) + \beta(A\mathscr{M})\ \beta(B\mathscr{M}) \\ &\leq \|B(S)\|\ \Phi(\beta(\mathscr{M})) + \|A(S)\|\ \Psi(\beta(\mathscr{M})) + \Phi(\beta(\mathscr{M}))\ \Psi(\beta(\mathscr{M})).\end{aligned}$$

This prove that Q is countably $\mathscr{D}$-set-contraction on S with $\mathscr{D}$-function φ given by

$$\varphi(r) = \|B(S)\|\ \Phi(r) + \|A(S)\|\ \Psi(r) + \Phi(r)\ \Psi(r), \quad \text{for } r > 0.$$

Let us demonstrate that Q is weakly sequentially continuous. Consider $(x_n)_n$ as a weakly convergent sequence contained in the set S to a point $x \in S$. Since A and B are strongly continuous, then

$$A(x_n) \to Ax \quad \text{and} \quad B(x_n) \to Bx.$$

This implies, in particular, that $Qx_n \rightharpoonup Qx$. Next, we will show that Q maps weakly compact sets into relatively weakly compact ones. To do this, let D be an arbitrary weakly compact subset of S and take a sequence $y_n \in Q(D)$. Then, there exists a sequence $(x_n)_n$ contained in the set D such that $y_n \in Q(x_n)$. From the above discussion, and by using the monotony property of β, it follows that

$$\begin{aligned}\beta(\{y_n,\ n \in \mathbb{N}\}) \leq \|B(\Omega)\|\ \Phi\left(\beta\left(\{x_n,\ n \in \mathbb{N}\}\right)\right) + \|A(\Omega)\|\ \Psi\left(\beta\left(\{x_n,\ n \in \mathbb{N}\}\right)\right) \\ +\Phi(\beta\left(\{x_n,\ n \in \mathbb{N}\}\right))\ \Psi\left(\beta\left(\{x_n,\ n \in \mathbb{N}\}\right)\right).\end{aligned}$$

The use of Theorem 1.3.29 leads to $Q(D)$ is relatively weakly compact, and when we apply Theorem 1.5.18, we reach the desired conclusion. Q.E.D.

Corollary 2.4.10 *Let S be a nonempty, convex, and closed subset of an RWC-Banach algebra X and let $A, B : S \longrightarrow X$ be two strongly continuous mappings such that*

(i) A is countably nonexpansive with respect to β,
(ii) B is weakly compact,
(iii) $A(S)$ and $B(S)$ are bounded such that $\|B(S)\| < 1$,
(iv) $Ax \cdot Bx \in S$, for each $x \in S$,

Then, the operator equation $x = Ax \cdot Bx$ has a solution in S.

Now, we consider the operator $\frac{I}{A}$ with the aim of removing the strong continuity of the operator B and weakening the Burton condition in the above result.

Theorem 2.4.11 *Let S be a nonempty, convex, and closed subset of an RWC-Banach algebra X, and let $A : X \longrightarrow X$ and $B : S \longrightarrow X$ be two mappings such that*

(i) A is ww-compact,
(ii) $A(S)$ and $B(S)$ are bounded,
(iii) A is $\mathscr{D}$-Lipschitzian with $\mathscr{D}$-function Φ,
(iv) B is countably $\mathscr{D}$-set-Lipschitzian with $\mathscr{D}$-function Ψ,
(v) $\left(\frac{I}{A}\right)^{-1}$ and B are weakly sequentially continuous,
(vi) If $x = Ax \cdot By$, for $y \in S$, then $x \in S$.

Then, the operator equation $x = Ax \cdot Bx$ has a solution in S provided that

$$\|B(S)\|\Phi(r) + \|A(S)\|\Psi(r) + \Phi(r)\Psi(r) < r, \quad \textit{for } r > 0.$$

Proof Let $z \in B(S)$ be fixed and define the mapping $Q_z : X \longrightarrow X$ by putting

$$Q_z(x) = Ax \cdot z.$$

From Theorem 1.2.26, we can show that Q_z has an unique fixed point in X and so, by using of assumption (v), we can define the operator $Q : S \longrightarrow S$ by taking

$$Qx = \left(\frac{I}{A}\right)^{-1} Bx.$$

Let M be a countable and bounded subset of S. If we consider the boundedness of $A(S)$ and $B(S)$ and deploy the inclusion

$$Q(M) \subset A(Q(M)) \cdot B(M),$$

we get $Q(M)$ is bounded and

$$\beta\left(Q(M)\right) \leq \beta\left(A(Q(M)) \cdot B(M)\right).$$

Using the fact that X is a RWC-Banach algebra, together with Lemma 2.4.8, we obtain

$$\begin{aligned}\beta\left(Q(M)\right) \leq \beta\left(A(Q(M))\right)\|B(M)\| + \beta(B(M))\|A(Q(M))\| \\ +\beta\left(A(Q(M))\right)\beta(B(M)).\end{aligned}$$

The use of assumption (i) as well as Proposition 1.5.7 leads to

$$\beta\left(A(Q(M))\right) \leq \Phi(\beta(Q(M))).$$

We infer that

$$\beta\left(Q(M)\right) \leq \|B(S)\|\Phi(\beta(M)) + \|A(S)\|\Psi(\beta(M)) + \Phi(\beta(M))\Psi(\beta(M)).$$

Proceeding essentially as Theorem 2.4.9, we can prove that $Q(D)$ is relatively weakly compact, for any weakly compact D of S. Moreover, it is easy to show that Q is weakly sequentially continuous, in light of assumption (v). Accordingly, an application of Theorem 1.5.18 yields that Q has a fixed point in S. Q.E.D.

Corollary 2.4.12 *Let S be a nonempty, bounded, convex, and closed subset of an RWC-Banach algebra X and let $A : X \longrightarrow X$ and $B : S \longrightarrow X$ be two operators such that*

(i) A is strongly continuous,
(ii) A is $\mathcal{D}$-Lipschitzian with $\mathcal{D}$-function Φ,
(iii) B is weakly sequentially continuous such that $B(S)$ is bounded,
(iv) B is countably $\mathcal{D}$-set-Lipschitzian with $\mathcal{D}$-function Ψ,
(v) If $x = Ax \cdot By$, for $y \in S$, then $x \in S$.

Then, the operator equation $x = Ax \cdot Bx$ has a solution in S provided that

$$\max\left(\|S\|, \|BS\|\right)\Phi(r) + \|AS\|\Psi(r) + \Phi(r)\Psi(r) < r, \quad \textit{for } r > 0.$$

Proof Following the same procedures as in the proof of Theorem 2.4.11, we can demonstrate that $\left(\frac{I}{A}\right)^{-1}$ exists on $B(S)$. From assumption (i), it follows that A is ww-compact. We show that $\left(\frac{I}{A}\right)^{-1}$ is weakly sequentially continuous. In order to prove this fact, assume that $(x_n)_n$ is a weakly convergent sequence contained in the set S to an element $x \in S$ and take a sequence $(y_n)_n$ in X such that

$$y_n = A(y_n) \cdot x_n.$$

The use of Proposition 1.5.7 as well as Lemma 2.4.6 leads to

$$\begin{aligned}\beta\left(\{Ay_n,\ n \in \mathbb{N}\} \cdot \{x_n,\ n \in \mathbb{N}\}\right) &\leq \|x_n\|\beta\left(\{Ay_n,\ n \in \mathbb{N}\}\right)\\ &\leq \|S\|\beta\left(\{Ay_n,\ n \in \mathbb{N}\}\right).\end{aligned}$$

The use of the inclusion

$$\{y_n,\ n \in \mathbb{N}\} \subset \{A(y_n),\ n \in \mathbb{N}\} \cdot \{x_n,\ n \in \mathbb{N}\},$$

together with Proposition 1.5.7 leads to

$$\beta(\{y_n,\ n \in \mathbb{N}\}) \leq \|S\|\Phi\left(\beta\left(\{y_n,\ n \in \mathbb{N}\}\right)\right).$$

This means, in particular, that the set $\{y_n,\ n \in \mathbb{N}\}$ is relatively weakly compact. Then, on account of Theorem 1.3.29, the sequence $(y_n)_n$ has a renamed weakly convergent subsequence to a point $y \in S$. After simple calculations, we obtain

$$f\left(A(y_n) \cdot x_n - Ay \cdot x\right) \leq f\left((A(y_n) - Ay) \cdot x_n\right) + f\left(Ay \cdot (x_n - x)\right),$$

whenever $f \in X^*$. Since A is strongly continuous, we deduce that

$$f\left((A(y_n) - Ay) \cdot x_n\right) \to 0, \quad \text{as } n \to \infty.$$

Using Lemma 1.3.57, we obtain that

$$f\left(Ay \cdot (x_n - x)\right) \to 0, \quad \text{as } n \to \infty.$$

We infer that

$$A(y_n) \cdot x_n \rightharpoonup Ay \cdot x, \quad \text{as } n \to \infty.$$

By uniqueness of limit, we deduce that $y = Ay \cdot x$, which achieves the weak sequential continuity of $\left(\frac{I}{A}\right)^{-1}$. Now, we check that the whole sequence $\left(\left(\frac{I}{A}\right)^{-1} x_n\right)_n$ converges weakly to $\left(\frac{I}{A}\right)^{-1}(x)$. If not the case, there exists a subsequence, say $\left(x_{n_j}\right)_j$, of $(x_n)_n$ and there exists a weak neighborhood V of $\left(\frac{I}{A}\right)^{-1}(x)$ with

$$\left(\frac{I}{A}\right)^{-1}\left(x_{n_j}\right) \notin V, \quad \text{for all } j \in \mathbb{N}.$$

Using an argument similar to that above, we deduce the existence of a subsequence, say $\left(x_{n_{j_k}}\right)_k$, of $\left(x_{n_j}\right)_j$, which satisfies

$$\left(\frac{I}{A}\right)^{-1} x_{n_{j_k}} \rightharpoonup \left(\frac{I}{A}\right)^{-1}(x),$$

which is a contradiction, and the claim is proved. The remained proof follows along the lines of Theorem 2.4.11. Q.E.D.

References

1. O'Regan, D., Precup, R.: Continuation theory for contractions on spaces with two vector-valued metrics. Appl. Anal. **82**, 131–144 (2003)
2. O'Regan, D., Shahzad, N., Agarwal, R.P.: Fixed point theory for generalized contractive maps on spaces with vector-valued metrics. Fixed Point Theory Appl. **6**, 143–149 (2007)
3. Perov, A.I., Kibenko, A.V.: On a certain general method for investigation of boundary value problems. Izv. Akad. Nauk SSSR Ser. Mat. **30**, 249–264 (1966). (in Russian)
4. Nieto, J.J., Ouahab, A., Rodríguez-López, R.: Fixed point theorems in generalized Banach algebras and applications. Fixed Point Theory. **19**, 707–732 (2018)
5. Ouahab, A.: Some Perov's and Krasnosel'skii type fixed point results and application. Comm. Appl. Nonlinear Anal. 623–642 (2015)
6. Petre, I.R.: A multi-valued version of Krasnosel'skii's theorem in generalized Banach spaces. An. Ştiinţ. Univ. "Ovidius" Constanţa Ser. Mat., **22**, 177–192 (2014)

7. Banaś, J., Olszowy, L.: On a class of measures of non-compactness in Banach algebras and their application to nonlinear integral equations. Z. Anal. Anwend. **28**, 475–498 (2009)
8. Ben Amar, A., Boumiza, M., O'Regan, D.: Hybrid fixed point theorems for multi-valued mappings in Banach algebras under a weak topology setting. J. Fixed Point Theory Appl. **18**, 327–350 (2016)
9. Ben Amar, A., Chouayekh, S., Jeribi, A.: New fixed point theorems in Banach algebras under weak topology features and applications to nonlinear integral equations. J. Funct. Anal. **259**, 2215–2237 (2010)
10. Ben Amar, A., Chouayekh, S., Jeribi, A.: Fixed point theory in a new class of Banach algebras and application. Afr. Mat. **24**, 705–724 (2013)
11. Ben Amar, A., Feki, I., Jeribi, A.: Leray-Schauder and Furi-Pera types fixed point theorems for the sum of two weakly sequentially continuous mappings and application to transport equation. Afr. Mat. **25**(3), 707–722 (2014)
12. Ben Amar, A., O'Regan, D.: Measures of weak non-compactness and fixed point theory in Banach algebras satisfying condition $(\mathscr{P})$. Fixed Point Theory **18**, 37–46 (2017)
13. Ben Amara, K., Kaddachi, N., Jeribi, A.: On existence results in fixed set theory and applications to self-similarity. Fixed Point Theory **23**(1), 85–104 (2022)
14. Ben Amara, K., Kaddachi, N., Jeribi, A.: Equivalence of some properties in the theory of Banach algebras and applications. J. Math. Anal. Appl. **520**, 126865 (2023)
15. Ben Amara, K., Kaddachi, N., Jeribi, A.: Existence results for nonexpansive multi-valued operators and nonlinear integral inclusions. Afr. Mat. **34**(3), Paper No. 46, 25 (2023)
16. Dhage, B.C., O'Regan, D.: A fixed point theorem in Banach algebras with applications to functional integral equations. Funct. Differ. Equ. **7**, 259–267 (2000)
17. Jeribi, A., Kaddachi, N., Laouar, Z.: Generalized form of fixed point theorems in generalized Banach algebra relative to the weak topology with an application. Filomat **36**(18), 6253–6268 (2022)
18. Jeribi, A., Kaddachi, N., Laouar, Z.: Fixed point theorems for weakly asymptotically regular mappings in Banach spaces with an application. Numer. Funct. Anal. Optim. **43**(1), 68–87 (2022)
19. Kaddachi, N., Jeribi, A., Krichen, B.: Fixed point theorems of block operator matrices on Banach algebras and an application to functional integral equations. Math. Methods Appl. Sci. **36**, 659–673 (2013)
20. Sadovskii, B.N.: On a fixed point theorem. Funktsional. Anal. i Prilozhen. **1**, 74–76 (1967)
21. Banas, J., Olszowy, L.: On the equivalence of some concepts in the theory of Banach algebras. Ann. Funct. Anal. **10**, 277–283 (2019)
22. Jeribi, A., Krichen, B.: Nonlinear Functional Analysis in Banach Spaces and Banach Algebras Fixed Point Theory under Weak Topology for Nonlinear Operators and Block Operator Matrices with Applications. Monographs and Research Notes in Mathematics. CRC Press, Boca Raton, FL (2016)
23. Diestel, J.: The Dunford-Pettis property. Contemp. Math. **2**, 15–60 (1980)
24. Talagrand, M.: La propriété de Dunford-Pettis dans $C(K, E)$ et $L^1(E)$. J. Funct. Anal. **44**, 317–321 (1983)
25. Banaś, J., Taoudi, M.A.: Fixed points and solutions of operator equations for the weak topology in Banach algebras. Taiwanese J. Math. **18**(3), 871–893 (2014)
26. Diestel, J.: Remarks on weak compactness in $L^1(\nu, X)$. Glasgow Math. J. **18**, 87–91 (1977)

Chapter 3
Fixed Point Theory for Block Operator Matrices

Motivated by the investigations of Theorem 1.2.33 and by some results proved in [1–11] and the references therein, concerning the existence results of the operator equation

$$Ax \cdot Bx + Cx = x,$$

where A, B, and C are, in general nonlinear multi-valued mapping, we have the idea to extend their results by proving, in this chapter, that under certain hypotheses on its nonlinear entries, that the following 2×2 block operator matrix

$$\begin{pmatrix} A & B \cdot B' \\ C & D \end{pmatrix}. \tag{3.0.1}$$

has, at least, one fixed point. We establish some existence results of the block operator matrix (3.0.1), with both single-valued and multi-valued inputs acting in suitable Banach algebra, by using the concepts of condensing property and $\mathscr{D}$-set-contraction property.

3.1 Block Operator Matrices with Single-Valued Inputs

The objective of this section is to examine the existence of fixed point results of the block operator matrix (3.0.1), with nonlinear entries acting on a product of Banach algebras.

A. Jeribi and N. Kaddachi, *Topological Fixed-Point Theory in Suitable Banach Algebras with Applications*, Infosys Science Foundation Series,
https://doi.org/10.1007/978-981-96-5542-7_3

3.1.1 Extension of Dhage's Type of Fixed Point

We state some new variants of fixed point of the block operator matrix (3.0.1) by blending Theorem 1.2.26 with that Theorem 1.2.33.

Theorem 3.1.1 *Let S be a nonempty, bounded, closed, and convex subset of a Banach algebra X and let A, B, B', C, $D : S \longrightarrow X$ be five operators such that*

(i) A, B, C and $(I - D)^{-1}$ are Lipschitzians with constants q_1, q_2, q_3 and δ, respectively,
(ii) B' is continuous and C is compact,
(iii) for any x, $y \in S$, we have $A(x) + B(I - D)^{-1}C(x) \cdot B'(I - D)^{-1}C(y) \in S$.

Then, the block operator matrix (3.0.1) *has a fixed point in $S \times S$ provided that*

$$q_1 + q_2 q_3 \delta M < 1,$$

where $M = \|B'(I - D)^{-1}C(S)\|$.

Proof Let $y \in S$ be any fixed point, then the map which assigns to each $x \in S$ the value

$$A(x) + B(I - D)^{-1}C(x) \cdot B'(I - D)^{-1}C(y)$$

define a contraction with a constant $q_1 + q_2 q_3 \delta M$. It follows from Theorem 1.2.4, that the operator equation

$$x = A(x) + B(I - D)^{-1}C(x) \cdot B'(I - D)^{-1}C(y) \tag{3.1.1}$$

has an unique solution $x \in S$. Define the mapping $N : S \longrightarrow S$ by putting $N(y) = x$, where x is the unique solution of the equation (3.1.1). Now, let us show that the operator N is continuous on S. Indeed, consider $(y_n)_n$ as a converging sequence in S to an element $y \in S$. Then,

$$\begin{aligned}\|N(y_n) - N(y)\| &\leq \|B(I - D)^{-1}C(x_n) \cdot B'(I - D)^{-1}C(y_n) \\ &\quad - B(I - D)^{-1}C(x) \cdot B'(I - D)^{-1}C(y_n)\| \\ &\quad + \|B(I - D)^{-1}C(x) \cdot B'(I - D)^{-1}C(y_n) \\ &\quad - B(I - D)^{-1}C(x) \cdot B'(I - D)^{-1}C(y)\| \\ &\quad + \|A(x_n) - A(x)\| \\ &\leq (q_1 + q_2 q_3 \delta M)\, \|x_n - x\| + \|B(I - D)^{-1}C(x)\| \\ &\quad \times \|B'(I - D)^{-1}C(y_n) - B'(I - D)^{-1}C(y)\|.\end{aligned}$$

In other words, we have

$$\begin{aligned}\|N(y_n) - N(y)\| \leq (q_1 + q_2 q_3 \delta M)\, \|N(y_n) - N(y)\| + \|B(I - D)^{-1}C(Ny)\| \\ \times \|B'(I - D)^{-1}C(y_n) - B'(I - D)^{-1}C(y)\|.\end{aligned}$$

This means, in particular, that N is a continuous operator on S. Our next task is to shows that N is compact on S. First, one can verify easily that, for each $z \in S$, we have

$$\begin{aligned}\|B(I-D)^{-1}C(z)\| &\leq \|B(I-D)^{-1}C(a)\| + Mq_2q_3\delta\|z-a\| \\ &\leq c,\end{aligned}$$

where $c = \|B(I-D)^{-1}C(a)\| + q_2q_3\delta M \operatorname{diam}(S)$, for some $a \in S$. Let $\varepsilon > 0$ be given. Taking into account that $C(S)$ is totally bounded, and using the continuity of the operator $B'(I-D)^{-1}$, we get $B'(I-D)^{-1}C(S)$ is a totally bounded subset of X. Consequently, there exists a subset $Y = \{y_1, \ldots, y_n\}$ of S, that satisfies

$$B'(I-D)^{-1}C(S) \subseteq \bigcup_{i=1}^{n} \mathscr{B}_X(w_i, r),$$

where $r = \left(\dfrac{1-(q_1+q_2q_3\delta M)}{c}\right)\varepsilon$ and $w_i = B'(I-D)^{-1}C(y)_i$. So, for any $y \in S$, we have

$$\|B'(I-D)^{-1}C(y_k) - B'(I-D)^{-1}C(y)\| < \left(\frac{1-(q_1+q_2q_3\delta M)}{c}\right)\varepsilon.$$

Then,

$$\|Ny_k - Ny\| \leq (q_1+q_2q_3\delta M)\|z_k - z\| + c\|B'(I-D)^{-1}C(y_k) - B'(I-D)^{-1}C(y)\|.$$

This implies that

$$\begin{aligned}\|Ny_k - Ny\| &\leq \left(\frac{c}{1-(q_1+q_2q_3\delta M)}\right)\|B'(I-D)^{-1}C(y_k) - B'(I-D)^{-1}C(y)\| \\ &< \varepsilon.\end{aligned}$$

Since $y \in S$ was arbitrary, then

$$N(S) \subseteq \bigcup_{k=1}^{n} \mathscr{B}_X(N(y_k), \varepsilon).$$

As a result, N is a completely continuous operator on S into itself, and when we apply Theorem 1.2.33, we reach that the operator N has a fixed point in S. That is,

$$x = A(x) + B(I-D)^{-1}C(x) \cdot B'(I-D)^{-1}C(x).$$

Finally, the use of the vector $y = (I-D)^{-1}Cx$ achieves the proof. Q.E.D.

As an immediately consequence of Theorem 3.1.1, we have the following.

Corollary 3.1.2 *Let S be a nonempty, bounded, closed, and convex subset of a Banach algebra X and let A, B, B', C, $D : S \longrightarrow X$ be five operators such that*

(i) A, B, and C are Lipschitzians with constants q_1, q_2, and q_3, respectively,
(ii) D is a contraction with a constant k,
(iii) C is compact and $C(S) \subseteq (I - D)(S)$,
(iv) B' is continuous,
(v) $A(x) + B(I - D)^{-1}C(x) \cdot B'(I - D)^{-1}C(z) \in S$, for all $x, z \in S$.

Then, the block operator matrix (3.0.1) *has a fixed point in $S \times S$ provided that*

$$q_1 + \frac{q_2 q_3}{1-k} M < 1,$$

where $M = \|B'(I - D)^{-1}C(S)\|$.

Corollary 3.1.3 *Let S be a nonempty, bounded, closed, and convex subset of a Banach algebra X and let A, B, B', C, $D : S \longrightarrow X$ be five operators such that*

(i) A, B, and C are $\mathcal{D}$-Lipschitzians with $\mathcal{D}$-functions ϕ_A, ϕ_B and ϕ_C, respectively,
(ii) $(I - D)^{-1}$ is $\mathcal{D}$-Lipschitzian on $C(S)$ with $\mathcal{D}$-function ϕ_ψ,
(iii) B' is continuous and C is compact,
(iv) $\left(\dfrac{I}{B(I-D)^{-1}C}\right)^{-1}$ exists on $B'(I - D)^{-1}C(S)$,
(v) $A(x) + B(I - D)^{-1}C(x) \cdot B'(I - D)^{-1}C(z) \in S$ for all $x, z \in S$.

Then, the block operator matrix (3.0.1) *has a fixed point in $S \times S$ provided that*

$$M\phi_B \circ \phi_\psi \circ \phi_C(r) + \phi_A(r) < r,$$

for all $r > 0$, where $M = \|B'(I - D)^{-1}C(S)\|$.

Remark 3.1.4 Let X be a Banach space and let C, $D : X \longrightarrow X$ be two operators such that

(i) D is a contraction with a constant k,
(ii) C is Lipschitzian with a Lipschitz constant γ,
(iii) $C(X) \subset (I - D)(X)$.

Then the operator $(I - D)^{-1}C$ is Lipschitzian with a Lipschitz constant $\frac{\gamma}{1-k}$.

Theorem 3.1.5 *Let S (resp. S') be a nonempty, bounded, closed, and convex subset of a Banach algebra X (resp. Y) and let $A : S \longrightarrow X$, B, $B' : S' \longrightarrow X$, $C : S \longrightarrow Y$ and $D : S' \longrightarrow S'$ be five operators such that*

(i) A and C are $\mathcal{D}$-Lipschitzians with $\mathcal{D}$-functions ϕ_A and ϕ_C, respectively,
(ii) B is Lipschitzian with a Lipschitz constant q,
(iii) D is a contraction with a constant k such that $C(S) \subseteq (I - D)(S')$,
(iv) B' is continuous and C is compact,
(v) $A(x) + B(I - D)^{-1}C(x) \cdot B'(I - D)^{-1}C(y) \in S$, for all $x, y \in S$.

Then, the block operator matrix (3.0.1) *has a fixed point in* $S \times S$ *provided that*

$$\frac{qM}{1-k}\phi_C(r) + \phi_A(r) < r, \quad \textit{for all } r > 0,$$

where $M = \|B'(I-D)^{-1}C(S)\|$.

Proof After simple calculations, we can see that $(I-D)^{-1}$ exists and continuous on $(I-D)(S')$. Taking into account that C is compact and deploys the continuity of the operator B', we get $B'(I-D)^{-1}C$ is completely continuous. Let $y \in S$ be fixed and define the mapping $\varphi_y : S \longrightarrow S$ by putting

$$\varphi_y(x) = A(x) + B(I-D)^{-1}C(x) \cdot B'(I-D)^{-1}C(y).$$

It is easy to verify that φ_y is nonlinear contraction on S with $\mathscr{D}$-function ϕ given by

$$\phi(r) = \phi_A(r) + \frac{\beta M}{1-k}\phi_C(r).$$

Then, by using Theorem 1.2.26, we arrive that there exists exactly one point $x^* \in S$ satisfying the equality

$$x^* = A(x^*) + B(I-D)^{-1}C(x^*) \cdot B'(I-D)^{-1}C(y). \tag{3.1.2}$$

Define the mapping $N : S \longrightarrow S$ by putting $N(y) = z$, where z is the unique solution of the equation (3.1.2). In the similar way as in the proof of Theorem 2.2.4, we can demonstrate that N is a completely continuous operator on S, and when we apply Theorem 1.2.33, we reach the desired conclusion.

Q.E.D.

3.1.2 Particular Case

In this subsection, we discuss some new variants of fixed point theorems of the block operator matrix (3.0.1) under some suitable hypotheses on the operator

$$\frac{I-A}{B(I-D)^{-1}C}.$$

The obtained theorems significantly generalize some works in the literature, see for example [12–14].

Theorem 3.1.6 *Let* X *and* Y *be two Banach algebras and let* S *and* S' *be two nonempty, bounded, closed, and convex subsets of* X *and* Y*, respectively.*
Set $R = S \bigcap S'$ *(assume nonempty) and suppose that* $A, B, C, D, B' : R \longrightarrow R$ *are five operators such that*

(i) A and C are $\mathcal{D}$-Lipschitzians with $\mathcal{D}$-functions ϕ_A and ϕ_C, respectively,
(ii) B is Lipschitzian with a Lipschitz constant q,
(iii) D is a contraction with a constant k and $C(R) \subseteq (I - D)(R)$,
(iv) B′ is continuous and C is compact,
(v) $\left(\dfrac{I}{B(I-D)^{-1}C}\right)^{-1}$ exists on $B'(I-D)^{-1}C(S)$,
(vi) $A(x) + B(I-D)^{-1}C(x) \cdot B'(I-D)^{-1}C(y) \in R$, for all $x, y \in R$.

Then, the block operator matrix (3.0.1) *has a fixed point in $R \times R$ provided that*

$$\left(\frac{qM}{1-k}\right)\phi_C(r) + \phi_A(r) < r, \quad \textit{for all } r > 0,$$

where $M = \|B'(I-D)^{-1}C(R)\|$.

Proof From Theorem 1.2.4, it follows that the inverse operator $(I - D)^{-1}$ is well defined on $C(R)$. Indeed, let y be an arbitrarily fixed element contained in the set R and define the mapping $\varphi_y : R \longrightarrow X$ by putting

$$\varphi_y(x) = Cy + Dx.$$

After simple calculations, we can verify that φ_y is a contraction mapping. Then, on account of Theorem 1.2.4, we infer that φ_y has exactly one fixed point x^*. In other words, for an arbitrarily fixed point $y \in R$, there exists an unique element $x^* \in R$, that verifies

$$Cy = (I - D)x^*.$$

This shows that $(I - D)^{-1}$ exists on $C(R)$. On the other hand, one can show that the operator $B(I - D)^{-1}C$ is $\mathcal{D}$-Lipschitzian with $\mathcal{D}$-function $\left(\frac{q}{1-k}\right)\phi_C$. Due to the fact that $\phi_A(r) < r$, for $r > 0$ and Theorem 1.2.26, we infer that the operator $(I - A)^{-1}$ exists and continuous on R. Let us fix an arbitrarily element y in the set R and define the mapping $N_y : R \longrightarrow R$ by putting

$$N_y(x) = A(x) + B(I-D)^{-1}C(x) \cdot B'(I-D)^{-1}C(y).$$

Then, in virtue of assumption (i), we get

$$\begin{aligned}\|N_y(x_1) - N_y(x_2)\| &\leq \|A(x_1) - A(x_2)\| + M\|B(I-D)^{-1}C(x_1) - B(I-D)^{-1}C(x_2)\| \\ &\leq \psi(\|x_1 - x_2\|),\end{aligned}$$

whenever $x_1, x_2 \in R$, where

$$\psi(r) = \frac{qM}{1-k}\phi_C(r) + \phi_A(r) < r.$$

Based on Theorem 1.2.26, we obtain there exists an unique element $x^* \in R$ such that $N_y(x^*) = x^*$ and so, we can define the mapping $N : R \longrightarrow R$ by putting

$$N(y) = \left(\frac{I - A}{B(I - D)^{-1}C}\right)^{-1} B'(I - D)^{-1}C(y).$$

We show that N is completely continuous. To this end, take an arbitrary sequence $(x_n)_n$ contained in the set R, which is convergent to a point x. Let us denote

$$\begin{cases} z_n = \left(\frac{I - A}{B(I - D)^{-1}C}\right)^{-1} B'(I - D)^{-1}C(x_n), \\ z = \left(\frac{I - A}{B(I - D)^{-1}C}\right)^{-1} B'(I - D)^{-1}C(x). \end{cases}$$

In other words, we have

$$\begin{cases} z_n = A(z_n) + B(I - D)^{-1}C(z_n) \cdot B'(I - D)^{-1}C(x_n) \\ z = A(z) + B(I - D)^{-1}C(z) \cdot B'(I - D)^{-1}C(x). \end{cases}$$

This implies that,

$$\|z_n - z\| \leq \psi(\|z_n - z\|) + M\|B'(I - D)^{-1}C(x_n) - B'(I - D)^{-1}C(x)\|.$$

By assumption (iv), we infer that N is a compact operator on R. The desired conclusion follows from Theore 1.2.33. Q.E.D.

Theorem 3.1.7 *Let S be a nonempty, bounded, closed, and convex subset of a Banach algebra X and let $A, C : S \longrightarrow X$ and $B, B', D : X \longrightarrow X$ be five operators such that*

(i) A, C and D are $\mathcal{D}$-Lipschitzians with $\mathcal{D}$-functions ϕ_A, ϕ_C and ϕ_D, respectively,
(ii) B is Lipschitzian with a Lipschitz constant q,
(iii) D is a linear bounded operator and $C(S) \subset (I - D)(S)$,
(iv) B' is continuous and $C(S)$ is relatively compact,
(v) for all $r > 0$, we have $\phi_D(r) \leq \|D\|r < r$, and
(vi) $A(x) + B(I - D)^{-1}C(x) \cdot B'(I - D)^{-1}C(y) \in S$, for all $x, y \in S$.

Then, the block operator matrix (3.0.1) *has a fixed point in $S \times S$ provided that*

$$\sum_{n \geq 0} Mq\phi_D^n \circ \phi_C(r) + \phi_A(r) < r, \quad \textit{for all } r > 0,$$

where $M = \|B'(I - D)^{-1}C(S)\| > 1$.

Proof The use of Theorem 1.2.26, together with assumption (v) leads to the operator $(I - D)^{-1}$ is continuous on $C(S)$, and

$$(I-D)^{-1}x = \sum_{n\geq 0} D^n x, \quad \text{for all } x \in C(S).$$

Therefore, the operator $B(I-D)^{-1}C$ is $\mathscr{D}$-Lipschitzian with $\mathscr{D}$-function Ψ, where

$$\Psi(r) = \sum_{n\geq 0} q\phi_D^n \phi_C(r).$$

In fact, for all $x, y \in S$, we have

$$\begin{aligned}
\|B(I-D)^{-1}Cx - B(I-D)^{-1}Cy\| &\leq q \left\|(I-D)^{-1}Cx - (I-D)^{-1}Cy\right\| \\
&= q \left\|\sum_{n\geq 0} D^n Cx - \sum_{n\geq 0} D^n Cy\right\| \\
&\leq q \sum_{n\geq 0} \left\|D^n Cx - D^n Cy\right\| \\
&\leq \sum_{n\geq 0} q\phi_D^n \circ \phi_C\left(\|x-y\|\right).
\end{aligned}$$

Let $y \in S$ be fixed and define the mapping $N_y : S \longrightarrow S$ by putting

$$N_y(x) = A(x) + B(I-D)^{-1}C(x) \cdot B'(I-D)^{-1}C(y).$$

After simple calculations, we can verify that N_y is $\mathscr{D}$-Lipschitzian with $\mathscr{D}$-function Φ, where

$$\Phi(r) = Mq \sum_{n\geq 0} \phi_D^n \circ \phi_C(r) + \phi_A(r).$$

Then, on account of Theorem 1.2.26, we obtain there exists an unique point $z \in S$ satisfying

$$z = Az + B(I-D)^{-1}C(z) \cdot B'(I-D)^{-1}C(y). \tag{3.1.3}$$

Let us define the function $N : S \longrightarrow S$ by putting $N(y) = z$, where z is the unique solution of the equation (3.1.3). Due to the fact that the function Φ is continuous, we get N is also continuous on S. Our next task is to prove that N is a compact operator on S. Indeed, for all $z \in S$, we have

$$\begin{aligned}
\|B(I-D)^{-1}C(z)\| &\leq \|B(I-D)^{-1}C(a)\| + \sum_{n\geq 0} q\phi_D^n \circ \phi_C(\|z-a\|) \\
&< \|B(I-D)^{-1}C(a)\| + \|z-a\| \\
&\leq c
\end{aligned}$$

where $c = \|B(I-D)^{-1}C(a)\| + diam(S)$, for some $a \in S$. Let us take two sequence $(y_n)_n$ and $(z_n)_n$ contained in the set S such that

$$z_n = N(y_n).$$

Since $B'(I-D)^{-1}C(S)$ is relatively compact, there exists a subsequence, say $(y_{n_k})_k$, of $(y_n)_n$ such that the sequence $\big(B'(I-D)^{-1}C(y_{n_k})\big)_k$ is convergent.
Therefore, for all $k_1, k_2 \in \mathbb{N}$, we have

$$\begin{aligned}\|z_{n_{k_1}} - z_{n_{k_2}}\| &\leq \left(\sum_{n\geq 0} q\phi_D^n \circ \phi_C + \phi_A\right)(\|z_{n_{k_1}} - z_{n_{k_2}}\|) \\ &\quad + c\|B'(I-D)^{-1}C(y_{n_{k_1}}) - B'(I-D)^{-1}C(y_{n_{k_2}})\|.\end{aligned}$$

This implies that

$$\limsup_{k_1,\, k_2 \to +\infty} \|z_{n_{k_1}} - z_{n_{k_2}}\| \leq \phi\left(\limsup_{k_1,\, k_2 \to +\infty} \|z_{n_{k_1}} - z_{n_{k_2}}\|\right).$$

This means, in particular, that

$$\limsup_{k_1,\, k_2 \to +\infty} \|z_{n_{k_1}} - z_{n_{k_2}}\| = 0.$$

Consequently, $(N(y_{n_k}))_k$ is a Cauchy sequence and so, it possesses a convergent subsequence, since X is complete. This proves that $N(S)$ is a sequentially relatively compact subset of S, and when we apply Theorem 1.3.29, we deduce that $N(S)$ is relatively compact. Thus, the use of Theorem 1.2.33 ends the proof. Q.E.D.

In [15], it has been mentioned that assumption (vi) of Theorem 3.1.6 is very strong and can be replaced with a mild one. We state the following result.

Theorem 3.1.8 *Let S be a nonempty, bounded, closed, and convex subset of a Banach algebra X and let $A, C : S \longrightarrow X$ and $B, B', D : X \longrightarrow X$ be five operators such that*

(i) A, B, and C are $\mathcal{D}$-Lipschitzians with $\mathcal{D}$-functions ϕ_A, ϕ_B and ϕ_C, respectively,
(ii) B' is completely continuous on S,
(iii) D is a contraction with a constant k and $C(S) \subset (I-D)(S)$,
(iv) $\left(\dfrac{I-A}{B(I-D)^{-1}C}\right)^{-1}$ exists on $B'(I-D)^{-1}C(S)$,
(v) $x = A(x) + B(I-D)^{-1}C(x) \cdot B'(I-D)^{-1}C(z)$, for $z \in S \Rightarrow x \in S$.

Then, the block operator matrix (3.0.1) *has a fixed point in $S \times S$ provided that*

$$M\phi_B \circ L_{\frac{1}{1-k}}(\phi_C(r)) + \phi_A(r) < r, \ \ \textit{for all } r > 0,$$

where $M = \|B'(I-D)^{-1}C(S)\|$ and $L_s(t) = st$, for all $s, t \in \mathbb{R}^+$.

Proof Define a mapping $N : S \longrightarrow X$ by putting

$$N(x) = \left(\frac{I - A}{B(I - D)^{-1}C}\right)^{-1} B'(I - D)^{-1}C(x).$$

In what follows, we prove that the operator N is continuous. In fact, let us assume that $(x_n)_n$ is a sequence contained in the set S, which is convergent to a point $x \in S$. Let us denote

$$\begin{cases} z_n = A(z_n) + B(I - D)^{-1}C(z_n) \cdot B'(I - D)^{-1}C(x_n) \\ z = A(z) + B(I - D)^{-1}C(z) \cdot B'(I - D)^{-1}C(x). \end{cases}$$

After simple calculations, we can see that

$$\begin{aligned} \|z_n - z\| \leq \|B(I - D)^{-1}C(z)\| \|B'(I - D)^{-1}C(x_n) - B'(I - D)^{-1}C(x)\| \\ + \psi(\|z_n - z\|), \end{aligned} \tag{3.1.4}$$

where $\psi(r) = M\phi_B \circ L_{\frac{1}{1-k}} \circ \phi_C(r) + \phi_A(r)$. Taking into account the upper limit in the inequality (3.1.4), we obtain N is continuous. Further, from assumption (v), we immediately have

$$N(S) \subseteq \left(\frac{I - A}{B(I - D)^{-1}C}\right)^{-1} B'(S).$$

If we consider the continuity of map $\left(\frac{I-A}{B(I-D)^{-1}C}\right)^{-1}$ and deploy assumption (ii), we get N is compact. The desired conclusion follows from Theorem 1.2.33. Q.E.D.

The following result is an immediate consequence of Theorem 3.1.8.

Corollary 3.1.9 *Let S be a nonempty, compact, and convex subset of a Banach algebra X and let A, $C : S \longrightarrow X$ and B, D, $B' : X \longrightarrow X$ be five operators such that*

(i) A, B, and C are $\mathcal{D}$-Lipschitzians with $\mathcal{D}$-functions ϕ_A, ϕ_B and ϕ_C, respectively,
(ii) D is a contraction with a constant k such that $C(S) \subseteq (I - D)(S)$,
(iii) B' is continuous,
(iv) $\left(\frac{I - A}{B(I - D)^{-1}C}\right)^{-1}$ exists on $B'(I - D)^{-1}C(S)$, and
(v) $x = A(x) + B(I - D)^{-1}C(x) \cdot B'(I - D)^{-1}C(z)$, $z \in S \Rightarrow x \in S$.

Then, the block operator matrix (3.0.1) *has a fixed point in $S \times S$ provided that*

$$\phi_B \circ \left(\frac{1}{1 - k}\phi_C\right)(r) + \phi_A(r) < r$$

3.1.3 Existence Results Under Weak Topology Setting

In this subsection, we prove some new fixed point theorems in a product of Banach algebras under a weak topology setting for the block operator matrix (3.0.1). The main condition is formulated in terms of axiomatic measure of weak non-compactness. The obtained results can be found in [16]. The first result in this subsection is formulated as follows.

Theorem 3.1.10 *Let S be a nonempty, bounded, convex, and closed subset of a Banach algebra X satisfying the sequential condition $(\mathscr{P})$ and let A, $C : S \longrightarrow X$ and B, B', $D : X \longrightarrow X$ be five operators such that*

(i) A, B, C, and B' are weakly sequentially continuous,
(ii) D is linear, bounded and D^p is separate contraction, for some $p \in \mathbb{N}^*$,
(iii) A and C are weakly compact and $C(S) \subset (I - D)(S)$,
(iv) $\big(A + B(I - D)^{-1}C \cdot B'(I - D)^{-1}C\big)(S) \subset S$.

Then, the block operator matrix (3.0.1) *has a fixed point in $S \times S$.*

Proof It follows from Proposition 1.2.45, that $(I - D^p)^{-1}$ as well as $(I - D)^{-1}$ is well defined and

$$(I - D)^{-1} = (I - D^p)^{-1}\sum_{k=0}^{p-1} D^k. \tag{3.1.5}$$

Since D is a linear and bounded mapping, then $(I - D)^{-1}$ is a weakly continuous operator. Define the mapping $F : S \longrightarrow X$ by putting

$$F(x) = A(x) + B(I - D)^{-1}C(x) \cdot B'(I - D)^{-1}C(x). \tag{3.1.6}$$

If we consider the weak compactness of sets $\overline{A(S)}^w$ and $\overline{C(S)}^w$ and deploy the following inclusion

$$F(S) \subseteq A(S) + B(I - D)^{-1}C(S) \cdot B'(I - D)^{-1}C(S),$$

we get $F(S)$ is relatively weakly compact. We complete the proof by showing that F is weakly sequentially continuous. Consider $(\xi_n)_n$ as a weakly convergent sequence contained in the set S to a point ξ. Making use of the weak continuity of the mapping $(I - D)^{-1}$, together with the weak compactness of the set $\overline{C(S)}^w$, we acquire that the sequence $(\xi_n)_n$ possesses a subsequence, say $(\xi_{n_k})_k$, such that

$$(I - D)^{-1}C(\xi_{n_k}) \rightharpoonup \gamma.$$

Taking into account the weak sequential continuity of the maps C and D, and using the following equality

$$(I - D)^{-1}C = C + D(I - D)^{-1}C, \tag{3.1.7}$$

we obtain $\gamma = (I - D)^{-1}C(\xi)$. Thus,

$$(I - D)^{-1}C(\xi_{n_k}) \rightharpoonup (I - D)^{-1}C(\xi).$$

We show that

$$(I - D)^{-1}C(\xi_n) \rightharpoonup (I - D)^{-1}C(\xi).$$

If not the case, there exists a weak neighborhood V^w of $(I - D)^{-1}C(\xi)$ and there exists a subsequence, say $(\xi_{n_j})_j$, of the sequence $(\xi_n)_n$ such that

$$(I - D)^{-1}C(\xi_{n_j}) \notin V^w, \quad \text{for all} \quad j \geq 1.$$

Since $(\xi_{n_j})_j$ converges weakly to ξ, and arguing as before, we find a subsequence, say $(\xi_{n_{j_k}})_k$, of $(\xi_{n_j})_j$ such that

$$(I - D)^{-1}C(\xi_{n_{j_k}}) \rightharpoonup (I - D)^{-1}C(\xi),$$

which is a contradiction. As a result, $(I - D)^{-1}C$ is weakly sequentially continuous. The condition $(\mathscr{P})$ guaranties that F is weakly sequentially continuous on S. The desired conclusion follows by a direct application of Theorem 1.3.44. Q.E.D.

In the following result, it would be interesting to investigate the case when D is nonlinear mapping.

Theorem 3.1.11 *Let S be a nonempty, bounded, convex, and closed subset of a Banach algebra X satisfying the sequential condition $(\mathscr{P})$ and let $A, C : S \longrightarrow X$ and $B, B', D : X \longrightarrow X$ be five weakly sequentially continuous operators such that*

(i) A is condensing with respect to β,
(ii) D is a nonlinear contraction with $\mathscr{D}$-function ϕ,
(iii) C is weakly compact and $(I - D)^{-1}C(S)$ is bounded,
(iv) $A(x) + B(I - D)^{-1}C(x) \cdot B'(I - D)^{-1}C(x) \in S$ for all $x \in S$.

Then, the block operator matrix (3.0.1) *has a fixed point in $S \times S$.*

Proof Since D is nonlinear contraction with $\mathscr{D}$-function ϕ, then $I - D$ is one-to-one and so, the operator $(I - D)^{-1}$ is well defined on X. Indeed, let us fix an arbitrary element $y \in C(S)$ and define the mapping $\varphi : X \longrightarrow X$ by putting

$$\varphi(x) = y + Dx.$$

Notice that φ is a nonlinear contraction mapping. It follows from Theorem 1.2.26, that there exists an unique element x^*, that satisfies

$$y = (I - D)x^*.$$

In other words, $(I - D)^{-1}$ exists. We show that $(I - D)^{-1}C$ is weakly compact. If we suppose the contrary, then $\beta\left((I - D)^{-1}C(S)\right) = d > 0$. The use of (3.1.7) as well as the weak compactness of $\overline{C(S)}^{w}$ yields that

$$\beta\left((I - D)^{-1}C(S)\right) \leq \beta\left(D(I - D)^{-1}C(S)\right).$$

Let $\varepsilon > 0$. By the definition of the De Blasi measure β, there exists $K \in \mathscr{W}(X)$ satisfying

$$(I - D)^{-1}C(S) \subseteq K + \overline{\mathscr{B}}_X(d + \varepsilon).$$

Using Lemma 1.5.9, we obtain

$$\beta\left((I - D)^{-1}C(S)\right) \leq k\beta\left((I - D)^{-1}C(S)\right),$$

which is a contradiction, and so $(I - D)^{-1}C$ as well as F are weakly sequentially continuous, where F is described in (3.1.6). By using Lemma 1.3.62, together with assumption (i), we can prove that F is condensing with respect to β, and when we apply Theorem 1.5.12, we reach the desired conclusion. Q.E.D.

We will combine Theorem 3.1.11 and Lemma 1.5.9, in order to get the following result.

Corollary 3.1.12 *Let S be a nonempty, bounded, convex, and closed subset of a Banach algebra X satisfying the sequential condition $(\mathscr{P})$ and let $A, C : S \longrightarrow X$ and $B, B', D : X \longrightarrow X$ be five weakly sequentially continuous operators such that*

(i) A is a contraction with a constant k,
(ii) D is nonlinear contraction with $\mathscr{D}$-function ϕ,
(iii) C is weakly compact and $(I - D)^{-1}C(S)$ is bounded,
(iv) $A(x) + B(I - D)^{-1}C(x) \cdot B'(I - D)^{-1}C(x) \in S$, for all $x \in S$.

Then, the block operator matrix (3.0.1) *has, at least, one fixed point in $S \times S$.*

Theorem 3.1.13 *Let S be a nonempty, bounded, convex, and closed subset of a Banach algebra X satisfying the sequential condition $(\mathscr{P})$ and let $A, C : S \longrightarrow X$ and $B, B', D : X \longrightarrow X$ be five weakly sequentially continuous operators such that*

(i) B and C are Lipschitzians with constants q_1 and q_2, respectively,
(ii) D is expansive with constant $h > q_2 + 1$,
(iii) $A(S)$ and $B'(S)$ are relatively weakly compact,
(iv) C is strongly continuous and $C(S) \subseteq (I - D)(S)$,
(v) $A(x) + B(I - D)^{-1}C(x) \cdot B'(I - D)^{-1}C(x) \in S$, for all $x \in S$.

Then, the block operator matrix (3.0.1) *has a fixed point in $S \times S$ provided that*

$$0 \leq q_1\|B'(S)\| < 1.$$

Proof It follows from Proposition 1.2.11, that $(I - D)^{-1}$ exists on $(I - D)(X)$ and for all $x, y \in (I - D)(X)$, we have

$$\|(I-D)^{-1}x-(I-D)^{-1}y\| \leq \frac{1}{h-1}\|x-y\|.$$

Then, $(I-D)^{-1}$ is continuous. After simple calculations, we can verify that the mapping $(I-D)^{-1}C$ is Lipschitzian with a Lipschitz constant $\frac{q_2}{h-1}$. Thus, $(I-D)^{-1}C(S)$ is bounded. The use of Theorem 1.3.29 ensures that $\overline{B'(I-D)^{-1}C(S)}^{w}$ is weakly compact in X. We will prove that the operator F as is described in (3.1.6) is condensing with respect to β. Using the subadditivity of the De Blasi measure β, we arrive at the following

$$\beta\,(F(D)) \leq \beta(A(D)) + \beta\left(B(I-D)^{-1}C(D)\cdot B'(D)\right),$$

for any subset D of S. Taking into account that $A(D)$ is relatively weakly compact, and using Lemma 1.5.9, we get

$$\beta\,(F(D)) \leq M\beta\left(B(I-D)^{-1}C(D)\right).$$

So, if $\beta(D) > 0$, we obtain

$$\begin{aligned}\beta\left((A+B(I-D)^{-1}C\cdot B'(I-D)^{-1}C)(D)\right) &\leq \frac{q_1q_2}{h-1}M\beta(D)\\ &< \beta(D).\end{aligned}$$

The desired conclusion follows by a direct application of Theorem 1.5.12. Q.E.D.

Theorem 3.1.14 *Let S be a nonempty, bounded, convex, and closed subset of a Banach algebra X satisfying the sequential condition $(\mathscr{P})$ and let A, $C : S \longrightarrow X$ and B, B', $D : X \longrightarrow X$ be five weakly sequentially continuous operators such that*

(i) B and C are Lipschitzians with constants q_1 and q_2, respectively,
(ii) C is weakly compact and $C(S) \subseteq (I-D)(S)$,
(iii) A and D are two contractions with constants k and k', respectively,
(iv) $A(x)+B(I-D)^{-1}C(x)\cdot B'(I-D)^{-1}C(x) \in S$ for all $x \in S$.

Then, the block operator matrix (3.0.1) *has a fixed point in $S \times S$ provided that*

$$q_1q_2M < (1-k)(1-k'),$$

where $M = \left\|\overline{B'(I-D)^{-1}C(S)}^{w}\right\| > 1.$

Proof The use of Theorem 1.2.4 allows us to have $(I-D)^{-1}$ is continuous and $(I-D)^{-1}C(S)$ is bounded. In order to apply Theorem 1.5.4, we have to verify the following claims. We first claim that $(I-D)^{-1}C$ is weakly compact on S. If not the case, we obtain $\beta\left((I-D)^{-1}C(S)\right) > 0$. By using of the equality (3.1.7), we can prove that

$$\beta((I-D)^{-1}C(S)) \leq \beta(D(I-D)^{-1}C(S)). \tag{3.1.8}$$

Let $r > \beta((I-D)^{-1}C(S))$ and K be a relatively weakly compact subset in X satisfying

$$(I-D)^{-1}C(S) \subseteq K + \overline{\mathscr{B}}_X(r).$$

Keeping in mind that D is a contraction with a constant k', we infer that

$$D(I-D)^{-1}C(S) \subseteq \overline{D(K)}^{w} + \overline{\mathscr{B}}_X(k'r).$$

The use of assumption (iii) and Lemma 1.5.9, we obtain

$$\beta(D(I-D)^{-1}C(S)) \leq k'r < r.$$

Going back to the inequality (3.1.8) shows that

$$\begin{aligned}\beta((I-D)^{-1}C(S)) &\leq \beta(D(I-D)^{-1}C(S)) \\ &\leq k'r.\end{aligned}$$

Letting $r \to \beta((I-D)^{-1}C(S))$, we get

$$\begin{aligned}\beta((I-D)^{-1}C(S)) &\leq k'\beta((I-D)^{-1}C(S)) \\ &< \beta((I-D)^{-1}C(S)),\end{aligned}$$

which is a contradiction, and the claim is proved. In the similar way as in the proof of Theorem 3.1.11, we can demonstrate that the operator F, as is described in (3.1.6), is weakly sequentially continuous. Therefore, the operator F is convex-power condensing. Indeed, one can verify easily that

$$F(S) \subset A(S) + B(I-D)^{-1}C(S) \cdot B'(I-D)^{-1}C(S).$$

Since $\overline{B'(I-D)^{-1}C(S)}^{w}$ is weakly compact, by the subadditivity of β, we know

$$\beta(F(S)) \leq \beta(A(S)) + \beta\left(B(I-D)^{-1}C(S) \cdot \overline{B'(I-D)^{-1}C(S)}^{w}\right).$$

The use of assumption (iii) as well as Lemma 1.5.10 leads to

$$\beta(F(S)) \leq k\beta(S) + \left\|\overline{B'(I-D)^{-1}C(S)}^{w}\right\| \beta\left(B(I-D)^{-1}C(S)\right).$$

Since the operator $B(I-D)^{-1}C$ is Lipschitzian with a Lipschitz constant $\frac{q_1q_2}{1-k'}$, then

$$\beta(F(S)) \leq \left(k + M\frac{q_1q_2}{1-k'}\right)\beta(S).$$

Letting $x_0 \in S$ and proceeding by induction, we get

$$\begin{aligned}\beta\left(F^{(n,x_0)}(S)\right) &= \beta\left(F\left(\overline{co}\left\{F^{(n-1,x_0)}(S)\bigcup\{x_0\}\right\}\right)\right)\\ &\leq \left(k+M\tfrac{q_1q_2}{1-k'}\right)\beta\left(\overline{co}\left\{F^{(n-1,x_0)}(S)\bigcup\{x_0\}\right\}\right)\\ &\leq \left(k+M\tfrac{q_1q_2}{1-k'}\right)^n\beta(S).\end{aligned}$$

Since $0 < k + M\frac{q_1q_2}{1-k'} < 1$, then F is a convex-power condensing mapping. Therefore, on account of Theorem 1.5.4, we can deduce that F has a fixed point x in S. Q.E.D.

We can modify some assumptions of Theorem 3.1.14, in order to obtain a similar approach.

Theorem 3.1.15 *Let S be a nonempty, convex, and closed subset of a Banach algebra X satisfying condition $(\mathscr{P})$ and let A, $C : S \longrightarrow X$ and B, B', $D : X \longrightarrow X$ be five weakly sequentially continuous operators such that*

(i) A, B, and C are $\mathscr{D}$-Lipschitzians with $\mathscr{D}$-functions ϕ_A, ϕ_B and ϕ_C, respectively,
(ii) $A(S)$ and $B(S)$ are bounded and $C(S) \subseteq (I - D)(S)$,
(iii) the set $B'(S)$ is relatively weakly compact,
(iv) D is a contraction with constant k,
(v) $A(x) + B(I - D)^{-1}C(x) \cdot B'(I - D)^{-1}C(x) \in S$, for all $x \in S$.

Then, the block operator matrix (3.2.6) has a fixed point in $S \times S$ provided that

$$\phi_A(r) + M\phi_B\left(\frac{1}{1-k}\phi_C\right)(r) < r, \quad \text{for all } r > 0,$$

where $M = \|B'(I - D)^{-1}C(S)\|$.

Proof Define the mapping $F : S \longrightarrow S$ by putting

$$F(x) = A(x) + B(I - D)^{-1}C(x) \cdot B'(I - D)^{-1}C(x).$$

Notice that $(I - D)^{-1}C$ as well as F is weakly sequentially continuous on S. In fact, consider $(x_n)_n$ as a weakly convergent sequence contained in the set S to a point x. Going back equality (3.1.7) and Lemma 1.5.9, we obtain

$$\begin{aligned}\beta(\{(I-D)^{-1}C(x_n),\ n\in\mathbb{N}\}) &\leq \beta(\{C(x_n),\ n\in\mathbb{N}\}) + \beta\left(\{D(I-D)^{-1}C(x_n),\ n\in\mathbb{N}\}\right)\\ &\leq k\beta\left(\{(I-D)^{-1}C(x_n),\ n\in\mathbb{N}\}\right).\end{aligned}$$

Then, the sequence $(x_n)_n$ possesses a subsequence, say $(x_{n_k})_k$, such that

$$(I - D)^{-1}C(x_{n_k}) \rightharpoonup y.$$

Thus, on account of equality (3.1.7), we get $y = (I - D)^{-1}C(x)$. Now, we show that

$$(I-D)^{-1}C(x_n) \rightharpoonup (I-D)^{-1}C(x).$$

Suppose that this is not the case, there exists a weakly open V^w of $(I-D)^{-1}C(x)$ and there exists a subsequence, say $(x_{n_j})_j$, of $(x_n)_n$ such that $(I-D)^{-1}C(x_{n_j}) \notin V^w$, for all $j \geq 1$. Since the sequence $(x_{n_j})_j$ converges weakly to x, we can find a subsequence, say $(x_{n_{j_k}})_k$, of $(x_{n_j})_j$, that satisfies

$$(I-D)^{-1}C(x_{n_{j_k}}) \rightharpoonup (I-D)^{-1}C(x),$$

which is a contradiction, since $(I-D)^{-1}C(x_{n_{j_k}}) \notin V^w$. As a result, $(I-D)^{-1}C$ is weakly sequentially continuous. Moreover, the condition $(\mathcal{P})$ guarantees the weak sequential continuity of the mapping F on S. Our next task is to show that F is condensing with respect to β. Indeed, for an arbitrary bounded subset $\mathcal{M}$ of S with $\beta(\mathcal{M}) > 0$, we get

$$\begin{aligned} F(\mathcal{M}) &\subset A(\mathcal{M}) + B(I-D)^{-1}C(\mathcal{M}) \cdot B'(I-D)^{-1}C(\mathcal{M}) \\ &\subset A(\mathcal{M}) + B(I-D)^{-1}C(S) \cdot B'(I-D)^{-1}C(S) \\ &\subset A(S) + B(S) \cdot B'(S). \end{aligned}$$

Based on assumption (ii), we deduce that $F(\mathcal{M})$ is bounded. Since the operator $(I-D)^{-1}C$ is $\mathcal{D}$-Lipschitzian with $\mathcal{D}$-function $\frac{1}{1-k}\phi_C$, we can prove that $(I-D)^{-1}C(\mathcal{M})$ is a bounded subset of S. Taking into account assumption (i), and using Lemma 1.5.9, we can show that

$$\begin{aligned} \beta(F(\mathcal{M})) &= \beta(A(\mathcal{M})) + \beta\left(B(I-D)^{-1}C(\mathcal{M}) \cdot B'(I-D)^{-1}C(\mathcal{M})\right) \\ &\leq \beta(A(\mathcal{M})) + \left\|B'(I-D)^{-1}C(\mathcal{M})\right\| \beta\left(B(I-D)^{-1}C(\mathcal{M})\right) \\ &\leq \phi_A(\beta(\mathcal{M})) + \frac{M}{1-k}\phi_C(\beta(\mathcal{M})) \\ &\leq \left(\phi_A + \frac{M}{1-k}\phi_C\right)\beta(\mathcal{M}). \end{aligned}$$

This means, in particular that F is a condensing with respect to β, and when we apply Theorem 1.5.18, we reach the desired conclusion. Q.E.D.

In the sequel, we exhibit a fixed point theorem for regular mapping in commutative Banach algebra satisfying the sequential condition $(\mathcal{P})$. In what follows, we will continue the study in the case where, the operator $B(I-D)^{-1}C$ is regular, that is, $B(I-D)^{-1}C$ maps S into the set of all invertible elements of X. Before stating, we would like to prove the following lemma.

Lemma 3.1.16 *Let S be a nonempty, convex, and closed subset of a Banach algebra X satisfying condition $(\mathcal{P})$ and let $A, C : S \longrightarrow X$ and $B, B', D : X \longrightarrow X$ be five operators such that*

(i) A, B, and C are $\mathcal{D}$-Lipschitzians with $\mathcal{D}$-functions ϕ_A, ϕ_B and ϕ_C, respectively,
(ii) D is a contraction with a constant k and $C(S) \subseteq (I - D)(S)$,
(iii) $B'(S)$ is bounded and $C(S) \subseteq (I - D)(S)$,
(iv) $B(I - D)^{-1}C$ is regular $B'(S)$,
(v) $A(x) + B(I - D)^{-1}C(x) \cdot B'(I - D)^{-1}C(y) \in S$, for all $x, y \in S$.

Then, the operator $\left(\frac{I-A}{B(I-D)^{-1}C}\right)^{-1}$ is well defined on $B'(I - D)^{-1}C(S)$ provided that

$$M\phi_B\left(\frac{1}{1-k}\phi_C\right)(r) + \phi_A(r) < r, \quad \text{for all } r > 0,$$

where $M = \|B'(I - D)^{-1}C(S)\|$.

Proof It is not hard to verify, by Theorem 1.2.4, that $(I - D)^{-1}$ exists on X and

$$\begin{aligned}\|B(I - D)^{-1}Cx - B(I - D)^{-1}Cy\| &\leq \phi_B\left(\|(I - D)^{-1}Cx - (I - D)^{-1}Cy\|\right)\\ &\leq \phi_B\left(\tfrac{1}{1-k}\phi_C\right)(\|x - y\|),\end{aligned}$$

whenever $x, y \in S$. This means that the operator $B(I - D)^{-1}C$ is $\mathcal{D}$-Lipschitzian with $\mathcal{D}$-function $\phi_B \circ \varphi$, where

$$\varphi(r) = \frac{1}{1-k}\phi_C(r), \quad \text{for } r > 0. \tag{3.1.9}$$

Let us fix an arbitrarily element $y \in S$ and define the mapping $T_y : S \longrightarrow S$ by putting

$$T_y(x) = A(x) + B(I - D)^{-1}C(x) \cdot B'(I - D)^{-1}C(y).$$

It is obvious, via simple calculations, that the mapping T_y is $\mathcal{D}$-Lipschitzian with $\mathcal{D}$-function ψ, where

$$\psi(r) = M\phi_B \circ \varphi(r) + \phi_A(r).$$

In view of Theorem 1.2.26, we deduce that T_y has an unique fixed point, say $x_y \in S$. In other words, the point x_y satisfies

$$A(x_y) + B(I - D)^{-1}C(x_y) \cdot B'(I - D)^{-1}C(y) = x_y.$$

Then,

$$\left(\frac{I - A}{B(I - D)^{-1}C}\right)(x_y) = B'(I - D)^{-1}C(y).$$

The desired result is deduced. Q.E.D.

By combining Theorem 1.5.12, together with Lemma 3.1.16, we obtain the following result.

Theorem 3.1.17 *Let S be a nonempty, bounded, convex, and closed subset of a Banach algebra X satisfying the sequential condition ($\mathcal{P}$) and let A, $C : S \longrightarrow X$ and B, B', $D : X \longrightarrow X$ be five mapping such that*

(i) A, B, and C are $\mathcal{D}$-Lipschitzians with $\mathcal{D}$-functions ϕ_A, ϕ_B and ϕ_C, respectively,
(ii) B' is Lipschitzian with a Lipschitz constant q_1,
(iii) D is a contraction with a constant k such that $k \in \left[0, \frac{3-\sqrt{5}}{2}\right[$,
(iv) $B(I-D)^{-1}C$ is regular on $B'(I-D)^{-1}C(S)$,
(v) C is strongly continuous such that $C(S) \subseteq (I-D)(S)$,
(vi) $x = A(x) + B(I-D)^{-1}C(x) \cdot B'(I-D)^{-1}C(y)$, for $y \in S \Rightarrow x \in S$.

Then, the block operator matrix (3.0.1) *has a fixed point in $S \times X$ provided that*

$$\max\{M\phi_B\varphi(r) + \phi_A(r),\ \rho q_1 \phi_C(r)\} \leq kr, \quad \text{for all } r > 0,$$

where φ is given in the equality (3.1.9) *and $M = \|B'(I-D)^{-1}C(S)\| > 0$ and*

$$\rho = \|B(I-D)^{-1}C(S)\| + \frac{k}{M} diam(S).$$

Proof It is obvious to show that $B(I-D)^{-1}C(S)$ and $B'(I-D)^{-1}C(S)$ are bounded. From Lemma 3.1.16, we can deduce that the operator

$$\left(\frac{I-A}{B(I-D)^{-1}C}\right)^{-1} \tag{3.1.10}$$

exists on $B'(I-D)^{-1}C(S)$. Define the mapping $N : S \longrightarrow X$ by putting

$$N(x) = \left(\frac{I-A}{B(I-D)^{-1}C}\right)^{-1} B'(I-D)^{-1}C(x). \tag{3.1.11}$$

In view of Theorem 1.5.4, the proof is composed of the following steps.
Step 1. We start by showing that N is a weakly sequentially continuous mapping, then we first prove the continuity of the operator described in (3.1.10). For this purpose, let us assume that $(x_n)_n$ is a sequence contained in the set $B'(I-D)^{-1}C(S)$, which is convergent to a point x. Let us denote

$$\begin{cases} y_n = \left(\dfrac{I-A}{B(I-D)^{-1}C}\right)^{-1}(x_n) \\ y = \left(\dfrac{I-A}{B(I-D)^{-1}C}\right)^{-1}(x). \end{cases}$$

In other words, we have

$$\begin{cases} y_n = A(y_n) + B(I-D)^{-1}C(y_n) \cdot x_n \\ y = A(y) + B(I-D)^{-1}C(y) \cdot x. \end{cases}$$

After simple calculations, we can verify that

$$\|y_n - y\| \leq \left(M\phi_B \circ \left(\frac{1}{1-k}\phi_C\right) + \phi_A\right)(\|y_n - y\|) + \|B(I-D)^{-1}C(y)\|\|x_n - x\|.$$

Taking into account that the operator $B(I-D)^{-1}C$ is $\mathscr{D}$-Lipschitzian with $\mathscr{D}$-function $\phi_B \circ \left(\frac{1}{1-k}\phi_C\right)$, and using the growth of the functions ϕ_B and ϕ_C, we get

$$\|B(I-D)^{-1}C(y)\| < \|B(I-D)^{-1}C(a)\| + \frac{k}{M}\|y - a\|,$$

for some $a \in S$. Consequently, we have

$$\limsup_n \|y_n - y\| \leq \left(M\phi_B \circ \left(\frac{1}{1-k}\phi_C\right) + \phi_A\right) \limsup_n \|y_n - y\|.$$

Thus,

$$\limsup_n \|y_n - y\| = 0.$$

On the other hand, by assumption (v) and via simple calculations, one can show that the mapping N is weakly sequentially continuous.

Step 2. We have to show that N is convex-power condensing with respect to β. Let $x_1, x_2 \in S$ and $y_1, y_2 \in X$ be such that $y_1 = Nx_1$ and $y_2 = Nx_2$. Then,

$$\begin{cases} y_1 = A(y_1) + B'(I-D)^{-1}C(y_1) \cdot B'(I-D)^{-1}C(x_1) \\ y_2 = A(y_2) + B'(I-D)^{-1}C(y_2) \cdot B'(I-D)^{-1}C(x_2). \end{cases}$$

This implies that

$$\begin{aligned} \|y_1 - y_2\| &\leq \phi_A(\|y_1 - y_2\|) + M\phi_B \circ \varphi\,(\|y_1 - y_2\|) + \rho q_1 \varphi\,(\|x_1 - x_2\|) \\ &\leq k\|y_1 - y_2\| + \tfrac{k}{1-k}\|x_1 - x_2\|, \end{aligned}$$

where the function φ as is described in equality (3.1.9) and so,

$$\|Nx_1 - Nx_2\| \leq \frac{k}{(1-k)^2}\|x_1 - x_2\|.$$

The use of Lemma 1.5.9 and Step 1 leads to the following inequality

$$\beta(N(S)) \leq \frac{k}{(1-k)^2}\beta(S).$$

This show that the operator N is convex-power condensing, and when we apply Theorem 1.5.4, we reach the desired conclusion. Q.E.D.

Remark 3.1.18 The assumption (vi) of Theorem 3.1.17 was introduced by Burton [15] instead of assuming that

$$A(x) + B(I - D)^{-1}C(x) \cdot B'(I - D)^{-1}C(y) \in S, \quad \text{for all } x,\ y \in S.$$

In the following result, we may combine Theorem 1.3.44 and Lemma 3.1.16.

Theorem 3.1.19 *Let S be a nonempty, convex, and weakly compact subset of a Banach algebra X satisfying the sequential condition $(\mathscr{P})$ and let $A,\ C : S \longrightarrow X$ and $B,\ B',\ D : X \longrightarrow X$ be five operators such that*

(i) A, B, and C are $\mathscr{D}$-Lipschitzians with $\mathscr{D}$-functions ϕ_A, ϕ_B and ϕ_C, respectively,
(ii) B' is continuous on S,
(iii) $B(I - D)^{-1}C$ is regular on $B'(S)$,
(iv) C is strongly continuous and $C(S) \subseteq (I - D)(S)$,
(v) D is a contraction with a constant k,
(vi) $x = A(x) + B(I - D)^{-1}C(x) \cdot B'(I - D)^{-1}C(y),\ y \in S \Rightarrow x \in S$.

Then, the block operator matrix (3.0.1) *has, at least, one fixed point provided that*

$$M\phi_B \circ \varphi(r) + \phi_A(r) < r, \quad \textit{for all } r > 0,$$

where φ is given in (3.1.9) *and* $M = \|B'(I - D)^{-1}C(S)\|$.

Proof According to Theorem 3.1.17, we infer that the operator N already defined in (3.1.11) is weakly sequentially continuous. Due to the fact that S is weakly compact and Theorem 1.3.29, we infer that the set $N(S)$ is relatively weakly compact. Now, the operator N satisfies the hypotheses of Theorem 1.3.46, so there exists an element $x^* \in S$ such that $Nx^* = x^*$. Q.E.D.

3.1.4 Weakly Asymptotically Regular Mappings

We are examining the existence of a fixed point for the following the block operator matrix

$$\begin{pmatrix} A & B \\ C & D \end{pmatrix}, \tag{3.1.12}$$

with nonlinear single-valued entries. This study has been based on the convexity of the unbounded domain and on the properties of the inputs (such as k-set-weakly contractive, weakly compact, weakly asymptotically regular,...). The results of this subsection can be found in [17]. Asymptotic regularity is usually used to obtain fixed point theorems by many researchers, see for example [18, 19]. Our first result, motivated in part by Ben Amar, O'Regan and Touati [20], involves k-set-weakly contractive operators in Banach space.

Theorem 3.1.20 *Let S be a nonempty, convex, and closed subset of a Banach space X and let A, B, C, $D : S \longrightarrow S$ be four continuous operators such that*

(i) A is k-set-weakly contractive, which satisfies $(\mathfrak{H}_1)$,
(ii) B is λ-set-weakly contractive,
(iii) If $x_n - y_n \rightharpoonup \theta$, then $A(x_n) - A(y_n) \rightharpoonup \theta$,
(iv) C is weakly compact which satisfies $(\mathfrak{H}_1)$ and $C(S) \subset (I - D)(S)$,
(v) D is linear, bounded and D^p is separate contraction, for some $p \in \mathbb{N}$,
(vi) A is weakly asymptotically regular with respect to $B(I - D)^{-1}C$,
(vii) $A(S) + B(I - D)^{-1}C(S)$ is a bounded subset in S,
$(viii)$ $B(I - D)^{-1}C$ has a weakly sequentially closed graph.

Then, the block operator matrix (3.1.12) *has a fixed point in $S \times S$.*

Proof According to Proposition 1.2.45, we obtain $I - D$ is a homeomorphism and the operator $(I - D)^{-1}$ is weakly sequentially continuous on $C(S)$. Due to the fact that $C(S)$ is relatively weakly compact, and using the subadditivity of the De Blasi measure β, we obtain

$$\begin{aligned}\beta(A^n(M) + B(I - D)^{-1}C(M)) &\leq \beta(A^n(M)) + \beta(B(I - D)^{-1}C(M))\\ &\leq k^n\beta(M) + \lambda\beta((I - D)^{-1}C(M)),\end{aligned}$$

for any bounded subset M of S with $\beta(M) > 0$. If we consider the continuity of the operator $(I - D)^{-1}$, and using the fact that $C(M)$ is relatively weakly compact, we can deduce that the operator $A^n + B(I - D)^{-1}C$ is condensing with respect to β. By using our assumptions, we deduce that the operator $A^n + B(I - D)^{-1}C$ satisfies $(\mathfrak{H}_1)$. Hence, an application of Theorem 1.5.25 shows that, there exists a bounded sequence $(x_n)_n$ contained in the set S that satisfies

$$A^n(x_n) + B(I - D)^{-1}C(x_n) = x_n. \tag{3.1.13}$$

Since A is weakly asymptotically regular with respect to $B(I - D)^{-1}C$, then

$$A^n(x_n) - A^{n-1}(x_n) + B(I - D)^{-1}C(x_n) \rightharpoonup \theta.$$

This implies that

$$x_n - A^{n-1}(x_n) \rightharpoonup \theta.$$

Therefore, $A(x_n) - A^n(x_n) \rightharpoonup \theta$, in view of assumption (iii). Note that

$$x_n - A(x_n) - B(I - D)^{-1}C(x_n) = x_n - A(x_n) + A^n(x_n) - A^n(x_n) - B(I - D)^{-1}C(x_n).$$

It is not hard to verify that,

$$x_n - Ax_n - B(I - D)^{-1}C(x_n) \rightharpoonup \theta \text{ as } n \to \infty. \tag{3.1.14}$$

Using equality (3.1.13) combined with the fact that $(I - D)^{-1}C$ is weakly compact, we obtain

$$\begin{aligned}\beta(\{x_n,\ n \in \mathbb{N}\}) &\le \beta\left(\{A^n(x_n),\ \ n \in \mathbb{N}\}\right) + \beta\left(\{B(I - D)^{-1}C(x_n),\ \ n \in \mathbb{N}\}\right) \\ &\le k\beta\left(\{A^{n-1}(x_n),\ \ n \in \mathbb{N}\}\right) + \lambda\beta\left(\{(I - D)^{-1}C(x_n),\ \ n \in \mathbb{N}\}\right) \\ &\le k^n\beta(\{x_n,\ n \in \mathbb{N}\}) \\ &< \beta(\{x_n,\ n \in \mathbb{N}\}).\end{aligned}$$

As a result, $(x_n)_n$ contains a weakly convergent subsequence, say $(x_{n_k})_k$, to an element $x \in S$. Taking into account the weak continuity of the operator $(I - D)^{-1}$ and assumption (iv), we obtain $(x_{n_k})_k$ has a subsequence, say $(x_{n_{k_s}})_s$, that satisfies

$$A(x_{n_{k_s}}) + B(I - D)^{-1}C(x_{n_{k_s}}) \rightharpoonup Ax + y, \text{ for some } y \in X.$$

Based on the equality (3.1.14), we get

$$x_{n_{k_s}} \rightharpoonup Ax + y \in S.$$

This means that the operator $A + B(I - D)^{-1}C$ has a fixed point. Q.E.D.

Theorem 3.1.21 *Let S be a nonempty, convex, and closed subset of a Banach space X and let $A, B, C, D : S \longrightarrow S$ be four operators such that*

(i) A, B and C are $\mathcal{D}$-Lipschitzians with $\mathcal{D}$-functions ϕ_A, ϕ_B and ϕ_C, respectively,
(ii) A, B and C are weakly sequentially continuous,
(iii) If $x_n - y_n \rightharpoonup \theta$, then $A(x_n) - A(y_n) \rightharpoonup \theta$,
(iv) D is linear, bounded and D^p is a separate contraction, for some $p \in \mathbb{N}$,
(v) A is weakly asymptotically regular with respect to $B(I - D)^{-1}C$,
(vi) $A(S) + B(I - D)^{-1}C(S)$ is bounded in S.

Then, the block operator matrix (3.1.12) *has a fixed point in $S \times S$ provided that*

$$\|D\| < 1 \ \textit{and}\ \ \phi_A^n(r) + \phi_B \circ \left(\frac{1}{1 - \|D\|}\phi_C\right)(r) < r, \ \ \textit{for all } r > 0.$$

Proof We show that the operator $A^n + B(I - D)^{-1}C$ is condensing with respect to β and it is weakly sequentially continuous, for each $n \in \mathbb{N}$. In order to prove this fact, assume that M is a bounded subset of S with $\beta(M) > 0$. Due to the fact that the operator A^n is weakly sequentially continuous and Lemma 1.5.9, we infer that

$$\begin{aligned}\beta\left(A^n(M) + B(I - D)^{-1}C(M)\right) &\le \beta\left(A^n(M)\right) + \beta\left(B(I - D)^{-1}C(M)\right) \\ &\le \phi_A\left[\beta\left(A^{n-1}(M)\right)\right] + \phi_B\left[\beta\left((I - D)^{-1}C(M)\right)\right].\end{aligned}$$

This implies that

$$\beta\left(A^n(M) + B(I-D)^{-1}C(M)\right) \leq \left[\phi_A^n + \phi_B \circ \left(\frac{1}{1-\|D\|}\phi_C\right)\right]\beta(M).$$

This means, in particular, that the map $A^n + B(I-D)^{-1}C$ is condensing with respect to β, and when we apply Theorem 1.5.18, we deduce the existence of a bounded sequence $(x_n)_n$ contained in the set S, that satisfies

$$A^n(x_n) + B(I-D)^{-1}C(x_n) = x_n, \text{ for each } n \in \mathbb{N}. \tag{3.1.15}$$

Since A is weakly asymptotically regular with respect to $B(I-D)^{-1}C$, then

$$A^n(x_n) - A^{n-1}(x_n) + B(I-D)^{-1}C(x_n) \rightharpoonup \theta.$$

This implies that $x_n - A^{n-1}(x_n) \rightharpoonup \theta$. It follows from assumption (iii), that

$$A(x_n) - A^n(x_n) \rightharpoonup \theta, \tag{3.1.16}$$

It is easy to see that

$$x_n - A(x_n) - B(I-D)^{-1}C(x_n) = x_n - A(x_n) + A^n(x_n) - A^n(x_n) - B(I-D)^{-1}C(x_n). \tag{3.1.17}$$

Using (3.1.15) and (3.1.16) in (3.1.17), we obtain

$$x_n - Ax_n - B(I-D)^{-1}C(x_n) \rightharpoonup \theta.$$

Our next task is to demonstrate that $(x_n)_n$ possesses a weakly convergent subsequence. If not the case, we get

$$\begin{aligned}
\beta(\{x_n,\ n \in \mathbb{N}\}) &\leq \beta\left(A^n(\{x_n,\ n \in \mathbb{N}\})\right) + \beta\left(B(I-D)^{-1}C(\{x_n,\ n \in \mathbb{N}\})\right) \\
&\leq \phi_A^n\left(\beta(\{x_n,\ n \in \mathbb{N}\})\right) + \left[\phi_B \circ \left(\frac{1}{1-\|D\|}\right) \circ \phi_C\right]\beta(\{x_n,\ n \in \mathbb{N}\}) \\
&< \beta(\{x_n,\ n \in \mathbb{N}\}).
\end{aligned}$$

which is a contradiction and so, we reach the result that $(x_n)_n$ possesses a weakly convergent subsequence, in view of Theorem 1.3.29. In other words, there exists subsequence, say $(x_{n_k})_k$, of $(x_n)_n$ that satisfies $x_{n_k} \rightharpoonup x \in S$. Using assumption (ii), we obtain

$$A(x_{n_k}) \rightharpoonup Ax \text{ and } B(I-D)^{-1}C(x_{n_k}) \rightharpoonup B(I-D)^{-1}C(x).$$

This completes the proof. Q.E.D.

By combining Theorem 3.1.21 and the Krasnoselskii-Daher fixed point theorem, proved in [21], we obtain the following.

Theorem 3.1.22 *Let S be a nonempty, bounded, convex and closed subset of a non-separable Banach space X with $\theta \in S$ and let $A, C : S \longrightarrow X$ and $B, D : X \longrightarrow X$ be four operators such that*

(i) A pseudo-contractive and strongly continuous,
(ii) B and C are weakly sequentially continuous and $C(S) \subset (I - D)(S)$,
(iii) A, B, and C are countably β-contractive with constants k_A, k_B and k_C, respectively,
(iv) D is linear, bounded and D^p is a separate contraction, for some $p \in \mathbb{N}^*$,
(v) $A(x) + B(I - D)^{-1}C(y) \in S$ for all $x, y \in S$.

Then, the block operator matrix (3.1.12) *has a fixed point in $S \times S$ provided that*

$$\|D\| < 1 \text{ and } \frac{k_B k_C}{(1 - \|D\|)(1 - k_A)} < 1.$$

Proof It is easy to verify that $I - A$ is one-to-one on S and, then the operator inverse $(I - A)^{-1}$ exists on $(I - A)(S)$. Indeed, let y be an arbitrarily fixed element contained in the set S and define the mapping $\varphi_y : S \longrightarrow S$ by putting

$$\varphi_y(x) = A(x) + B(I - D)^{-1}C(y).$$

The assumption (i) guaranties that the operator $I - \varphi_y$ is accretive. Then, on account of Lemma 1.2.51, there exists an element $x^* \in S$ satisfying

$$(I - \varphi_y)(x^*) = \theta.$$

In other words, we have

$$B(I - D)^{-1}C(y) = (I - A)(x^*) \in (I - A)(S).$$

Define the mapping $T : S \longrightarrow S$ by putting

$$T(x) = (I - A)^{-1}B(I - D)^{-1}C(x).$$

By using Theorem 1.5.18, we only demonstrate that T is weakly sequentially continuous and it is countably condensing with respect to β. To prove the above-announced fact, let $(x_n)_n$ be a weakly convergent sequence contained in the set S to an element $x \in S$. From the following equality

$$T = B(I - D)^{-1}C + AT, \tag{3.1.18}$$

we derive the estimate

$$\begin{aligned}\beta(\{T(x_n),\ n\in\mathbb{N}\}) &\le \beta\left(\{B(I-D)^{-1}C(x_n),\ n\in\mathbb{N}\}\right)+\beta(\{AT(x_n),\ n\in\mathbb{N}\})\\ &\le k_A\beta(\{T(x_n),\ n\in\mathbb{N}\}).\end{aligned}$$

From the above established facts, we deduce that $(T(x_n))_n$ possesses a weakly convergent subsequence, say $(T(x_{n_k}))_k$, to a point z. Making use of equality (3.1.18), we obtain $z = T(x)$. Now, we claim that

$$T(x_n) \rightharpoonup Tx.$$

If not the case, there exists a subsequence, say $(x_{n_j})_j$, of $(x_n)_n$ and there exists a weak neighborhood V of $T(x)$, that satisfies $T(x_{n_j}) \notin V$, for all $j \ge 0$ and so, there exists a subsequence of $(x_{n_j})_j$, say $(x_{n_{j_k}})_k$, that satisfies

$$T(x_{n_{j_k}}) \rightharpoonup T(x),$$

which is a contradiction and the claim is proved. Let us take a countable subset M of S with $\beta(M) > 0$. Going back to equality (3.1.18), we get

$$\begin{aligned}\beta(T(M)) &\le \beta\left(B(I-D)^{-1}C(M)\right)+\beta(AT(M))\\ &\le \frac{k_Bk_C}{1-\|D\|}\beta(M)+k_A\beta(T(M)).\end{aligned}$$

This implies that

$$\beta(T(M)) \le \frac{k_Bk_C}{(1-\|D\|)(1-k_A)}\beta(M).$$

The use of Theorem 1.5.18 achieves the proof. Q.E.D.

Theorem 3.1.23 *Let S be a nonempty, bounded, convex, and closed subset of a non-separable Banach space X with $\theta \in S$ and let A, B, C, $D : S \longrightarrow X$ be four operators such that*

(i) A is pseudo-contractive, continuous and weakly countably 1*-set-contractive,*
(ii) B and C are weakly sequentially continuous,
(iii) $C(S)$ is relatively weakly compact,
(iv) $(I-D)^{-1}$ is weakly sequentially continuous on $C(S)$,
(v) $\lambda A(x) + B(I-D)^{-1}C(y) \in S$, for all $x, y \in S$ and all $\lambda \in (0, 1)$.

Then, the block operator matrix (3.1.12) *has a fixed point in $S \times S$.*

Proof Via simple calculations, one can show that the operator λA is countably condensing with respect to β, for any $\lambda \in (0, 1)$. Since A is pseudo-contractive, then $I - A$ is accretive, that is,

$$Re\langle (I-A)(x)-(I-A)(y), x-y\rangle \ge 0, \quad \text{for all } x,\ y\in S,$$

and so, there exists $j \in J(x - y)$ such that

$$j((I-A)(x)-(I-A)(y)) \geq 0$$

where J is the normalized duality mapping which is given in (1.2.3). It should be noted that the operator λA is also pseudo-contractive on S. Indeed, for all $x, y \in S$, we have

$$\begin{aligned}
Re\langle (I-\lambda A)(x)-(I-\lambda A)(y), x-y\rangle &\geq j((I-\lambda A)(x)-(I-\lambda A)(y)) \\
&= j(\lambda x-\lambda A(x)-\lambda y+\lambda A(y)) \\
&\quad +j((1-\lambda)(x-y)) \\
&= \lambda j((I-A)(x)-(I-A)(y)) \\
&\quad +(1-\lambda)j(x-y) \\
&\geq (1-\lambda)\|x-y\|^2.
\end{aligned}$$

Then, the operator $I-\lambda A$ is ψ-expansive with function $\psi(t)=(1-\lambda)t$. In the similar way as in the proof of Theorem 3.1.22, we can show that there exists an element $x_\lambda \in S$ such that

$$x_\lambda = \lambda A(x_\lambda) + B(I-D)^{-1}C(x_\lambda).$$

Let $(\lambda_n)_n$ be a sequence contained in the set $[0,1]$, which is convergent to 1 and take $(x_n)_n$ the corresponding sequence. Then,

$$\begin{aligned}
\beta(\{x_n,\ n\in\mathbb{N}\}) &\leq \lambda_n\beta(\{A(x_n),\ n\in\mathbb{N}\}) + \beta\left(\{B(I-D)^{-1}C(x_n),\ n\in\mathbb{N}\}\right) \\
&\leq \lambda_n\beta(\{A(x_n),\ n\in\mathbb{N}\}) + \beta\left(\overline{\{B(I-D)^{-1}C(x_n),\ n\in\mathbb{N}\}}^{w}\right) \\
&\leq \lambda_n\beta(\{x_n,\ n\in\mathbb{N}\}).
\end{aligned}$$

This means, in particular, that $(x_n)_n$ possesses a subsequence, say $(x_{n_k})_k$, with $\beta\left(\{x_{n_k},\ k\in\mathbb{N}\}\right)=0$. This completes the proof. Q.E.D.

Theorem 3.1.24 *Let S be a nonempty, bounded, convex and closed subset of a non-separable Banach space X with $\theta \in S$ and let $A, B, C, D : S \longrightarrow X$ be four operators such that*

(i) A is pseudo-contractive and continuous such that C is weakly compact,
(ii) C is weakly compact and $C(S) \subset (I-D)(S)$,
(iii) A, B, and C are weakly sequentially continuous,
(iv) B and C are $\mathcal{D}$-Lipschitzians with $\mathcal{D}$-functions ϕ_B and ϕ_C, respectively,
(v) D is linear, bounded and D^p is a separate contraction, for some $p \in \mathbb{N}^*$,
(vi) $I-A$ is ψ-expansive and $B(I-D)^{-1}C$ is weakly demi-compact, and
(vii) $A(x)+B(I-D)^{-1}C(y) \in S$, for all $x, y \in S$.

Then, the block operator matrix (3.1.12) *has a fixed point in $S \times S$ provided that*

$$\|D\| < 1 \ \text{ and } \ \phi_B\left(\frac{1}{1-\|D\|}\phi_C\right)(r) \leq r, \ \text{ for } r > 0.$$

Proof It is sufficient to demonstrate that the map, which assigns to each x in S, the value

$$T(x) = (I - A)^{-1}B(I - D)^{-1}C(x) \in S$$

meets all the conditions of Theorem 1.5.21. In what follows, we prove that the operator T is weakly countably 1-set-contraction with respect to β. Let V be a countably subset contained in the set S with $\beta(V) > 0$. Taking into account that $T(V) \subseteq S$, together with the weak compactness of the operator A and deploy the equality (3.1.18), we get

$$\begin{aligned}\beta\left(T(V)\right) &\leq \beta\left(B(I - D)^{-1}C(V) + AT(V)\right)\\ &\leq \beta\left(B(I - D)^{-1}C(V)\right) + \beta(AT(V))\\ &\leq \beta(B(I - D)^{-1}C(V)).\end{aligned}$$

This implies that

$$\beta\left(T(V)\right) \leq \left[\phi_B \circ \left(\frac{1}{1 - \|D\|}\phi_C\right)\right](\beta(V)).$$

On the other hand, let us assume that $(x_n)_n$ is a sequence contained in the set S, which is weakly convergent to a point x. Taking into account the weak compactness of the set $\overline{C(S)}^w$, it is shown that $(x_n)_n$ possesses a subsequence, say $(x_{n_k})_k$, that satisfies

$$AT(x_{n_k}) \rightharpoonup z, \quad \text{for some } z \in S.$$

Going back to equality (3.1.18), we obtain

$$T(x_{n_k}) \rightharpoonup B(I - D)^{-1}C(x) + z.$$

Then, in view of assumption (iii), we infer that

$$(I - A)T(x_{n_k}) \rightharpoonup B(I - D)^{-1}C(x).$$

Thus, $T(x) = B(I - D)^{-1}C(x) + z$. Q.E.D.

3.2 Critical-Type of Fixed Point Theorems

The strategy used in the above-mentioned sections consists in giving some hypotheses, which guarantee the reversibility of the operator $I - A$, in order to process with the operator $(I - A)^{-1}B(I - D)^{-1}C$. Hence, it will be important to verify, under weak topology features, the case of critical-type of fixed point. The assumptions imposed on the mentioned operators are formulated with help of weak topology and measures of weak non-compactness. Let us mention that results obtained in this section, which can be found in [17], generalize a few ones contained in [22, 23].

3.2.1 Fixed Point Theorems in Banach Spaces

Theorem 3.2.1 *Let S be a nonempty, convex, and closed subset of a Banach space X and let $A, C : S \longrightarrow X$ and $B, D : X \longrightarrow X$ be four operators such that*

(i) A, B and C are $\mathcal{D}$-Lipschitzians with $\mathcal{D}$-functions ϕ_A, ϕ_B and ϕ_C, respectively,
(ii) A, B, and C are weakly sequentially continuous,
(iii) D is linear, bounded and D^p is separate contraction with $\|D\| < 1$,
(iv) $A(S)$ and $(I - D)^{-1}C(S)$ are bounded.

Then, the block operator matrix (3.1.12) *has a fixed point in $S \times S$ provided that*

$$\phi_B\left(\frac{1}{1-\|D\|}\phi_C\right)(r) + \phi_A(r) < r, \;\; \textit{for } r > 0.$$

Proof In view of Proposition 1.2.45, we infer that $I - D$ is a homeomorphism and the operator $(I - D)^{-1}C$ is weakly sequentially continuous. Define the mapping $T : S \longrightarrow S$ by putting

$$T(x) = A(x) + B(I - D)^{-1}C(x). \tag{3.2.1}$$

We show that T is condensing with respect to β. Indeed, let M be an arbitrary bounded subset of S with $\beta(M) > 0$. Then,

$$\begin{aligned}\beta(TM) &\le \beta\left(AM + B(I - D)^{-1}CM\right)\\ &\le \beta(AM) + \beta(B(I - D)^{-1}CM).\end{aligned}$$

The use of Lemma 1.5.9 allows us to have

$$\beta(TM) \le \left[\phi_A + \phi_B \circ \left(\tfrac{1}{1-\|D\|}\phi_C\right)\right]\beta(M).$$

This means, in particular, that the operator T is condensing with respect to β, and when we apply Theorem 1.5.12, we reach the desired conclusion. Q.E.D.

Let us observe that, if we put S as a bounded subset of X and the operator A is linear, we obtain a similar approach of Theorem 3.2.1.

Theorem 3.2.2 *Let S be a nonempty, bounded, closed, and convex subset of a Banach space X and let $A, C : S \longrightarrow X$ and $B, D : X \longrightarrow X$ be four operators such that*

(i) A is linear and bounded such that A^p is separate contraction, for some $p \in \mathbb{N}^*$,
(ii) B and C are $\mathcal{D}$-Lipschitzians with $\mathcal{D}$-functions ϕ_B and ϕ_C, respectively,

(iii) B and C are weakly sequentially continuous,
(iv) D is linear, bounded and D^q *is separate contraction, for some* $q \in \mathbb{N}^*$,
(v) $\|A\| < 1$, $\|D\| < 1$ *and* $C(S) \subset (I - D)(S)$,
(vi) $B(I - D)^{-1}C(S) \subset (I - A)(S)$.

Then, the block operator matrix (3.1.12) *has a fixed point in* $S \times S$ *provided that*

$$\left(\frac{1}{1 - \|A\|}\phi_B\right) \circ \left(\frac{1}{1 - \|D\|}\phi_C\right)(r) < r, \quad for \ r > 0.$$

Proof After simple calculations, we can see that $I - D$ and $I - A$ are two homeomorphism and $(I - D)^{-1}$ and $(I - A)^{-1}$ are weakly continuous. Define the mapping $F : S \longrightarrow X$ by putting

$$F(x) = (I - A)^{-1}B(I - D)^{-1}C(x). \tag{3.2.2}$$

We prove that F is condensing with respect to β. To this end, let M be a subset of S with $\beta(M) > 0$. Going back to the following equality

$$F = AF + B(I - D)^{-1}C, \tag{3.2.3}$$

we obtain

$$\beta(FM) \leq \left(\frac{1}{1 - \|A\|}\phi_B\right) \circ \left(\frac{1}{1 - \|D\|}\phi_C\right)(\beta(M)).$$

The desired conclusion follows by a direct application of Theorem 1.5.12. Q.E.D.

Let us point out an immediate consequence of the result contained in Theorem 3.2.2.

Corollary 3.2.3 *Let* S *be a nonempty, bounded, closed, and convex subset of a Banach space* X *and let* A, $C : S \longrightarrow X$ *and* B, $D : X \longrightarrow X$ *be four operators such that*

(i) A is linear and bounded such that A^p *is separate contraction, for some* $p \in \mathbb{N}$,
(ii) C is weakly compact and $C(S) \subset (I - D)(S)$,
(iii) B and C are weakly sequentially continuous,
(iv) D is linear, bounded and D^q *is separate contraction, for some* $q \in \mathbb{N}$,
(v) $(I - A)^{-1}B(I - D)^{-1}C(S)$ *is bounded in* S.

Then, the block operator matrix (3.1.12) *has a fixed point in* $S \times S$.

The proof of Theorems 3.2.1 and 3.2.2 are based on the fact that A or D is linear. So, it is important to treat the case, when the inputs of the block operator matrix (3.1.12) are all nonlinear functional.

Theorem 3.2.4 *Let* S *be a nonempty, bounded, closed, and convex subset of a Banach space* X *and let* A, $C : S \longrightarrow X$ *and* B, $D : X \longrightarrow X$ *be four weakly sequentially continuous operators such that*

(i) A is nonlinear contraction with $\mathscr{D}$-function ϕ_A,
(ii) C is weakly compact and $C(S) \subset (I - D)(S)$,
(iii) D is a contraction with a constant k,
(iv) $A(x) + B(I - D)^{-1}C(x) \in S$, for all $x \in S$.

Then, the block operator matrix (3.1.12) *has a fixed point in $S \times S$.*

Proof Taking into account that $C(S)$ is relatively weakly compact, and using the subadditivity of the De Blasi measure β, we get

$$\begin{aligned}\beta\left((I - D)^{-1}C(S)\right) &\le \beta\left(\overline{C(S)}^{w}\right) + \beta\left(D(I - D)^{-1}C(S)\right)\\ &\le \beta\left(D(I - D)^{-1}C(S)\right).\end{aligned}$$

Let $\varepsilon > \beta\left((I - D)^{-1}C(S)\right)$. Then, there exist $0 \le \varepsilon_0 < \varepsilon$ and there exists a subset $K \in \mathscr{W}(S)$ such that the following inclusion

$$D(I - D)^{-1}C(S) \subseteq D(K) + \overline{\mathscr{B}}_X(k\varepsilon_0)$$

is satisfied. It follows from Lemma 1.5.9, that

$$\beta(D(I - D)^{-1}C(S)) < k\varepsilon.$$

This yields that

$$\begin{aligned}\beta\left((I - D)^{-1}C(S)\right) &\le \beta\left(D(I - D)^{-1}C(S)\right)\\ &< \beta\left((I - D)^{-1}C(S)\right).\end{aligned}$$

This means that $(I - D)^{-1}C(S)$ is a relatively weakly compact subset of S. In the similar way as in Theorem 3.1.20, we can prove that $(I - D)^{-1}C$ as well as T are weakly sequentially continuous, where T is given in the equality (3.2.1). In view of Theorem 1.5.12, it is sufficient to demonstrate that T is condensing with respect to β. In order to prove this fact, let us take a subset M of S with $\beta(M) > 0$. Then, $T(M)$ is bounded and

$$\begin{aligned}\beta(T(M)) &\le \beta\left[A(M) + B(I - D)^{-1}C(M)\right]\\ &\le \beta(A(M)) + \beta(B(I - D)^{-1}C(M))\\ &\le \phi_A(\beta(M)).\end{aligned}$$

The desired conclusion follows from Theorem 1.5.12. Q.E.D.

3.2.2 Critical-Type of Fixed Point Theory in Banach Space

The mentioned conditions S is bounded and $I - A$ is invertible were used in the above demonstrations. So, it is important to treat the case when the operator $I - A$ is not invertible and S is an unbounded subset of a Banach space. Therefore, we will establish a critical type of fixed point result of the block operator matrix (3.1.12). The obtained results, which can be found in [24], generalized and extended the counterparts in [25]. The first theorem of this section extends Theorem 1.5.34.

Theorem 3.2.5 *Let S be a nonempty, convex, and closed subset of a Banach space X and let A, B, C $D : X \longrightarrow X$ be four operators such that*

(i) A is condensing with respect to β with a weakly sequentially closed graph,
(ii) B, C and $(I - D)^{-1}$ are weakly sequentially continuous,
(iii) $C(S)$ is relatively weakly compact and $C(S) \subset (I - D)(S)$,
(iv) the set $\{x \in X \ : \ (I - A)(x) = y\} \in \mathscr{P}_{cv}(X)$, for each $y \in B(I - D)^{-1}C(S)$, and
(v) If $x = A(x) + B(I - D)^{-1}C(y)$, for $y \in S$, then $x \in S$.

Then, the block operator matrix (3.1.12) *has a fixed point in $S \times S$.*

Proof At first, let us assume that the operator $I - A$ is invertible. In order to prove our theorem, we will apply the result contained in Theorem 1.3.44. Then, we are going to show that, the operator F, as is described in the equality (3.2.2), is weakly sequentially continuous such that the set $\overline{F(S)}^{w}$ is weakly compact. If not the case, we have $\beta(F(S)) > 0$. Going back to the equality (3.2.3), we obtain

$$\begin{aligned}\beta(F(S)) &\leq \beta\left(B(I - D)^{-1}C(S)\right) + \beta\left(AF(S)\right)\\ &\leq \beta\left(AF(S)\right).\end{aligned}$$

According to the assumption (i), we get

$$\begin{aligned}\beta(F(S)) &\leq \beta\left(AF(S)\right)\\ &< \beta\left(F(S)\right)\end{aligned}$$

which is a contradiction, and so $\beta\left(F(S)\right) = 0$. It is not hard, via simple calculations, to verify the weak sequential continuity of the mapping F and then, the desired result is valid by applying Theorem 1.3.44. Suppose that the operator $I - A$ is not invertible. Then, we can define the multi-valued mapping $G : S \longrightarrow \mathscr{P}(S)$ by putting

$$G(x) = (I - A)^{-1}B(I - D)^{-1}C(x). \tag{3.2.4}$$

It follows from assumption (iv), that $G(x) \in \mathscr{P}_{cv}(S)$, for any $x \in S$. We demonstrate that the graph of G is sequentially closed with respect to the weak topology. To this end, let $(x_n)_n$ be a sequence contained in the set S, which is weakly convergent to a point x and take $y_n \in G(x_n)$ such that $y_n \rightharpoonup y$. This yields

$$(I - A)(y_n) = B(I - D)^{-1}C(x_n).$$

This implies that

$$A(y_n) \rightharpoonup y + B(I - D)^{-1}C(x).$$

Since A has a weakly sequentially closed graph, it follows that

$$(I - A)(y) = B(I - D)^{-1}C(x).$$

By using the definition of G, we can verify that

$$G(S) \subseteq AG(S) + B(I - D)^{-1}C(S).$$

Taking into account that $B(I - D)^{-1}C(S)$ is relatively weakly compact, and using the subadditivity of the De Blasi measure β, we get

$$\begin{aligned} \beta(G(S)) &\leq \beta\left(B(I - D)^{-1}C(S) + AG(S)\right) \\ &\leq \beta\left(\overline{B(I - D)^{-1}C(S)}^{w}\right) + \beta\left(AG(S)\right) \\ &\leq \beta\left(AG(S)\right). \end{aligned}$$

This means that $\overline{G(S)}^{w}$ is weakly compact. The other conditions of Theorem 1.5.34 are satisfied and hence, the multi-valued mapping G has a fixed point. Q.E.D.

As consequence of Theorem 3.2.5, we obtain the following result.

Corollary 3.2.6 *Let S be a nonempty, bounded, convex, and closed subset of a Banach space X and let A, $C : S \longrightarrow X$ and B, $D : X \longrightarrow X$ be four operators such that*

(i) A is nonlinear contraction with $\mathcal{D}$-function ϕ_A,
(ii) A has a weakly sequentially closed graph,
(iii) B, C and $(I - D)^{-1}$ are weakly sequentially continuous,
(iv) C is weakly compact and $C(S) \subset (I - D)(S)$,
(v) $B(I - D)^{-1}C(S) \subset (I - A)(S)$.

Then, the block operator matrix (3.1.12) *has a fixed point in $S \times S$.*

Proof Let y be an arbitrary fixed element contained in the set S and define the mapping $\varphi_y : X \longrightarrow X$ by putting

$$\varphi_y(x) = A(x) + B(I - D)^{-1}C(y).$$

Making full use of the estimate $\phi_A(r) < r$, and according to Theorem 1.2.26, we acquire that the operator φ_y has an unique fixed point z_y. In other words, we have

$$z_y = A(z_y) + B(I - D)^{-1}C(y). \tag{3.2.5}$$

Define now a mapping N on S by putting $N(y) = z_y$, where z_y is as defined in (3.2.5). Therefore, we finish the proof in view of Theorem 3.2.5. Q.E.D.

In the next result, we employ and present a generalization of that containing in Theorem 1.4.46.

Theorem 3.2.7 *Let S be a nonempty, convex, and closed subset of a Banach space X, and let $A, C : S \longrightarrow X$ and $B, D : X \longrightarrow X$ be four operators such that*

(i) for each $K \in \mathscr{W}(S)$, the restriction mapping $A : K \longrightarrow X$ has a weakly sequentially closed graph,
(ii) $C(S) \subset (I - D)(S)$,
(iii) for any $M \in \mathscr{B}(S)$ with $\beta(M) > 0$, the following inequality is satisfied

$$\beta\left(A(M) + B(I - D)^{-1}C(M)\right) < \beta(M),$$

(iv) B, $(I - D)^{-1}$ and C are weakly sequentially continuous on K,
(v) $\{x \in X,\ (I - A)(x) = y\} \in \mathscr{P}_{cv}(S)$, for each $y \in B(I - D)^{-1}C(S)$,
(vi) $x = A(x) + B(I - D)^{-1}C(y),\ y \in S \Rightarrow x \in S$,
(vii) $(I - A)^{-1}\left(B(I - D)^{-1}C(S)\right)$ is bounded.

Then, the block operator matrix (3.1.12) *has a fixed point in $S \times S$.*

Proof It is assumed that the operator $I - A$ is non-invertible. Then, $(I - A)^{-1}$ may be seen as a multi-valued mapping. Choose $x_0 \in S$ and define the set

$$\Sigma = \{M \subset S,\ x_0 \in M,\ M \text{ is bounded, convex and closed and } G(M) \subseteq M\},$$

where, G is as described in the equality (3.2.4). It is not hard to verify that

$$\overline{co}(G(S) \bigcup \{x_0\}) \in \Sigma.$$

Besides, if we put $M \in \Sigma$, then

$$\begin{aligned} G(M) &\subseteq AG(M) + B(I - D)^{-1}C(M) \\ &\subseteq A(M) + B(I - D)^{-1}C(M). \end{aligned}$$

This means that G is condensing with respect to β, in view of assumption (iii). Choose $M_0 = \bigcap\limits_{M \in \Sigma} M$. After simple calculations, one can show that $M_0 \in \Sigma$, since

$$\begin{cases} x_0 \in M_0 \\ M_0 \text{ is a convex and closed subset of } S \\ G(M_0) \subseteq M_0. \end{cases}$$

Due to the fact that $\overline{co}\left(G(M_0) \bigcup \{x_0\}\right) \subseteq M_0$, we infer that

$$\begin{aligned} G\left(\overline{co}\left(G(M_0) \bigcup \{x_0\}\right)\right) &\subseteq G(M_0) \\ &\subseteq \overline{co}(G(M_0) \bigcup \{x_0\}), \end{aligned}$$

which implies $\overline{co}(G(M_0) \bigcup \{x_0\}) \in \Sigma$. Hence,

$$
\begin{aligned}
\beta(M_0) &\leq \beta\left(\overline{co}(G(M_0)\bigcup\{x_0\})\right) \\
&\leq \beta\left(G(M_0)\bigcup\{x_0\}\right) \\
&\leq \beta(G(M_0)).
\end{aligned}
$$

Consequently $M_0 \in \mathscr{W}(S)$, since G is condensing with respect to β. Therefore, the multi-valued mapping $G : M_0 \longrightarrow \mathscr{P}_{cv}(M_0)$ is well defined. Now, let $(x_n)_n$ be a weakly convergent sequence contained in the set M_0 to an element x and take a weakly convergent sequence $y_n \in G(x_n)$ to an element y. Then,

$$y_n \in M_0 \bigcap \left\{z \in S \text{ such that } z = A(z) + B(I - D)^{-1}C(x_n)\right\}.$$

This implies that

$$A(y_n) \rightharpoonup y - B(I - D)^{-1}C(x).$$

From assumption (i), we get

$$y \in M_0 \bigcap \left\{z \in S \text{ such that } z = A(z) + B(I - D)^{-1}C(x)\right\}.$$

In our next step, we show that G is weakly upper semi-continuous (G is upper semi-continuous with respect to the weak topologies). To do so, let us fix an arbitrary weakly closed subset S_1 in M_0 and choose an element $u \in \overline{G^{-1}(S_1)}^{w}$. Then, there exists a sequence $(u_n)_n$ contained in the set $G^{-1}(S_1)$ such that $u_n \rightharpoonup u$ and so, for any sequence $v_n \in G(u_n)\bigcap S_1 \subseteq M_0$, we can easy prove that $v_n \rightharpoonup v$, for some $v \in S_1$. This yields

$$A(v_n) \rightharpoonup v - B(I - D)^{-1}C(u).$$

The use of assumption (i) leads to $v \in G(u)\bigcap S_1$. In other words, $u \in G^{-1}(S_1)$. Thus, $\overline{G^{-1}(S_1)}^{w} = G^{-1}(S_1)$. Since X endowed with its weak topology is a Hausdorff locally convex space, we apply Theorem 1.4.46 to guarantee that G has a fixed point in S. Q.E.D.

Remark 3.2.8 Linking the proof of Theorem 3.2.7, we can see that one of advantages of this last is that instead of demanding that $(I - A)^{-1}$ be a weakly continuous mapping, we necessitate only that $G : M_0 \longrightarrow M_0$ be weakly sequentially continuous. The weak compactness of $B(I - D)^{-1}C$ is freed, which is an additional benefit. Our results weaken the weak continuity of A.

In what follows in this subsection, we shall denote by $\Upsilon_S^{(A,B,C,D)}$ the following set

$$\Upsilon_S^{(A,B,C,D)} = \{x \in S,\ x = A(x) + B(I - D)^{-1}C(y), \quad \text{for some } y \in S\}.$$

We state and prove the following result.

Theorem 3.2.9 *Let S be a nonempty, convex, and closed subset of a Banach space X, and let A, $C : S \longrightarrow X$ and B, $D : X \longrightarrow X$ be four operators such that*

(i) $\Upsilon_S^{(A,B,C,D)}$ is a bounded subset of S,
(ii) A is condensing with respect to β,
(iii) A has a weakly sequentially closed graph on $\Upsilon_S^{(A,B,C,D)}$,
(iv) B, $(I-D)^{-1}$ and C are weakly sequentially continuous,
(v) $C(S)$ is relatively weakly compact and $C(S) \subset (I-D)(S)$,
(vi) the set $\{x \in X \,:\, (I-A)(x) = y\} \in \mathscr{P}_{cv}(X)$, for any $y \in B(I-D)^{-1}C(S)$, and
(vii) $x = A(x) + B(I-D)^{-1}C(y)$, for $y \in S \Rightarrow x \in S$.

Then, the block operator matrix (3.1.12) *has a fixed point in $S \times S$.*

Proof Assume that $I - A$ is non-invertible. Let $(\theta_n)_n$ be a sequence contained in the set S and take a sequence $\xi_n \in G(\theta_n)$ where, G is as described in (3.2.4). Then,

$$(I-A)(\xi_n) = B(I-D)^{-1}C(\theta_n).$$

Since $C(S)$ is relatively weakly compact, it follows that $(\theta_n)_n$ possesses a subsequence, say $(\theta_{n_k})_k$, such that

$$B(I-D)^{-1}C(\theta_{n_k}) \rightharpoonup B(I-D)^{-1}(y).$$

Using the subadditivity of the De Blasi measure, we arrive at

$$\begin{aligned}\beta\left(\{\xi_{n_k},\ k \in \mathbb{N}\}\right) &\leq \beta\left(\{(I-A)(\xi_{n_k}),\ k \in \mathbb{N}\}\right) + \beta\left(\{A(\xi_{n_k}),\ k \in \mathbb{N}\}\right)\\ &\leq \beta\left(\{B(I-D)^{-1}C(\theta_{n_k}),\ k \in \mathbb{N}\}\right) + \beta\left(\{A(\xi_{n_k}),\ k \in \mathbb{N}\}\right)\\ &\leq \beta\left(\{A(\xi_{n_k}),\ k \in \mathbb{N}\}\right).\end{aligned}$$

It follows from Theorem 1.3.29, that the set $\{\xi_{n_k},\ k \in \mathbb{N}\}$ as well as the set $G(S)$ is relatively weakly compact. Consider $(x_n)_n$ as a weakly convergent sequence contained in the set S to an element x and take a sequence $y_n \in G(x_n)$ such that $y_n \rightharpoonup y$. This implies that $y_n \in \Upsilon_S^{(A,B,C,D)}$ and consequently,

$$A(y_n) \rightharpoonup y - B(I-D)^{-1}C(x).$$

Using assumption (iii), we get $y \in G(\theta)$. In other words, the graph of G is weakly sequentially closed. Now, the multi-valued mapping G satisfies all conditions of Theorem 1.5.34, so there exists $x \in X$ such that $x \in G(x)$. Q.E.D.

A similar approach of Theorem 3.2.9, will be adopted to study the case $\Upsilon_S^{(A,B,C,D)}$ is relatively weakly compact.

Theorem 3.2.10 *Let S be a nonempty, closed, and convex subset of a Banach space X and let A, $C : S \longrightarrow X$ and B, $D : X \longrightarrow X$ be four operators such that*

(i) A is sequentially weakly closed on S,
(ii) $C(S) \subset (I - D)(S)$,
(iii) $\Upsilon_S^{(A,B,C,D)}$ *is relatively weakly compact,*
(iv) B, $(I - D)^{-1}$ *and* C *are weakly sequentially continuous,*
(v) $\{x \in X,\ (I - A)(x) = y\} \in \mathscr{P}_{cv}(S)$, *for any* $y \in B(I - D)^{-1}C(S)$,
(vi) $x = A(x) + B(I - D)^{-1}C(y),\ y \in S \Rightarrow x \in S$.

Then, the block operator matrix (3.1.12) *has a fixed point in* $S \times S$.

Proof Consider an arbitrary sequence $(\xi_n)_n$ of $G(S)$, where G is described in (3.2.4). By the definition of G, we infer that

$$(I - A)(\xi_n) = B(I - D)^{-1}C(\theta_n), \quad \text{for some } \theta_n \in S.$$

Then, $\xi_n \in \Upsilon_S^{(A,B,C,D)}$ and consequently, the sequence $(\xi_n)_n$ possesses a weakly convergent subsequence, say $(\xi_{n_k})_k$. Thus, on account of Theorem 1.3.29, we get the weak compactness of the set $\overline{G(S)}^w$. We are going to show that, the graph of the multi-valued mapping G is weakly sequentially closed. In order to prove this fact, assume that $(\theta_n)_n$ is a weakly convergent sequence of S to an element θ and take a weakly convergent sequence $\xi_n \in G(\theta_n)$ to a point ξ. From the above discussion, we can extract a subsequence, say $(\xi_{n_k})_k$, of $(\xi_n)_n$ such that

$$A(\xi_{n_k}) \rightharpoonup \xi - B(I - D)^{-1}C(\theta).$$

Since A is sequentially weakly closed, then $\xi \in G(\theta)$. Q.E.D.

3.2.3 Critical-Type in Banach Algebras Satisfying Condition ($\mathscr{P}$)

The fundamental aim of this section is to provide the critical-type of fixed points of the following block operator matrices

$$\begin{pmatrix} A & B \cdot B' \\ C & D \end{pmatrix}, \tag{3.2.6}$$

where the inputs are in general nonlinear operators acting in a product of Banach algebra satisfying the sequential condition ($\mathscr{P}$). Based on the previous extension, we are planning to extend these results by proving that, under certain hypotheses associated with its nonlinear entries, the block operator matrix (3.2.6) has, at least, one fixed point. This discussion is based on absence of the invertibility of the operator

$$\frac{I - A}{B(I - D)^{-1}C}. \tag{3.2.7}$$

The obtained results may be found in [8]. The first result is formulated as follows.

Theorem 3.2.11 *Let S be a nonempty, closed, and convex subset of a Banach algebra X satisfying the sequential condition $(\mathcal{P})$ and let $A, B, C, D, B' : S \longrightarrow X$ be five operators such that*

(i) $(I - D)^{-1}$ is well defined on $C(S)$ and $B(I - D)^{-1}C$ is regular,
(ii) $B'(I - D)^{-1}C(S)$ is relatively weakly compact,
(iii) If the sequence $\left(\left(\frac{I-A}{B(I-D)^{-1}C}\right)(x_n)\right)_n$ is weakly convergent, then the sequence $(x_n)_n$ possesses a weakly convergent subsequence,
(iv) $\left(\frac{I-A}{B(I-D)^{-1}C}\right)^{-1} B'(I - D)^{-1}C(x) \in \mathcal{P}_{cv}(X)$, for each $x \in S$,
(v) $B'(I - D)^{-1}C$ and $\frac{I-A}{B(I-D)^{-1}C}$ are weakly sequentially continuous,
(vi) $x = A(x) + B(I - D)^{-1}C(x) \cdot B'(I - D)^{-1}C(y)$, for $y \in S \Rightarrow x \in S$.

Then, the block operator matrix (3.2.6) has a fixed point in $S \times S$.

Proof There are two cases as follows. At first, assume that the operator (3.2.7) is invertible. We will apply the result contained in Theorem 1.3.44. Then, we are going to verify the weak continuity and the weak compactness of the operator, which assigns to any $x \in S$, the value

$$T(x) = \left(\frac{I - A}{B(I - D)^{-1}C}\right)^{-1} B'(I - D)^{-1}C(x).$$

Let $(y_n)_n$ be any sequence contained in the set $T(S)$. Then, we can find a sequence $(x_n)_n$ contained in the set S such that

$$\left(\frac{I - A}{B(I - D)^{-1}C}\right)(y_n) = B'(I - D)^{-1}C(x_n),$$

Based on assumption (ii), we can show that $(x_n)_n$ possesses a subsequence, say $(x_{n_k})_k$, that satisfies

$$B'(I - D)^{-1}C(x_{n_k}) \rightharpoonup z, \quad \text{for some } z \in S.$$

Then, the sequence $(y_n)_n$ possesses a weakly convergent subsequence, in light of assumption (iii). Consequently, the set $T(S)$ is relatively weakly sequentially compact. Thus, on account of Theorem 1.3.29, we deduce that $T(S)$ is relatively weakly compact. Let us assume that $(x_n)_n$ is a weakly convergent sequence contained in the set S to a point $x \in S$. It is clear that $(x_n)_n$ possesses a subsequence, say $(x_{\varphi_1(n)})_n$, which satisfies

$$T(x_{\varphi_1(n)}) \rightharpoonup y, \quad \text{for some } y \in S.$$

The use of the following equality

$$\left(\frac{A-I}{B(I-D)^{-1}C}+I\right)T=-B'(I-D)^{-1}C+T,$$

leads to

$$\left(\frac{A-I}{B(I-D)^{-1}C}+I\right)T(x_{\varphi_1(n)})\rightharpoonup -B'(I-D)^{-1}C(x)+y.$$

Since $\left(\frac{I-A}{B(I-D)^{-1}C}\right)$ is weakly sequentially continuous, we get

$$\left(\frac{A-I}{B(I-D)^{-1}C}+I\right)(y)=-B'(I-D)^{-1}C(x)+y.$$

Now, a standard argument shows that

$$T(x_n)\rightharpoonup T(x).$$

Suppose the contrary, then there would exist a weak neighborhood V^w of $T(x)$ and a subsequence, say $(x_{n_k})_k$, of $(x_n)_n$ such that

$$T(x_{n_k})\notin V^w,\quad \text{for all } k\geq 1.$$

Similarly, the sequence $(x_{n_k})_k$ possesses a subsequence, say $(x_{n_{k_j}})_j$, such that

$$T\left(x_{n_{k_j}}\right)\rightharpoonup y,$$

which is absurd and hence, T is weakly sequentially continuous. Suppose that the operator that appears in (3.2.7) is non-invertible and define the multi-valued mapping $G:S\longrightarrow \mathscr{P}(X)$ by putting

$$G(x)=\left(\frac{I-A}{B(I-D)^{-1}C}\right)^{-1}B'(I-D)^{-1}C(x). \tag{3.2.8}$$

It follows from assumption (iv), that $G(x)\in\mathscr{P}_{cv}(S)$, for all $x\in S$ and the set $G(S)$ is relatively weakly compact. Besides, we can prove that the graph of G is weakly sequentially closed. To prove the above-announced fact, let us assume that $(x_n)_n$ is a sequence contained in the set S, which is weakly convergent to a point $x\in S$ and take a sequence $y_n\in G(x_n)$, which is weakly convergent to a point y. Then,

$$\left(\frac{I-A}{B(I-D)^{-1}C}\right)(y_n)=B'(I-D)^{-1}C(x_n).$$

From assumption (v), it follows that $y\in G(x)$. Consequently, the operator G satisfies all assumptions of Theorem 1.5.34 and so, G has a fixed point. Q.E.D.

As an immediate consequence of Theorem 3.2.11, we derive the following result.

Corollary 3.2.12 *Let S be a nonempty, closed, and convex subset of a Banach algebra X satisfying condition $(\mathscr{P})$ and let A, $C : S \longrightarrow X$ and B, B', $D : X \longrightarrow X$ be five operators such that*

(i) A, B, and C are $\mathscr{D}$-Lipschitzians with $\mathscr{D}$-functions ϕ_A, ϕ_B and ϕ_C, respectively,
(ii) D is a contraction mapping with constant k,
(iii) $C(S) \subseteq (I-D)(S)$ and $(I-D)^{-1}C(S)$ is relatively weakly compact,
(iv) $B'(I-D)^{-1}C$ and $\frac{I-A}{B(I-D)^{-1}C}$ are weakly sequentially continuous,
(v) $\left(\frac{I-A}{B(I-D)^{-1}C}\right)^{-1} B'(I-D)^{-1}C$ is weakly compact,
(vi) If $x = A(x) + B(I-D)^{-1}C(x) \cdot B'(I-D)^{-1}C(y)$, for $y \in S$, then $x \in S$.

Then, the block operator matrix (3.2.6) has a fixed point in $S \times S$ provided that

$$M\phi_B \circ \varphi(r) + \phi_A(r) < r, \quad \textit{for all } r > 0,$$

where φ is given in the equality (3.1.9) and $M = \|B'(I-D)^{-1}C(S)\|$.

Theorem 3.2.13 *Let S be a nonempty, closed, and convex subset of a Banach algebra X satisfying condition $(\mathscr{P})$ and let A, $C : S \longrightarrow X$ and B, B', $D : X \longrightarrow X$ be five operators such that*

(i) A, B, and C are $\mathscr{D}$-Lipschitzians with $\mathscr{D}$-functions ϕ_A, ϕ_B and ϕ_C, respectively,
(ii) D is a contraction with a constant k and $C(S) \subseteq (I-D)(S)$,
(iii) $B'(I-D)^{-1}C$ and $\frac{I}{B(I-D)^{-1}C}$ are weakly sequentially continuous,
(iv) the set $B'(I-D)^{-1}C(S)$ is relatively weakly compact,
(v) $A(x) + B(I-D)^{-1}C(x) \cdot B'(I-D)^{-1}C(y) \in S$, for all $x, y \in S$.

Then, the block operator matrix (3.2.6) has a fixed point in $S \times S$ provided that

$$M\phi_B \circ \varphi(r) + \phi_A(r) < r, \quad \textit{for all } r > 0,$$

where φ is given in the equality (3.1.9) and $M = \|B'(I-D)^{-1}C(S)\|$.

Proof An application of Theorem 1.2.26 ensures the existence of $(I-D)^{-1}$ such that $(I-D)^{-1}C$ is $\mathscr{D}$-Lipschitzian with $\mathscr{D}$-function φ, where φ is given in the equality (3.1.9). Let y be an arbitrarily fixed element of the set $B'(I-D)^{-1}C(S)$ and define the mapping $T_y : S \longrightarrow S$ by putting

$$T_y(x) = A(x) + B(I-D)^{-1}C(x) \cdot y.$$

After simple calculations, we can show that T_y is nonlinear contraction with $\mathscr{D}$-function ψ, where

$$\psi(r) = M\phi_B \circ \varphi(r) + \phi_A(r).$$

Then, on account of Theorem 1.2.26, there exists an unique point $x_y \in S$ such that $T_y(x_y) = x_y$. In other words, we have

$$x_y = A(x_y) + B(I - D)^{-1}C(x_y) \cdot y. \tag{3.2.9}$$

Let us consider the function $N : S \longrightarrow S$ by putting $N(y) = x_y$, where x_y is the unique solution of the Eq. (3.2.9). In the similar way as in the proof of Theorem 3.2.11, we can demonstrate the weak sequential continuity of N. Then, by using Theorem 1.5.34, there exists $x \in S$ such that $N(x) = x$. Q.E.D.

Now, we can modify some assumptions of Theorem 3.2.13, in order to obtain a similar approach.

Theorem 3.2.14 *Let S be a nonempty, closed, and convex subset of a Banach algebra X satisfying condition $(\mathcal{P})$ and let A, B, C, B' : $S \longrightarrow X$ be four weakly sequentially continuous operators and $D : X \longrightarrow X$ be an operator such that*

(i) D is linear, bounded and D^p is separate contraction, for some $p \in \mathbb{N}^*$,
(ii) $C(S) \subseteq (I - D)(S)$ and $B(I - D)^{-1}C$ is regular,
(iii) $\Gamma_y = \{x \in S,\ x = A(x) + B(I - D)^{-1}C(x) \cdot y\} \in \mathcal{P}_{cv}(X)$, for $y \in B(I - D)^{-1}C(S)$,
(iv) the set $\left(\frac{I-A}{B(I-D)^{-1}C}\right)^{-1} B'(I - D)^{-1}C(S)$ is relatively weakly compact,
(v) $B'(I - D)^{-1}C(S) \subset \left(\dfrac{I - A}{B(I - D)^{-1}C}\right)(S)$.

Then, the block operator matrix (3.2.6) has a fixed point in $S \times S$.

Proof Making use of assumption (i), together with equality (3.1.5) enables us to have the weak continuity of the operator $(I - D)^{-1}C$. Consider $(\theta_n)_n$ as a sequence contained in the set S, which is weakly convergent to an element $\theta \in S$ and take a sequence $\xi_n \in G(\theta_n)$ such that $\xi_n \rightharpoonup \xi$, where G is as described in (3.2.8). Then, the following inequality

$$\xi_n = A(\xi_n) + B(I - D)^{-1}C(\xi_n) \cdot B'(I - D)^{-1}C(\theta_n)$$

is satisfied. The sequential condition $(\mathcal{P})$ guaranties that $\xi \in G(\theta)$. In other words, the operator G admits a weakly sequentially closed graph. The desired result follows from Theorem 1.5.34. Q.E.D.

As a consequence of Theorem 3.2.14, we derive the following fixed point theorem.

Corollary 3.2.15 *Let S be a nonempty, closed, and convex subset of a Banach algebra X satisfying condition $(\mathcal{P})$ and let A, B, C, B' : $S \longrightarrow X$ be four weakly sequentially continuous operators and $D : X \longrightarrow X$ such that*

(i) D is linear, bounded and D^p is separate contraction, for some $p \in \mathbb{N}^*$,
(ii) $B(I - D)^{-1}C$ is regular on $B'(I - D)^{-1}C(S)$,
(iii) $\Gamma_y = \{x \in S,\ \ x = A(x) + B(I - D)^{-1}C(x) \cdot y\}$ is a convex subset of X, for each $y \in B'(I - D)^{-1}C(S)$,

(iv) $A(S)$ and $C(S)$ are relatively weakly compact,
(v) $B'(I-D)^{-1}C(S) \subset \left(\frac{I-A}{B(I-D)^{-1}C}\right)(S)$.

Then, the block operator matrix (3.2.6) has a fixed point in $S \times S$.

Proof Let us consider a sequence $(\xi_n)_n$ contained in the set S and take a sequence $\theta_n \in G(\xi_n)$, where G is as described in the equality (3.2.8). Then,

$$(I-A)(\theta_n) = B(I-D)^{-1}C(\theta_n) \cdot B'(I-D)^{-1}C(\xi_n).$$

It follows from assumption (iv), that the sequence $(\theta_n)_n$ possesses two subsequences, say $(\theta_{\varphi_1(n)})_n$ and $(\theta_{\varphi_2(n)})_n$ and there exists a subsequence, say $(\xi_{\varphi_3(n)})_n$ of $(\xi_n)_n$ such that

$$A\left(\theta_{\varphi_1(n)}\right) \rightharpoonup y \text{ and } C\left(\theta_{\varphi_2(n)}\right) \rightharpoonup z \text{ and } C\left(\xi_{\varphi_3(n)}\right) \rightharpoonup u,$$

for some $y, z, u \in X$. Due to the fact that $(I-D)^{-1}$ is weakly sequentially continuous, we get

$$B(I-D)^{-1}C\left(\theta_{\varphi_2(n)}\right) \rightharpoonup B(I-D)^{-1}(z) \text{ and } B'(I-D)^{-1}C\left(\xi_{\varphi_3(n)}\right) \rightharpoonup B'(I-D)^{-1}(u).$$

Then, there exists a subsequence, say $(\theta_{\varphi_4(n)})_n$ of $(\theta_n)_n$ such that

$$\theta_{\varphi_4(n)} \rightharpoonup y + B(I-D)^{-1}(z) \cdot B'(I-D)^{-1}(u).$$

Thus, $G(S)$ is relatively weakly compact, in view of Theorem 1.3.29. The desired conclusion follows by a direct application of Theorem 3.2.14. Q.E.D.

Now, we prove a critical-type fixed point theorem for 1-set-weakly contractive in Banach algebra satisfying condition $(\mathscr{P})$.

Theorem 3.2.16 *Let S be a nonempty, bounded, closed, and convex subset of a Banach algebra X satisfying the sequential condition $(\mathscr{P})$ and let $A, C : S \longrightarrow X$ and $B, B', D : X \longrightarrow X$ be five weakly sequentially continuous mapping such that*

(i) A is 1-set-weakly contractive such that $\lambda A(S) \subset A(S)$, for each $\lambda \in (0,1)$,
(ii) $A(S)$, $B(S)$ and $B'(S)$ are bounded,
(iii) $C(S) \subset (I-D)(S)$ and $(I-D)^{-1}C$ is weakly compact,
(iv) $B(I-D)^{-1}C$ is regular on $B'(S)$,
(v) $\left\{x \in X,\ x = \lambda A(x) + B(I-D)^{-1}C(x) \cdot B'(I-D)^{-1}C(y)\right\} \in \mathscr{P}_{cv}(X)$, for each $y \in S$,
(vi) $B'(I-D)^{-1}C(S) \subset \left(\frac{I-\lambda A}{B(I-D)^{-1}C}\right)(S)$.

Then, the block operator matrix (3.2.6) has a fixed point in $S \times S$.

Proof At first, suppose that the mapping F_λ defined by

$$F_\lambda = \frac{I - \lambda A}{B(I - D)^{-1}C} \tag{3.2.10}$$

is invertible. Let y be an arbitrarily fixed element of the set S and consider $\lambda \in (0, 1)$. Since F_λ is invertible, there exists an unique element $x_y^\lambda \in X$ such that the following equality

$$A(x_y^\lambda) + B(I - D)^{-1}C(x_y^\lambda) \cdot B'(I - D)^{-1}C(y) = x_y^\lambda$$

is satisfied. This yields $x_y^\lambda \in S$, in view of assumption (vi) and so, we can define the mapping $T_\lambda : S \longrightarrow S$ by putting

$$T_\lambda(x) = F_\lambda^{-1} \circ B'(I - D)^{-1}C(x).$$

Due to the fact that $(I - D)^{-1}C(S)$ is relatively weakly compact, and using the following equality

$$T_\lambda = B(I - D)^{-1}CT_\lambda \cdot B'(I - D)^{-1}C + \lambda AT_\lambda, \tag{3.2.11}$$

we deduce that $T_\lambda(D)$ is bounded, for any subset D of S and

$$\begin{aligned}\beta(T_\lambda(D)) &\leq \beta\left(\lambda AT_\lambda(D)\right) + \beta\left(B(I - D)^{-1}CT_\lambda(D) \cdot B'(I - D)^{-1}C(D)\right) \\ &\leq \lambda\beta\left(T_\lambda(D)\right) \\ &< \beta\left(T_\lambda(D)\right).\end{aligned}$$

This means that T_λ is a weakly compact operator. Next, let $(x_n)_n$ be a sequence contained in the set S, which is weakly convergent to an element $x \in S$. After simple calculations, one can show that the sequence $(x_n)_n$ possesses a subsequence, say $(x_{\varphi_1(n)})_n$, that satisfies

$$T_\lambda(x_{\varphi_1(n)}) \rightharpoonup y,$$

for some $y \in S$. Going back to the equality (3.1.7) yields that the sequence $(x_n)_n$ possesses a subsequence, say $(x_{\varphi_2(n)})_n$, that satisfies

$$(I - D)^{-1}C(x_{\varphi_2(n)}) \rightharpoonup C(x) + D(z), \quad \text{for some } z \in S.$$

Thus, $z = C(x) + D(z)$. We show that

$$(I - D)^{-1}C(x_n) \rightharpoonup (I - D)^{-1}C(x).$$

If this is not the case, there exists a subsequence, say $(x_{n_k})_k$, and there exists a weak neighborhood V^w of $(I - D)^{-1}C(x)$ such that

$$(I - D)^{-1}C(x_{n_k}) \notin V^w.$$

Therefore, we can extract a subsequence, say $(x_{n_{k_j}})_j$, of $(x_{n_k})_k$ such that

$$(I - D)^{-1}C(x_{n_{k_j}}) \rightharpoonup z_1,$$

which is absurd. If we consider the equality (3.2.11), together with assumptions on maps A, B and B', we can show

$$T_\lambda(x_{\varphi_1(n)}) \rightharpoonup \lambda A(y) + B(I - D)^{-1}C(y) \cdot B'(I - D)^{-1}C(x).$$

This implies that

$$y = \lambda A(y) + B(I - D)^{-1}C(y) \cdot B'(I - D)^{-1}C(x).$$

Thus, the operator T_λ satisfies all conditions of Theorem 1.3.44, so there exists an element $x^\lambda \in S$ such that

$$x^\lambda = T_\lambda(x^\lambda).$$

Choose a sequence $(\lambda_n)_n$ contained in the interval $[0, 1]$, which is convergent to 1 and take $(x_n)_n$ the corresponding sequence in S. Then,

$$x_n = \lambda_n A(x_n) + B(I - D)^{-1}C(x_n) \cdot B'(I - D)^{-1}C(x_n).$$

After simple calculations, we can see that the sequence $(x_n)_n$ has a weakly convergent renamed subsequence to an element x. This implies that

$$x = A(x) + B(I - D)^{-1}C(x) \cdot B'(I - D)^{-1}C(x).$$

Second, we study the case when F_λ is not invertible. Then, the operator F_λ could be seen as a multi-valued mapping. Define the mapping $G_\lambda : S \longrightarrow \mathscr{P}(S)$ by putting

$$G_\lambda(x) = (F_\lambda)^{-1}\, B'(I - D)^{-1}C(x).$$

The assumption (v) guaranties that $G_\lambda(x) \in \mathscr{P}_{cv}(S)$, for all $x \in S$. On the other hand, via simple calculations, one can prove that

$$\begin{aligned} G_\lambda(S) &\subseteq \lambda A G_\lambda(S) + B(I - D)^{-1}CG_\lambda(S) \cdot B'(I - D)^{-1}C(S) \\ &\subseteq A(S) + B(I - D)^{-1}C(S) \cdot B'(I - D)^{-1}C(S). \end{aligned}$$

Using the subadditivity of the De Blasi measure β, we arrive at the following

$$\begin{aligned} \beta(G_\lambda(S)) &\leq \beta\left(\lambda A G_\lambda(S)\right) + \beta\left(B(I - D)^{-1}CG_\lambda(S) \cdot B'(I - D)^{-1}C(S)\right) \\ &\leq \lambda\beta\left(G_\lambda(S)\right), \end{aligned}$$

which is absurd and consequently, $G_\lambda(S)$ is relatively weakly compact. Let us assume that $(x_n)_n$ is a sequence contained in the set S, which is weakly convergent to an element $x \in S$ and take a sequence $y_n \in G_\lambda(x_n)$, which is weakly converging to an element y. Then,

$$F_\lambda(y_n) = B'(I - D)^{-1}C(x_n).$$

The use of assumptions on maps λA and $(I - D)^{-1}C$ allows us to have $y \in G_\lambda(x)$. In other words, the graph of the multi-valued mapping G_λ is weakly sequentially closed. The desired conclusion follows by an application of Theorem 1.5.34. Q.E.D.

As an immediate consequence of Theorem 3.2.16, we derive the following.

Corollary 3.2.17 *Let S be a nonempty, bounded, closed, and convex subset of a Banach algebra X satisfying the sequential condition $(\mathcal{P})$ and let $A, C : S \longrightarrow X$ and $B, B', D : X \longrightarrow X$ be five weakly sequentially continuous operators such that*

(i) $A(S)$ is relatively weakly compact,
(ii) B, C, B' and D are Lipschitzians with constants k_1, k_2, k_3 and k_4, respectively,
(iii) $x = D(x) + C(y), y \in S \Rightarrow x \in S$,
(iv) F_λ^{-1} exists on $B'(I - D)^{-1}C(S)$, where F_λ is given in (3.2.10),
(v) $x = \lambda A(x) + B(I - D)^{-1}C(x) \cdot B'(I - D)^{-1}C(y)$, for $y \in S \Rightarrow x \in S$.

Then, the block operator matrix (3.2.6) has fixed point in $S \times S$ provided that

$$\max\left\{k_4 + \|B'(S)\|k_1k_2, \frac{\|B(S)\|k_2k_3(1 - k_4) + k_1k_2^2k_3\beta(S)}{(1 - k_4)(1 - (k_4 + \|B'(S)\|k_1k_2))}\right\} < 1.$$

Proof Using arguments similar to those used in the proof of Theorem 3.2.16, we can prove that there exists an element $x_y^\lambda \in X$ such that

$$x_y^\lambda = \lambda A(x_y^\lambda) + B(I - D)^{-1}C(x_y^\lambda) \cdot B'(I - D)^{-1}C(y), \quad \text{for each } y \in S.$$

Thus, we can define a mapping $T_\lambda : S \longrightarrow S$ by putting

$$T_\lambda(x) = F_\lambda^{-1}B'(I - D)^{-1}C(x).$$

Let us define the iterative sequence $(S_n)_n$, consisting of subsets of the set S, as

$$\begin{cases} S_1 = S, \\ S_{n+1} = \overline{co}(T_\lambda(S_n)) \end{cases}$$

Obviously all sets of the sequence $(S_n)_n$ are nonempty, bounded, convex and closed in S. Going back to equality (3.2.11) yields that

$$\begin{aligned} T_\lambda(S_n) &\subseteq \lambda AT_\lambda(S_n) + B(I - D)^{-1}CT_\lambda(S_n) \cdot B'(I - D)^{-1}C(S_n) \\ &\subseteq \lambda A(S) + B(I - D)^{-1}CT_\lambda(S_n) \cdot B'(I - D)^{-1}C(S_n). \end{aligned}$$

The use of Lemma 1.5.11 as well as Lemma 1.5.9 leads to

$$\begin{aligned}
\beta(S_{n+1}) &\leq \|B'(I-D)^{-1}C(S_n)\| \left(\frac{k_1k_2}{1-k_4}\right) \beta(T_\lambda(S_n)) \\
&\quad + \|B(I-D)^{-1}C(T_\lambda(S_n))\| \left(\frac{k_2k_3}{1-k_4}\right) \beta(S_n) \\
&\quad + \left(\frac{k_1k_2^2k_3}{(1-k_4)^2}\right) \beta(T_\lambda(S_n))\beta(S_n) \\
&\leq \left(\frac{\|B'(S)\|k_1k_2}{1-k_4}\right) \beta(S_{n+1}) + \left(\frac{\|B(S)\|k_2k_3}{1-k_4}\right) \beta(S_n) \\
&\quad + \left(\frac{k_1k_2^2k_3}{(1-k_4)^2}\right) \beta(S)\beta(S_n),
\end{aligned}$$

which yields that

$$\beta(S_{n+1}) \leq \left(\frac{\|B(S)\|k_2k_3(1-k_4) + k_1k_2^2k_3\beta(S)}{(1-k_4)(1-(k_4+\|B'(S)\|k_1k_2))}\right) \beta(S_n).$$

Using the induction, we infer that

$$\beta(S_n) \leq \left(\frac{\|B(S)\|k_2k_3(1-k_4) + k_1k_2^2k_3\beta(S)}{(1-k_4)(1-(k_4+\|B'(S)\|k_1k_2))}\right)^n \beta(S). \qquad (3.2.12)$$

Linking our assumption with estimate (3.2.12), we infer that the set S_∞ given by

$$S_\infty = \bigcap_{n\geq 1} S_n$$

is weakly compact. Besides, it is easy to verify that $T_\lambda(S_\infty) \subseteq S_\infty$. It follows from Theorem 3.2.16, that the operator $T_\lambda : S_\infty \longrightarrow S_\infty$ is weakly sequentially continuous. The desired conclusion follows by a direct application of Theorem 1.3.44. Q.E.D.

3.3 Multi-Valued Fixed Point Theory

In this section, several theorems are proved for nonlinear multi-valued mappings in Banach algebras. The obtained results, which can be found in [26], generalize and improve results for single-valued mapping in [3, 5, 8–10].

3.3.1 Fixed Point Theory for Block Operator Matrices

We begin this subsection by proving some results providing necessary assumptions on the inputs, so that the block operator matrix (3.2.6) has a fixed point.

Theorem 3.3.1 *Let S be a bounded, closed, and convex subset of a Banach algebra X, and let $A : S \longrightarrow \mathscr{P}_{bd,cl,cv}(X)$, $B : X \longrightarrow \mathscr{P}_{bd,cl,cv}(X)$, $B' : X \longrightarrow \mathscr{P}_{cp,cv}(X)$, $C : S \longrightarrow X$ and $D : X \longrightarrow X$ be five operators such that*

(i) A, B, and C are Lipschitzians with constants q_1, q_2 and q_3, respectively,
(ii) D is a contraction with a constant k such that $C(S) \subseteq (I - D)(S)$,
(iii) B' is lower semi-continuous and totally bounded,
(iv) $A(x) + B(I - D)^{-1}C(x) \cdot B'(I - D)^{-1}C(y) \in \mathscr{P}_{cl,cv}(S)$, for all x, $y \in S$.

Then, the block operator matrix (3.2.6) *has a fixed point in $S \times S$ provided that*

$$\frac{q_2 q_3}{1-k} M + q_1 < 1,$$

where $M = \|B'(I - D)^{-1}C(S)\|$.

Proof It is clear, in virtue of Theorem 1.2.4, that $I - D$ is a homeomorphism and the operator $(I - D)^{-1}C$ is Lipschitzian with a Lipschitz constant $\frac{q_3}{1-k}$. Let us take an arbitrary fixed element y of the set S and define the mapping $T_y : S \longrightarrow \mathscr{P}_{cl,cv}(S)$ by putting

$$T_y(x) = A(x) + B(I - D)^{-1}C(x) \cdot B'(I - D)^{-1}C(y).$$

It should be noted that T_y is a contraction mapping, since

$$\begin{aligned} \text{dist}_H(T_y(x_1), T_y(x_2)) &\leq \text{dist}_H(A(x_1), A(x_2)) + \\ &\quad \text{dist}_H(B(I-D)^{-1}C(x_1) \cdot B'(I-D)^{-1}C(y), B(I-D)^{-1}C(x_2) \cdot B'(I-D)^{-1}C(y)) \\ &\leq \text{dist}_H(A(x_1), A(x_2)) \ + M\text{dist}_H(B(I-D)^{-1}C(x_1), B(I-D)^{-1}C(x_2)) \\ &\leq \left(q_1 + M\frac{q_2 q_3}{1-k}\right)\|x_1 - x_2\|, \end{aligned}$$

whenever $x_1, x_2 \in S$. An application of Theorem 1.4.40 yields that, there exists an element $x_y \in X$ such that

$$x_y \in A(x_y) + B(I - D)^{-1}C(x_y) \cdot B'(I - D)^{-1}C(y).$$

Taking into account assumption (iii), and using Theorem 1.4.47, we obtain there exists a continuous function $f : S \times S \longrightarrow S$ such that

$$f(x, y) \in A(f(x, y)) + B(I - D)^{-1}C(f(x, y)) \cdot B'(I - D)^{-1}C(y).$$

Define two mapping $G : S \longrightarrow \mathscr{P}_{cl}(S)$ and $g : S \longrightarrow S$ by putting

$$Gy = \mathscr{F}_{T_y} \quad \text{and} \quad g(x) = f(x, x).$$

After simple calculations, we can see that g is continuous and

$$g(x) = f(x, x) \in A(f(x, x)) + B(I - D)^{-1}C(f(x, x)) \cdot B'(I - D)^{-1}C(x),$$

for any $x \in S$. Let z be an arbitrarily fixed element of the set S. Then,

$$\begin{aligned}\|B(I-D)^{-1}C(z)\|_{\mathscr{P}} &\leq \|B(I-D)^{-1}C(a)\|_{\mathscr{P}} + \frac{q_2q_3}{1-k}\|z-a\| \\ &\leq \|B(I-D)^{-1}C(a)\|_{\mathscr{P}} + \frac{\|z-a\|}{M} \\ &\leq \eta,\end{aligned}$$

for some a in S, where $\|B(I-D)^{-1}C(a)\|_{\mathscr{P}} = \mathrm{dist}_H\left(B(I-D)^{-1}C(a), \{0\}\right)$ and

$$\eta = \|B(I-D)^{-1}C(a)\|_{\mathscr{P}} + \frac{diam(S)}{M}. \tag{3.3.1}$$

We show that the operator $B'(I-D)^{-1}C$ is totally bounded. Indeed, taking into account that $B'(I-D)^{-1}C(S) \subset B'(S)$, and using assumption (iii), it is shown that, for a fixed $\varepsilon > 0$, there exists a set $Y = \{y_1, \ldots, y_n\} \subset S$ such that the following inclusion is satisfied

$$B'(I-D)^{-1}C(S) \subset \{w_1, \ldots, w_n\} + \mathscr{B}_X\left(\frac{(1-q_1)\varepsilon}{\eta} - \frac{Mq_2q_3}{\eta(1-k)}\varepsilon\right),$$

where, $w_i \in B'(I-D)^{-1}C(y_i)$. This yields that, for an arbitrarily fixed element y in the set S, we obtain

$$\mathrm{dist}_H\left(B'(I-D)^{-1}C(y), B'(I-D)^{-1}C(y_k)\right) < \frac{(1-q_1)\varepsilon}{\eta} - \frac{Mq_2q_3}{\eta(1-k)}\varepsilon,$$

for some $y_k \in Y$. Therefore,

$$\begin{aligned}\mathrm{dist}_H(G(y), G(y_k)) &= \mathrm{dist}_H(\mathscr{F}_{T_y}, \mathscr{F}_{T_{y_k}}) \\ &\leq \frac{1}{1-\left(q_1 + \frac{Mq_2q_3}{1-k}\right)} \sup_{x\in S}\{\mathrm{dist}_H\left(T_y(x), T_{y_k}(x)\right)\} \\ &\leq \frac{1}{1-\left(q_1 + \frac{Mq_2q_3}{1-k}\right)} \|B(I-D)^{-1}C(x)\|_{\mathscr{P}}\, \mathrm{dist}_H\left(B'(I-D)^{-1}C(y), B'(I-D)^{-1}C(y_k)\right) \\ &< \frac{\eta}{1-\left(q_1 + \frac{Mq_2q_3}{1-k}\right)}\left(\frac{(1-q_1)\varepsilon}{\eta} - \frac{Mq_2q_3}{\eta(1-k)}\varepsilon\right) \\ &\leq \varepsilon.\end{aligned}$$

Hence, for each $u \in G(y)$, there exists $u_k \in G(y_k)$ such that

$$\|u - u_k\| < \varepsilon.$$

Therefore, we infer that

$$G(y) \subseteq \bigcup_{k=1}^{n} \mathscr{B}_X(u_k, \varepsilon),$$

where $u_i \in G(y_i)$, $i = 1, \ldots, n$. This implies that

$$g(S) \subset G(S) \subseteq \bigcup_{k=1}^{n} \mathscr{B}_X(u_k, \varepsilon)$$

and so g has, at least, one fixed point in S. We finish the proof. Q.E.D.

Similar to Theorem 3.3.1 and under a weaker assumption, we prove the following result.

Theorem 3.3.2 *Let S be a bounded, closed, and convex subset of a Banach algebra X, and let $A, B : X \longrightarrow \mathscr{P}_{bd,cl,cv}(X)$, $B' : X \longrightarrow \mathscr{P}_{cp,cv}(X)$ and $C, D : S \longrightarrow X$ be five multi-valued mappings such that*

(i) A, B, and C are Lipschitzians with constants q_1, q_2 and q_3, respectively,
(ii) D is a contraction with a constant k such that $C(S) \subseteq (I - D)(X)$,
(iii) B' is lower semi-continuous such that $B'(S)$ is relatively compact,
(iv) $A(x) + B(I - D)^{-1}C(x) \cdot B'(I - D)^{-1}C(y) \in \mathscr{P}_{cl,cv}(X)$, for all x, $y \in S$,
(v) $x \in A(x) + B(I - D)^{-1}C(x) \cdot B'(I - D)^{-1}C(y)$, $y \in S \Rightarrow x \in S$.

Then, the block operator matrix (3.2.6) *has a fixed point in $S \times S$ provided that*

$$\frac{q_2 q_3}{1-k} M + q_1 < 1,$$

where $M = \|B'(I - D)^{-1}C(S)\|_{\mathscr{P}}$.

Proof Define the multi-valued mapping $T : X \times S \longrightarrow \mathscr{P}_{cl,cv}(X)$ by putting

$$T(x, y) = A(x) + B(I - D)^{-1}C(x) \cdot B'(I - D)^{-1}C(y).$$

In the similar way as in the proof of Theorem 3.3.1, we can show that, for each $y \in S$, there exists $x_y \in S$ such that

$$x_y \in A(x_y) + B(I - D)^{-1}C(x_y) \cdot B'(I - D)^{-1}C(y).$$

Let us denote by $\mathscr{F}_{T_y}$ the set

$$\mathscr{F}_{T_y} = \left\{x \in X \text{ such that } x \in A(x) + B(I - D)^{-1}C(x) \cdot B'(I - D)^{-1}C(y)\right\}.$$

Then, $\mathscr{F}_{T_y}$ is a nonempty and closed subset of S, in view of Theorem 1.4.40. Define the two mapping $G : S \longrightarrow \mathscr{P}_{cl}(S)$ and $g : S \longrightarrow S$ by putting

$$Gy = \mathscr{F}_{T_y} \quad \text{and} \quad g(y) = f(y, y).$$

After simple calculations, we can show that the operator g is continuous such that the following equality

$$g(y) = f(y, y) \in A(f(y, y)) + B(I - D)^{-1}C(f(y, y)) \cdot B'(I - D)^{-1}C(y)$$

is satisfied, for each $y \in S$. Hence, there exists $u \in S$ such that

$$u \in A(u) + B(I - D)^{-1}C(u) \cdot B'(I - D)^{-1}C(u).$$

This completes the proof. Q.E.D.

In what follows, we are in position to formulate the following result by combining Theorem 1.5.38 with Lemma 1.5.30.

Theorem 3.3.3 *Let S be a nonempty, bounded, closed, and convex subset of a Banach algebra X, and let A, B, $B' : S \longrightarrow \mathscr{P}_{bd,cv}(X)$ and C, $D : S \longrightarrow X$ be five operators such that*

(i) A and B are $\mathscr{D}$-Lipschitzians with $\mathscr{D}$-functions ψ_1 and ψ_2, respectively,
(ii) C and D are Lipschitzians with constants q and k such that $q + k < 1$,
(iii) B' is lower semi-continuous such that $B'(S)$ is totally bounded,
(iv) $C(S) \subseteq (I - D)(S)$,
(v) $A(x) + B(I - D)^{-1}C(x) \cdot B'(I - D)^{-1}C(x) \in \mathscr{P}_{cl,cv}(S)$, for all $x \in S$.

Then, the block operator matrix (3.2.6) has a fixed point in $S \times S$ provided that

$$\psi_1(2r) + \|B'(S)\|_{\mathscr{P}} \psi_2(2r) < r, \ for\ r > 0.$$

Proof Define the mapping $T : S \longrightarrow \mathscr{P}_{cl,cv}(S)$ by putting

$$T(x) = A(x) + B(I - D)^{-1}C(x) \cdot B'(I - D)^{-1}C(x).$$

Based on Lemma 1.5.30, we infer that

$$\begin{aligned}
\mu(T(x)) &= \mu\left(A(x) + B(I - D)^{-1}C(x) \cdot B'(I - D)^{-1}C(x)\right) \\
&\leq \mu(A(x)) + \mu\left(B(I - D)^{-1}C(x) \cdot B'(I - D)^{-1}C(x)\right) \\
&\leq \mu(A(x)) + \mu\left(B(I - D)^{-1}C(x)\right) \|B'(I - D)^{-1}C(x)\|_{\mathscr{P}} \\
&\leq \alpha(A(x)) + \alpha\left(B(I - D)^{-1}C(x)\right) \|B'(I - D)^{-1}C(x)\|_{\mathscr{P}} \\
&\leq \psi_1\left(2\mu(\{x\})\right) + \psi_2\left(2\mu\left(\left\{\frac{q}{1-k}x\right\}\right)\right) \|B'(I - D)^{-1}C(x)\|_{\mathscr{P}},
\end{aligned}$$

whenever $x \in S$. This yields that $T : S \longrightarrow \mathscr{P}_{cp,cv}(S)$ is well defined. We demonstrate that T is closed. To prove the above-announced fact, let us assume that $(x_n)_n$ is a sequence contained in the set S, which is convergent to a point x and take a sequence $y_n \in T(x_n)$ such that $y_n \to y$. After simple calculations, we can show that

$$D(y, T(x)) \leq \text{dist}(y, y_n) + \text{dist}(y_n, T(x_n)) + \text{dist}_H(T(x_n), T(x)). \quad (3.3.2)$$

For arbitrary $x, z \in S$, we get

$$\begin{aligned}
\mathrm{dist}_H\,(T(x), T(z)) &\le \mathrm{dist}_H\,(A(x), A(z)) \\
&\quad + \mathrm{dist}_H\left(B(I-D)^{-1}C(x)\cdot B'(I-D)^{-1}C(x), B(I-D)^{-1}C(z)\cdot B'(I-D)^{-1}C(z)\right) \\
&\le \psi_1(\|x-z\|) + \mathrm{dist}_H\left(B(I-D)^{-1}C(x), B(I-D)^{-1}C(z)\right)\mathrm{dist}_H(\{0\}, B'(I-D)^{-1}C(x)) \\
&\quad + \mathrm{dist}_H(0, B(I-D)^{-1}C(z))\mathrm{dist}_H\left(B'(I-D)^{-1}C(x), B'(I-D)^{-1}C(z)\right) \\
&\le \psi_1(\|x-z\|) + M\psi_2\left(\frac{q}{1-k}\|x-z\|\right) + \eta\,\mathrm{dist}_H(B'(I-D)^{-1}C(x), B'(I-D)^{-1}C(z)) < \\
&\le (\psi_1 + M\psi_2)(\|x-z\|) + \eta\,\mathrm{dist}_H(B'(I-D)^{-1}C(x>), B'(I-D)^{-1}C(z)),
\end{aligned}$$

where, η is as described in (3.3.1). From assumption (iii), it follows that

$$\lim_{n\to+\infty} \mathrm{dist}_H\left(B'(I-D)^{-1}C(x_n), B'(I-D)^{-1}C(x)\right) = 0.$$

This implies that

$$\lim_{n\to+\infty} \mathrm{dist}_H\,(T(x_n), T(x)) = 0.$$

Due to the fact that $T(x) \in \mathscr{P}_{cl}(S)$, for any $x \in S$ and using the estimate (3.3.2), we conclude that $y \in T(x)$ and so $T : S \longrightarrow \mathscr{P}_{cp,cv}(S)$ is a closed multi-valued mapping. Finally, let us take a subset $S_1 \subset S$ with $\mu(S_1) > 0$. Based on Lemma 1.4.37, we can prove that $A(S_1)$ and $(I-D)^{-1}C(S_1)$ are bounded. Since $B'(S_1)$ is relatively compact, it follows that $B'(I-D)^{-1}C(S_1)$ is bounded. The use of the sub-linearity of the Hausdorff measure of non-compactness μ, together with the following inclusion

$$T(S_1) \subset A(S_1) + B(I-D)^{-1}C(S_1)\cdot B'(I-D)^{-1}C(S_1),$$

we get

$$\begin{aligned}
\mu(T(S_1)) &\le \mu(A(S_1)) + \mu\left(B(I-D)^{-1}C(S_1)\cdot B'(I-D)^{-1}C(S_1)\right) \\
&\le \mu(A(S_1)) + \mu\left(B(I-D)^{-1}C(S_1)\right)\|B'(I-D)^{-1}C(S_1)\|_{\mathscr{P}} \\
&\quad + \mu\left(B'(I-D)^{-1}C(S_1)\right)\|B(I-D)^{-1}C(S_1)\|_{\mathscr{P}} \\
&\le \alpha(A(S_1)) + \alpha\left(B(I-D)^{-1}C(S_1)\right)\|B'(I-D)^{-1}C(S_1)\|_{\mathscr{P}} \\
&\quad + \alpha\left(B'(I-D)^{-1}C(S_1)\right)\|B(I-D)^{-1}C(S_1)\|_{\mathscr{P}} \\
&\le (\psi_1 + M\psi_2)(\alpha(S_1)) \\
&\le (\psi_1 + M\psi_2)(2\mu(S_1)).
\end{aligned}$$

This prove that T is nonlinear $\mathscr{D}$-set-contraction multi-valued mapping. The desired conclusion follows from Theorem 1.5.38. Q.E.D.

The next result presents a generalization of that containing in Theorem 3.3.3.

Corollary 3.3.4 *Let X and Y be two Banach algebras and let S and S' be two nonempty bounded, convex and closed subsets of X and Y, respectively.*
Let $A : S \longrightarrow X$, $B : S' \longrightarrow X$, $C : S \longrightarrow Y$, $D : S' \longrightarrow Y$ and $B' : S' \longrightarrow \mathscr{P}_{cp,cv}(X)$ be five multi-valued mappings such that

(i) A, B, and C are Lipschitzians with constants q_1, q_2 and q_3, respectively,
(ii) D is a contraction with a constant k and $C(S) \subseteq (I-D)(S')$,

(iii) B' *is upper semi-continuous such that* $B'(S)$ *is relatively compact,*
(iv) for all $x \in S$, *we have* $A(x) + B(I - D)^{-1}C(x) \cdot B'(I - D)^{-1}C(x) \subseteq S$.

Then, the block operator matrix (3.2.6) *has a fixed point in* $S \times S$ *provided that*

$$\frac{q_2 q_3}{1-k} \|B'(I - D)^{-1}C(S)\|_{\mathscr{P}} + q_1 < \frac{1}{2}.$$

By combining Theorem 3.3.3, together with Theorem 1.4.46, we obtain the following fixed point result.

Theorem 3.3.5 *Let S be a nonempty, convex, and closed subset of a Banach algebra* X, *and let* A, B, $B' : X \longrightarrow \mathscr{P}_{cp}(X)$, $C : S \longrightarrow X$ *and* $D : X \longrightarrow X$ *be five multi-valued mappings such that*

(i) A, B, *and* C *are Lipschitzians with constants* q_1, q_2 *and* q_3, *respectively,*
(ii) A *and* C *are totally bounded,*
(iii) B *and* B' *are upper semi-continuous,*
(iv) D *is a contraction with a constant* $k \in [0, \frac{1}{2}[$,
(v) $A(x) + B(I - D)^{-1}C(x) \cdot B'(I - D)^{-1}C(x) \in \mathscr{P}_{cp,cv}(S)$, *for all* $x \in S$.

Then, the operator matrix (3.2.6) *has, at least, one fixed point in* $S \times S$ *provided that*

$$\frac{q_2 q_3}{1-k} M + q_1 < 1,$$

where $M = \|B'(I - D)^{-1}C(S)\|_{\mathscr{P}}$.

Proof Let S_1 be a nonempty and bounded subset contained in the set S with $\mu(S_1) > 0$. We show that the set $(I - D)^{-1}C(S_1)$ is relatively compact. Suppose the contrary, then $d = \beta((I - D)^{-1}C(S_1)) > 0$. If we consider the estimate (3.1.7) and deploy Remark 1.5.32 , it is shown that

$$\begin{aligned}
\mu\left((I - D)^{-1}C(S_1)\right) &\leq \mu(\overline{C(S_1)}) + \mu\left(D(I - D)^{-1}C(S_1)\right) \\
&\leq k\alpha\left((I - D)^{-1}C(S_1)\right) \\
&\leq 2k\mu\left((I - D)^{-1}C(S_1)\right) \\
&< \mu\left((I - D)^{-1}C(S_1)\right),
\end{aligned}$$

which is absurd and consequently, $(I - D)^{-1}C(S_1)$ is relatively compact. Define now the mapping $T : S \longrightarrow \mathscr{P}_{cl,cv}(S)$ by putting

$$T(x) = A(x) + B(I - D)^{-1}C(x) \cdot B'(I - D)^{-1}C(x).$$

It follows from assumption (ii), that $T(S)$ is relatively compact. Again, when we proceed with the same arguments as in Theorem 3.3.3, we can show that T is upper semi-continuous. The desired conclusion follows from Theorem 1.4.46. Q.E.D.

3.3.2 Multi-Valued Maps with a Weakly Sequentially Closed Graph

We may establish new fixed point theorem of the block operator matrix (3.2.6), by imposing some conditions on its inputs, which are assumed to be nonlinear multi-valued mapping acting on a product of a Banach algebra satisfying the sequential condition ($\mathscr{P}$). The obtained results generalize a few ones contained in [16].

Theorem 3.3.6 *Let S be a nonempty, closed, and convex subset of a Banach algebra X satisfying the sequential condition ($\mathscr{P}$) and let A, B, $B' : S \longrightarrow \mathscr{P}(X)$ be three multi-valued mappings with weakly sequentially closed graphs, $C : S \longrightarrow X$ and $D : X \longrightarrow X$ be two weakly sequentially continuous operators such that*

(i) A, B and B' are $\mathscr{D}$-Lipschitzians with $\mathscr{D}$-functions ϕ_A, ϕ_B and $\phi_{B'}$, respectively,
(ii) A, B and B' are sequentially weakly upper semi-compact,
(iii) C is $\mathscr{D}$-Lipschitzian with $\mathscr{D}$-function ϕ_C,
(iv) D is a contraction with a constant k such that $C(S) \subset (I - D)(S)$,
(v) $A(S)$, $B(S)$ and $B'(S)$ are bounded,
(vi) $A(x) + B(I - D)^{-1}C(x) \cdot B'(I - D)^{-1}C(x) \in \mathscr{P}_{cv}(S)$, for all $x \in S$.

Then, the block operator matrix (3.2.6) *has a fixed point in $S \times S$ provided that*

$$\phi_A(r) + \phi_B\,(2\phi_C)\,(r)\phi_{B'}\,(2\phi_C)\,(r) + \left[\|B'(S)\|_{\mathscr{P}}\phi_B + \|B(S)\|_{\mathscr{P}}\phi_{B'}\right](2\phi_C)\,(r) < r.$$

Proof Let $(\theta_n)_n$ be a sequence contained in the set S, which is weakly convergent to an element θ. Then, the set $\{\theta_n,\ n \in \mathbb{N}\}$ is relatively weakly compact. By using the equality (3.1.7) and Lemma 1.5.9, together with assumption on maps C and D, we obtain

$$\begin{aligned}\beta\left(\{(I - D)^{-1}C(\theta_n),\ n \in \mathbb{N}\}\right) &\le \beta\left(\{D(I - D)^{-1}C(\theta_n),\ n \in \mathbb{N}\}\right)\\ &\le k\beta(\{(I - D)^{-1}C(\theta_n),\ n \in \mathbb{N}\}).\end{aligned}$$

This implies that

$$\beta(\{(I - D)^{-1}C(\theta_n),\ n \in \mathbb{N}\}) = 0.$$

Then, the sequence $\left((I - D)^{-1}C(\theta_n)\right)_n$ possesses a weakly convergent subsequence to a point γ. Taking into account the weak sequential continuity of D, and using equality (3.1.7), we obtain

$$\gamma = C\theta + D\gamma.$$

We show that

$$(I - D)^{-1}C(\theta_n) \rightharpoonup \gamma.$$

If not the case, let us observe that the sequence $(\theta_n)_n$ has a subsequence, say $(\theta_{n_j})_j$, that satisfies

$$(I - D)^{-1}C(\theta_{n_j}) \notin V^w,$$

for some weak neighborhood V^w of $(I-D)^{-1}C(\theta)$. Similarly, we can find a subsequence, say $(\theta_{n_{j_k}})_k$, of $(\theta_{n_j})_j$ that satisfies

$$(I-D)^{-1}C(\theta_{n_{j_k}}) \rightharpoonup (I-D)^{-1}C(\theta),$$

which is a contradiction. This means that the operator $(I-D)^{-1}C$ is weakly sequentially continuous. Let M be a nonempty and bounded subset contained in the set S with $\beta(M) > 0$. It follows from Lemma 1.5.9, that

$$\beta\left((I-D)^{-1}C(M)\right) \leq \frac{1}{1-k}\phi_C(\beta(M)).$$

Define the multi-valued mapping $G : S \longrightarrow \mathscr{P}_{cv}(S)$ by putting

$$G(x) = A(x) + B(I-D)^{-1}C(x) \cdot B'(I-D)^{-1}C(x). \tag{3.3.3}$$

Notice that G is sequentially weakly upper semi-compact. In fact, let us assume that $(\theta_n)_n$ is a sequence contained in S, which is weakly convergent to an element θ and take a sequence $z_n \in G(\theta_n)$. Then, there exist three weakly convergent sequences $a_n \in A(\theta_n)$, $b_n \in B(I-D)^{-1}C(\theta_n)$ and $b'_n \in B'(I-D)^{-1}C(\theta_n)$ to some a, b and b', respectively such that the following equality

$$z_n = a_n + b_n \cdot b'_n$$

is satisfied. The use of assumption (ii) as well as condition $(\mathscr{P})$ leads to $(z_n)_n$ has a renamed weakly convergent subsequence to a point $z \in G(\theta)$. This yields that G is a sequentially weakly upper semi-compact multi-valued mapping and so, the graph of G is weakly sequentially closed. However, by taking M as a bounded subset contained in the set S with $\beta(M) > 0$, we can verify that

$$\begin{aligned} G(M) &\subseteq A(M) + B(I-D)^{-1}C(M) \cdot B'(I-D)^{-1}C(M) \\ &\subseteq A(S) + B(S) \cdot B'(S), \end{aligned}$$

which yields that $G(M)$ is bounded and

$$\begin{aligned} \beta(G(M)) &\leq \beta\left(A(M) + B(I-D)^{-1}C(M) \cdot B'(I-D)^{-1}C(M)\right) \\ &\leq \|B(S)\|_{\mathscr{P}}\beta\left(B'(I-D)^{-1}C(M)\right) + \|B'(S)\|_{\mathscr{P}}\beta\left(B(I-D)^{-1}C(M)\right) \\ &\quad +\beta\left(B(I-D)^{-1}C(M)\right)\beta\left(B'(I-D)^{-1}C(M)\right) + \beta(A(M)) \\ &\leq \left[\|B(S)\|_{\mathscr{P}}\phi'_B\left(2\phi_C\right) + \|B'(S)\|_{\mathscr{P}}\phi_B\left(2\phi_C\right) + \phi_A\right](\beta(M)) \\ &\quad +\phi_B(2\phi_C)(\beta(M))\phi_{B'}(2\phi_C)(\beta(M)). \end{aligned}$$

This means, in particular, that G is condensing with respect to β, and when we apply Theorem 1.5.36, we reach the desired conclusion. Q.E.D.

Let us notice that the following results are a particular case of Theorem 3.3.6, if we put A, B, and B' as three single-valued mappings. We omit their proofs.

Corollary 3.3.7 *Let S be a nonempty, closed, and convex subset of a Banach algebra X satisfying the sequential condition $(\mathscr{P})$ and let $A, B, C, D, B' : S \longrightarrow X$ be five weakly sequentially continuous operators such that*

(i) A, B, C, and B' are $\mathscr{D}$-set-Lipschitzians with $\mathscr{D}$-functions ϕ_A, ϕ_B, ϕ_C and $\phi_{B'}$, respectively,
(ii) D is a contraction with a constant k and $C(S) \subset (I-D)(S)$,
(iii) $A(S)$, $B(S)$ and $B'(S)$ are bounded,
(iv) for all $x \in S$, we have $A(x) + B(I-D)^{-1}C(x) \cdot B'(I-D)^{-1}C(x) \in S$.

Then, the block operator matrix (3.2.6) has a fixed point in $S \times S$ provided that

$$\phi_A(r) + (\phi_B \cdot \phi_{B'})\left(\frac{1}{1-k}\phi_C(r)\right) + \left(\|B'(S)\|_{\mathscr{P}}\,\phi_B + \|B(S)\|_{\mathscr{P}}\,\phi_{B'}\right)\left(\frac{1}{1-k}\phi_C(r)\right) < r.$$

Corollary 3.3.8 *Let S be a nonempty, closed, and convex subset of a Banach algebra X satisfying the sequential condition $(\mathscr{P})$ and let $A, B, C, D, B' : S \longrightarrow X$ be five weakly sequentially continuous mappings such that*

(i) A, B, and C are $\mathscr{D}$-Lipschitzians with $\mathscr{D}$-functions ϕ_A, ϕ_B and ϕ_C, respectively,
(ii) D is a contraction with a constant k and $C(S) \subseteq (I-D)(S)$,
(iii) B' maps bounded sets into relatively weakly compact sets,
(iv) $A(S)$, $B(S)$ and $B'(S)$ are bounded,
(v) For all $x \in S$, we have $A(x) + B(I-D)^{-1}C(x) \cdot B'(I-D)^{-1}C(x) \in S$.

Then, the block operator matrix (3.2.6) has a fixed point in $S \times S$ provided that

$$\|B'(S)\|_{\mathscr{P}}\,\phi_B \circ \left(\frac{1}{1-k}\phi_C\right)(r) + \phi_A(r) < r, \text{ for } r > 0.$$

Remark 3.3.9 Notice that Corollary 3.3.8 presents a generalization of that containing in Theorem 3.1.14, since every weakly sequentially continuous and Lipschitzian operator with constant q is $\mathscr{D}$-set-Lipschitzian mapping with $\mathscr{D}$-function $\phi(r) = qr$.

In the special case when, we put $C = \mathbf{1}_X$ and $D = \mathbf{0}_X$ in Corollary 3.3.8, we obtain the following.

Corollary 3.3.10 *Let S be a nonempty, closed, and convex subset of a Banach algebra X satisfying the sequential condition $(\mathscr{P})$ and let $A, B : X \longrightarrow X$ and $B' : S \longrightarrow X$ be three weakly sequentially continuous operators such that*

(i) A and B are $\mathscr{D}$-Lipschitzians with $\mathscr{D}$-functions ϕ_A and ϕ_B, respectively,
(ii) B is regular and B' is weakly compact,
(iii) $A(S)$, $B(S)$ and $B'(S)$ are bounded,
(iv) $B'(S) \subset \left(\frac{I-A}{B}\right)(S)$.

Then, the operator inclusion $x \in Ax + Bx \cdot B'x$ has a solution in S provided that

$$\|B'(S)\|_{\mathcal{P}} \phi_B(r) + \phi_A(r) < r, \text{ for } r > 0.$$

Proof Let y be an arbitrarily fixed element of the set S and define the mapping $B'_y : X \longrightarrow X$ by putting

$$B'_y(x) = Ax + Bx \cdot B'y.$$

Then, in virtue of assumptions (i), we get

$$\begin{aligned}\|B'_y(x_1) - B'_y(x_2)\| &\leq \|Ax_1 - Ax_2\| + \|Bx_1 \cdot B'y - Bx_2 \cdot B'y\| \\ &\leq \|Ax_1 - Ax_2\| + \|Bx_1 - Bx_2\| \|B'y\| \\ &\leq \phi_A(\|x_1 - x_2\|) + \|B'(S)\| \phi_B(\|x_1 - x_2\|),\end{aligned}$$

for all $x_1, x_2 \in X$. It follows from Theorem 1.2.26, that there exists an unique $x_y \in X$ such that $B'_y(x_y) = x_y$. In other words, we have

$$x_y = Ax_y + Bx_y \cdot B'y.$$

Then, we can define the mapping $T : S \longrightarrow X$ by putting

$$Tx = \left(\frac{I - A}{B}\right)^{-1} B'x.$$

The assumption (iv) guaranties that $T(S) \subset S$. Going back to the following equality

$$T = AT + BTB',$$

and keeping in mind assumption (iii), we infer that $T(S)$ is bounded, since

$$\begin{aligned}T(S) &\subset A(T(S)) + B(T(S)) \cdot B'(S) \\ &\subset A(S) + B(S) \cdot B'(S).\end{aligned}$$

Using Lemma 1.5.9, we get

$$\begin{aligned}\beta(T(S)) &\leq \beta(A(T(S)) + B(T(S)) \cdot B'(S)) \\ &\leq \beta(AT(S)) + \beta(BT(S) \cdot B'(S)) \\ &\leq \phi_A(\beta(T(S))) + \|B'(S)\| \phi_B(\beta(T(S))),\end{aligned}$$

which is absurd and so, the set $T(S)$ is relatively weakly compact. On the other hand, via simple calculations, one can show the weak sequential continuity of the mapping T. Hence, the mapping T satisfies all conditions of Theorem 1.3.44, so there exists a point $x \in S$ such that $x = Tx$. This completes the proof. Q.E.D.

Remark 3.3.11 Corollary 3.3.10 presents a generalization and extension of that containing in Theorem 2.1.1.

Remark 3.3.12 Under hypotheses of Theorem 3.3.6, we can prove that the set of fixed point $\mathscr{F}_G$, where G is given in (3.3.3), is relatively weakly compact

Theorem 3.3.13 *Let S be a nonempty, closed, and convex subset of a Banach algebra X satisfying the sequential condition ($\mathscr{P}$) and let A, B, $B' : S \longrightarrow \mathscr{P}_{bd,cl,cv}(X)$, where $\mathscr{P}_{bd,cl,cv}(X)$ denote the family of nonempty, bounded, closed, and convex subsets of X, be three multi-valued mappings with weakly sequentially closed graphs, $C : S \longrightarrow X$ and $D : X \longrightarrow X$ be two weakly sequentially continuous single-valued mapping such that*

(i) A and B are $\mathscr{D}$-set-Lipschitzians with $\mathscr{D}$-functions ϕ_A and ϕ_B, respectively,
(ii) A and B are sequentially weakly upper semi-compact,
(iii) C is $\mathscr{D}$-Lipschitzian with $\mathscr{D}$-function ϕ_C,
(iv) D is a contraction with k and $C(S) \subseteq (I - D)(S)$,
(v) B' maps bounded sets into relatively weakly compact sets,
(vi) for all $x \in S$, we have $A(x) + B(I - D)^{-1}C(x) \cdot B'(I - D)^{-1}C(x) \in \mathscr{P}_{cv}(S)$.

Then, the block operator matrix (3.2.6) *has a fixed point in $S \times S$ provided that*

$$\|B'(S)\|_{\mathscr{P}} \phi_B \circ \varphi(r) + \phi_A(r) < r, \ for\ r > 0,$$

where the function φ is given in (3.1.9).

Proof In the similar way as in the proof of Theorem 3.3.6, we can show that the operator $(I - D)^{-1}C$ is weakly sequentially continuous. Let us assume that $(x_n)_n$ is a sequence contained in the set S, which is weakly convergent to an element $x \in S$ and take a sequence $z_n \in B'(I - D)^{-1}C(x_n)$. The assumption (v) guarantees that $(z_n)_n$ possesses a weakly convergent subsequence. This proves that the multi-valued mapping $B'(I - D)^{-1}C$ is sequentially weakly upper semi-compact. We show that the multi-valued mapping G, where G is as described in the equality (3.3.3), is condensing with respect to β. In order to prove this fact, let us take a nonempty and bounded subset M of S with $\beta(M) > 0$. Then, the set $(I - D)^{-1}C(M)$ is bounded with bound

$$\delta = \varphi(diam\ M) + \frac{1}{1-k}\|(I - D)^{-1}C(a)\|,$$

for some $a \in M$, where φ is as described in (3.1.9) and

$$\begin{aligned}\beta(G(M)) &\leq \beta\left(A(M) + B(I - D)^{-1}C(M) \cdot B'(I - D)^{-1}C(M)\right)\\ &\leq \beta(A(M)) + \|B(S)\|_{\mathscr{P}}\beta\left(B'(I - D)^{-1}C(M)\right)\\ &\quad + \|B'(S)\|_{\mathscr{P}}\beta\left(B(I - D)^{-1}C(M)\right)\\ &\leq \left(\|B'(S)\|_{\mathscr{P}}\phi_B \circ \varphi + \phi_A\right)(\beta(M)).\end{aligned}$$

This means, in particular, that G is a condensing multi-valued mapping with respect to β, and when we apply Theorem 1.5.36, we reach the desired conclusion. Q.E.D.

Theorem 3.3.14 *Let S be a nonempty, closed, and convex subset of a Banach algebra X satisfying condition $(\mathcal{P})$ and let A, B, $B' : S \longrightarrow \mathcal{P}_{bd,cl,cv}(X)$ be three multi-valued mappings with weakly sequentially closed graphs and let $C : S \longrightarrow X$ and $D : X \longrightarrow X$ be two weakly sequentially continuous operators such that*

(i) A, B, C and B' are $\mathcal{D}$-Lipschitzians with $\mathcal{D}$-functions ϕ_A, ϕ_B, ϕ_C and $\phi_{B'}$, respectively,
(ii) A, B and B' are sequentially weakly upper semi-compact,
(iii) $A(S)$, $B(S)$ and $B'(S)$ are bounded,
(iv) C is strongly continuous and $C(S) \subseteq (I - D)(S)$,
(v) D is a contraction with a constant $k \in [0, \frac{1}{2})$,
(v) for all $x \in S$, the set $A(x) + B(I - D)^{-1}C(x) \cdot B'(I - D)^{-1}C(x) \in \mathcal{P}_{cv}(S)$.

Then, the block operator matrix (3.2.6) *has a fixed point in $S \times S$ provided that*

$$\|B(S)\|_{\mathcal{P}}\phi_{B'}(2\phi_C(r)) + \|B'(S)\|_{\mathcal{P}}\phi_B(2\phi_C(r)) + \phi_A(r) + \phi_B(2\phi_C(r))\phi_{B'}(2\phi_C(r)) < r.$$

Proof Define the multi-valued mapping $G : S \longrightarrow \mathcal{P}_{cv}(S)$ by putting

$$G(x) = A(x) + B(I - D)^{-1}C(x) \cdot B'(I - D)^{-1}C(x).$$

Notice that G is sequentially weakly upper semi-compact. Indeed, let $(\theta_n)_n$ be a weakly convergent sequence of S to an element θ and take $z_n \in G(\theta_n)$. Then,

$$z_n = a_n + b_n \cdot b'_n,$$

for some sequences $a_n \in A(\theta_n)$, $b_n \in B(I - D)^{-1}C(\theta_n)$ and $b'_n \in B'(I - D)^{-1}C(\theta_n)$. Based on assumption (ii), we obtain

$$\begin{cases} a_n \rightharpoonup a, \\ b_n \rightharpoonup b, \text{ and} \\ b'_n \rightharpoonup b', \end{cases}$$

for some $a, b, b' \in X$. The sequential condition $(\mathcal{P})$ guaranties that

$$z_n = a_n + b_n \cdot b'_n \rightharpoonup z = a + b \cdot b' \in G(\theta).$$

This means that the multi-valued mapping G is sequentially weakly upper semi-compact and so, G has a weakly sequentially closed graph. Next, we demonstrate that G is condensing with respect to β. In order to prove this fact, let M be a nonempty and bounded subset contained in the set S with $r > \beta(M) > 0$. After simple calculations, one can prove that

$$\begin{aligned} G(M) &\subseteq A(M) + B(I - D)^{-1}C(M) \cdot B'(I - D)^{-1}C(M) \\ &\subseteq A(S) + B(S) \cdot B'(S). \end{aligned}$$

Then, $G(M)$ is a bounded subset, in light of assumption (iii). Moreover, for an arbitrarily fixed element $z \in M$, we have

$$\begin{aligned} \|B(I-D)^{-1}C(z)\|_{\mathscr{P}} &\leq \phi_B \circ \varphi(\|z-m\|) + \|B(I-D)^{-1}C(m)\|_{\mathscr{P}} \\ &< \frac{\|z-m\|}{\sup\left\{\|B'x\|_{\mathscr{P}},\ x \in M\right\}} + \|B(I-D)^{-1}C(m)\|_{\mathscr{P}} \\ &\leq \delta_1, \end{aligned}$$

where $\delta_1 = \dfrac{diam(M)}{\sup\left\{\|B'x\|_{\mathscr{P}},\ x \in M\right\}} + \|B(I-D)^{-1}C(m)\|_{\mathscr{P}}$, for some $m \in M$. Similarly, we obtain

$$\begin{cases} \|Az\|_{\mathscr{P}} < diam(M) + \|Aa\|_{\mathscr{P}}, \\ \|B'(I-D)^{-1}C(z)\|_{\mathscr{P}} \leq \dfrac{diam(M)}{\sup\{\|Bx\|_{\mathscr{P}},\ x \in M\}} + \|B'(I-D)^{-1}C(b)\|_{\mathscr{P}}, \end{cases}$$

for some $a, b \in M$. By the definition of β, there exists a subset $K \in \mathscr{W}(X)$ satisfying

$$M \subseteq K + \overline{\mathscr{B}}_X(r).$$

It follows from assumption (i), that

$$\begin{cases} A(M) \subseteq \overline{A(K)}^w + \overline{\mathscr{B}}_X(\phi_A(r)) \\ B(I-D)^{-1}C(M) \subseteq \overline{B(I-D)^{-1}C(K)}^w + \overline{\mathscr{B}}_X\,(\phi_B \circ \varphi(r)) \\ B'(I-D)^{-1}C(M) \subseteq \overline{B'(I-D)^{-1}C(K)}^w + \overline{\mathscr{B}}_X\left(\phi'_B \circ \varphi(r)\right). \end{cases}$$

Taking into account Lemma 1.5.9, and using the subadditivity of the De Blasi measure β, we get

$$\begin{cases} \beta(A(M)) \leq \phi_A(\beta(M)), \\ \beta(B(I-D)^{-1}C(M)) \leq \phi_B \circ \varphi(r), \\ \beta(B'(I-D)^{-1}C(M)) \leq \phi_{B'} \circ \varphi(r). \end{cases}$$

This implies that

$$\begin{aligned} \beta(G(M)) &\leq \beta\left(A(M) + B(I-D)^{-1}C(M) \cdot B'(I-D)^{-1}C(M)\right) \\ &\leq \|B(S)\|_{\mathscr{P}}\beta\left(B'(I-D)^{-1}C(M)\right) + \|B'(S)\|_{\mathscr{P}}\beta\left(B(I-D)^{-1}C(M)\right) \\ &\quad +\beta\left(B(I-D)^{-1}C(M)\right)\beta\left(B'(I-D)^{-1}C(M)\right) + \beta(A(M)) \\ &\leq \left[\|B(S)\|_{\mathscr{P}}\phi_{B'}\,(2\phi_C) + \|B'(S)\|_{\mathscr{P}}\phi_B\,(2\phi_C) + \phi_A\right](\beta(M)) \\ &\quad +\phi_B(2\phi_C)(\beta(M))\phi_{B'}(2\phi_C)(\beta(M)). \end{aligned}$$

Thus, G is condensing with respect to β. The desired conclusion of the theorem follows by a direct application of Theorem 1.5.36. Q.E.D.

If A, B, C and D are single-valued mapping, then Theorem 3.3.14 reduces to the following.

Corollary 3.3.15 *Let S be a nonempty, closed, and convex subset of a Banach algebra X satisfying condition $(\mathscr{P})$ and let $B' : S \longrightarrow \mathscr{P}_{bd,cl,cv}(X)$ be a multi-valued mapping with weakly sequentially closed graph and let $A,\ B,\ C : S \longrightarrow X$ and $D : X \longrightarrow X$ be four weakly sequentially continuous single-valued mapping such that*

(i) A, B, C and B' are $\mathscr{D}$-Lipschitzians with $\mathscr{D}$-functions ϕ_A, ϕ_B, ϕ_C and $\phi_{B'}$, respectively,
(ii) D is a contraction with a constant k and $C(S) \subseteq (I - D)(S)$,
(iii) B' is sequentially weakly upper semi-compact,
(iv) for all $x \in S$, the set $A(x) + B(I - D)^{-1}C(x) \cdot B'(I - D)^{-1}C(x) \in \mathscr{P}_{cv}(S)$.

Then, the block operator matrix (3.2.6) has a fixed point in $S \times S$ provided that

$$\|B(S)\|_{\mathscr{P}}\phi_{B'} \circ \varphi(r) + \|B'(S)\|_{\mathscr{P}}\phi_B \circ \varphi(r) + \phi_A(r) < r,$$

for all $r > 0$, where φ is given in (3.1.9).

The following result improve and generalize Theorem 3.3.4 relative to weak topology features.

Theorem 3.3.16 *Let S be a nonempty, closed, and convex subset of a Banach algebra X satisfying the sequential condition $(\mathscr{P})$ and let $A,\ B,\ B' : S \longrightarrow \mathscr{P}(X)$ be three multi-valued mappings with weakly sequentially closed graphs and $C : S \longrightarrow X$ and $D : X \longrightarrow X$ be two weakly sequentially continuous operators such that*

(i) A and B are $\mathscr{D}$-set-Lipschitzians with $\mathscr{D}$-functions ϕ_A and ϕ_B,
(ii) A and B are sequentially weakly upper semi-compact,
(iii) C is $\mathscr{D}$-Lipschitzian with $\mathscr{D}$-function ϕ_C,
(iii) D is a contraction with a constant k and $C(S) \subseteq (I - D)(S)$,
(iv) B' maps bounded sets into relatively weakly compact sets,
(v) $A(S)$, $B(S)$ and $B'(S)$ are bounded,
(vi) for all $x \in S$, we have $A(x) + B(I - D)^{-1}C(x) \cdot B'(I - D)^{-1}C(x) \in \mathscr{P}_{cv}(S)$.

Then, the block operator matrix (3.2.6) has a fixed point in $S \times S$ provided that

$$\|B'(S)\|\phi_B \circ \varphi(r) + \phi_A(r) < r, \text{ for } r > 0,$$

where φ is given in (3.1.9).

Proof By using Theorem 3.3.6, we can prove that the multi-valued mapping G, where G is as described in equality (3.3.3), has a weakly sequentially closed graph. We show that G is condensing with respect to β. Take an arbitrary bounded subset M contained in the set S with $\beta(M) > 0$. Since the operator B' is weakly compact and $(I - D)^{-1}C(M)$ is bounded, we get

$$\begin{aligned}\beta(G(M)) &\leq \|B'(I - D)^{-1}C(M)\|_{\mathscr{P}}\beta\left(B(I - D)^{-1}C(M)\right) + \beta(A(M)) \\ &\leq \left[\|B'(I - D)^{-1}C(M)\|_{\mathscr{P}}\phi_B \circ \varphi + \phi_A\right](\beta(M)).\end{aligned}$$

The desired conclusion follows by an application of Theorem 1.5.36. Q.E.D.

If A, B, C and D are weakly sequentially continuous single-valued mappings, then Theorem 3.3.16 reduces to the following.

Corollary 3.3.17 *Let S be a nonempty, closed, and convex subset of a Banach algebra X satisfying the sequential condition $(\mathscr{P})$ and let A, B, $C : S \longrightarrow X$ and $D : X \longrightarrow X$ be four weakly sequentially continuous single-valued mapping and let $B' : S \longrightarrow \mathscr{P}_{cv}(X)$ be a multi-valued mapping with weakly sequentially closed graph such that*

(i) A, B, and C are $\mathscr{D}$-Lipschitzians with $\mathscr{D}$-functions ϕ_A, ϕ_B and ϕ_C, respectively,
(ii) $A(S)$, $B(S)$ and $B'(S)$ are bounded,
(iii) D is a contraction with a constant k such that $C(S) \subseteq (I - D)(S)$,
(iv) B' maps bounded sets into relatively weakly compact sets,
(v) for all $x \in S$, we have $A(x) + B(I - D)^{-1}C(x) \cdot B'(I - D)^{-1}C(x) \in \mathscr{P}_{cv}(S)$.

Then, the block operator matrix (3.2.6) *has a fixed point in $S \times S$ provided that*

$$\|B'(S)\|\phi_B \circ \varphi(r) + \phi_A(r) < r, \text{ for } r > 0,$$

where φ is given in the equality (3.1.9).

Proof We show that $G(x) \in \mathscr{P}_{cv}(S)$, for all $x \in S$, where G is given in (3.3.3). Indeed, for any u, $u' \in Gx$, there exist two element v, $v' \in B'(I - D)^{-1}C(x)$ satisfying

$$\begin{cases} u = A(x) + B(I - D)^{-1}C(x) \cdot v \\ u' = A(x) + B(I - D)^{-1}C(x) \cdot v'. \end{cases}$$

This implies that

$$\lambda u + (1 - \lambda)u' = A(x) + B(I - D)^{-1}C(x) \cdot (\lambda v + (1 - \lambda)v'), \ \lambda \in [0, 1].$$

This yields that

$$\lambda u + (1 - \lambda)u' \in A(x) + B(I - D)^{-1}C(x) \cdot B'(I - D)^{-1}C(x).$$

This means, in particular, that $G : S \longrightarrow \mathscr{P}_{cv}(S)$ is well defined, and when we apply Theorem 3.3.16, we reach the desired conclusion. Q.E.D.

Theorem 3.3.18 *Let S be a nonempty, bounded, closed, and convex subset of a Banach algebra X satisfying condition $(\mathscr{P})$ and let A, B, $B' : S \longrightarrow \mathscr{P}_{bd,cl,cv}(X)$ be three multi-valued mappings with weakly sequentially closed graphs and let $C : S \longrightarrow X$ and $D : X \longrightarrow X$ be two weakly sequentially continuous single-valued mapping such that*

(i) A, B, C and B' are Lipschitzians with constants q_1, q_2, q_3 and q_4, respectively,
(ii) D is a contraction with a constant k such that $C(S) \subseteq (I - D)(S)$,

(iii) A, B and B' are sequentially weakly upper semi-compact,
(iv) for all $x \in S$, we have $A(x) + B(I-D)^{-1}C(x) \cdot B'(I-D)^{-1}C(x) \in \mathscr{P}_{cv}(S)$.

Then, the block operator matrix (3.2.6) *has a fixed point in $S \times S$ provided that*

$$\|B'(S)\|_{\mathscr{P}} \frac{q_2q_3}{1-k} + \|B(S)\|_{\mathscr{P}} \frac{q_3q_4}{1-k} + \frac{q_2q_3^2q_4}{(1-k)^2} diam(S) + q_1 < 1.$$

Proof After simple calculations, we can see that $B(I-D)^{-1}C$ and $B'(I-D)^{-1}C$ are Lipschitzians with constants $\frac{q_2q_3}{1-k}$ and $\frac{q_3q_4}{1-k}$, respectively. The sequential condition ($\mathscr{P}$) guarantees that, the operator G is sequentially weakly upper semi-compact, where G is described in the equality (3.3.3). Our next task is to show that G is condensing with respect to β. In order to prove this fact, take a subset $M \subset S$ with $\beta(M) > 0$. Then,

$$\begin{aligned}
\beta(G(M)) &\le \beta\left(A(M) + B(I-D)^{-1}C(M) \cdot B'(I-D)^{-1}C(M)\right) \\
&\le \|B'(S)\|_{\mathscr{P}} \beta\left(B(I-D)^{-1}C(M)\right) + \|B(S)\|_{\mathscr{P}} \beta\left(B'(I-D)^{-1}C(M)\right) \\
&\quad + \beta\left(B(I-D)^{-1}C(M)\right)\beta\left(B'(I-D)^{-1}C(M)\right) + \beta(A(M)) \\
&\le \left[\|B'(S)\|_{\mathscr{P}} \frac{q_2q_3}{1-k} + \|B(S)\|_{\mathscr{P}} \frac{q_3q_4}{1-k} + q_1\right]\beta(M) + \frac{q_2q_3^2q_4}{(1-k)^2}\beta(M)^2 \\
&\le \left[\|B'(S)\|_{\mathscr{P}} \frac{q_2q_3}{1-k} + \|B(S)\|_{\mathscr{P}} \frac{q_3q_4}{1-k} + \frac{q_2q_3^2q_4}{(1-k)^2} diam(S) + q_1\right]\beta(M).
\end{aligned}$$

The desired conclusion follows by an application of Theorem 1.5.36. Q.E.D.

As a consequence of Theorem 3.3.18, we derive the following fixed point theorem involving three nonlinear multi-valued mappings.

Corollary 3.3.19 *Let S be a nonempty, bounded, closed, and convex subset of a Banach algebra X satisfying condition ($\mathscr{P}$) and let A, B, $B' : S \longrightarrow \mathscr{P}_{bd,cl,cv}(S)$ be three multi-valued mappings with weakly sequentially closed graphs such that*

(i) A and B are Lipschitzians with constants q_1 and q_2, respectively,
(ii) A and B are sequentially weakly upper semi-compact,
(iii) B' is weakly completely continuous,
(iv) for all $x \in S$, the set $Ax + Bx \cdot B'x \in \mathscr{P}_{cv}(S)$.

Then, the operator inclusion $x \in Ax + Bx \cdot B'x$ has a solution in S provided that

$$q_1 + Mq_2 < 1,$$

where $M = \|B'(S)\|$.

3.3.3 New Class of Multi-Valued Mapping

In what follows, we shall denote X the given Banach algebra satisfying the condition $(\mathscr{P})$. Before displaying our main results, we first prove the following lemma.

Lemma 3.3.20 *Let S be a nonempty, bounded, closed, and convex subset of a Banach algebra X satisfying condition $(\mathscr{P})$ and let A, B, $B' : S \longrightarrow \mathscr{P}_{cv}(X)$ be three multi-valued mappings, $C : S \longrightarrow X$ and $D : X \longrightarrow X$ be two weakly sequentially continuous single-valued mapping such that*

(i) A and B are $\mathscr{D}$-set-Lipschitzians with $\mathscr{D}$-functions ϕ_A and ϕ_B, respectively,
(ii) $C(S) \subseteq (I - D)(S)$,
(iii) A and $B(I - D)^{-1}C$ are sequentially weakly upper semi-compact,
(iv) A and $B(I - D)^{-1}C$ have a weakly sequentially closed graph,
(v) C and D are two contractions with constants q and k such that $q + k < 1$,
(vi) $A(x) + B(I - D)^{-1}C(x) \cdot B'(I - D)^{-1}C(y) \subseteq S$, for all $x, y \in S$,
(vii) For any $r > 0$, the following inequality

$$\frac{Mq}{1-k}\phi_B(r) + \phi_A(r) < r, \quad for\ r > 0$$

is satisfied, where $M = \|B'(I - D)^{-1}C(S)\|$.

Then, the multi-valued mapping $\left(\dfrac{I - A}{B(I - D)^{-1}C}\right)^{-1}$ exists on $B'(I - D)^{-1}C(S)$.

Proof Let z be an arbitrarily fixed element of the set $B'(I - D)^{-1}C(S)$ and define the mapping $G_z : S \longrightarrow \mathscr{P}_{cv}(S)$ by putting

$$G_z(x) = A(x) + B(I - D)^{-1}C(x) \cdot z.$$

Let us assume that $(u_n)_n$ is a weakly converging sequence contained in the set S to an element u and take a sequence $v_n \in G_z(u_n)$ with $v_n \rightharpoonup v$. Then, there exist two sequences $\theta_n \in A(u_n)$ and $\xi_n \in B(I - D)^{-1}C(u_n)$ satisfying

$$v_n = \theta_n + \xi_n \cdot z.$$

Thus, in view of assumption (iii), the sequence $(\theta_n)_n$ possesses a renamed weakly converging subsequence to an element $\theta \in A(u)$ and the sequence $(\xi_n)_n$ possesses a renamed weakly converging subsequence to an element $\xi \in B(I - D)^{-1}C(u)$. The sequential condition $(\mathscr{P})$ guaranties that

$$v_n = \theta_n + \xi_n \cdot z \rightharpoonup \theta + \xi \cdot z \in G_z(u).$$

This means that the operator G_z has a weakly sequentially closed graph. Let M be an arbitrary bounded subset contained in the set S with $\beta(M) > 0$. Going back to equality (3.1.7), we obtain

$$\begin{aligned}\beta((I-D)^{-1}C(M)) &\leq \beta(C(M)) + \beta\left(D(I-D)^{-1}C(M)\right)\\ &\leq q\beta(M) + k\beta((I-D)^{-1}C(M)).\end{aligned}$$

This implies that

$$\beta((I-D)^{-1}C(M)) \leq \frac{q}{1-k}\beta(M).$$

On the other hand, via simple calculations, one can show that the set $G_z(M)$ is bounded and

$$\begin{aligned}\beta(G_z(M)) &\leq \beta(A(M)) + \beta(B(I-D)^{-1}C(M)\cdot z)\\ &\leq \left[\phi_A + \frac{qM}{1-k}\phi_B\right](\beta(M))\\ &< \beta(M).\end{aligned}$$

Then, G_z is condensing with respect to β. Thus, on account of Theorem 1.5.36, one has there exists a point $x \in S$ such that

$$x \in A(x) + B(I-D)^{-1}C(x)\cdot z.$$

In other words,

$$z \in \left(\frac{I-A}{B(I-D)^{-1}C}\right)(x).$$

This implies that

$$\left(\frac{I-A}{B(I-D)^{-1}C}\right)(x) \bigcap B'(I-D)^{-1}C(y) \neq \emptyset.$$

This completes the proof. Q.E.D.

Now, we state and prove our first main result by combining Theorem 3.3.6 together with Lemma 3.3.20.

Theorem 3.3.21 *Let S be a nonempty, closed, and convex subset of a Banach algebra X satisfying condition $(\mathscr{P})$ and let A, B, $B' : S \longrightarrow \mathscr{P}(X)$ be three multi-valued mappings with weakly sequentially closed graphs, $C : S \longrightarrow X$ and $D : X \longrightarrow X$ be two weakly sequentially continuous operators such that*

(i) A and B are $\mathscr{D}$-set-Lipschitzians with $\mathscr{D}$-functions ϕ_A and ϕ_B, respectively,
(ii) C is $\mathscr{D}$-Lipschitzian with $\mathscr{D}$-function ϕ_C,
(iii) D is a contraction with a constant k and $C(S) \subseteq (I-D)(S)$,
(iv) $A(S)$, $B(S)$ and $B'(S)$ are bounded such that the following inequality

$$M\phi_B\left(\frac{1}{1-k}\phi_C\right)(r) + \phi_A(r) < r, \quad \textit{for } r > 0$$

is satisfied, where $M = \|B'(I-D)^{-1}C(S)\|$.
(v) A, B *and* B' *are sequentially weakly upper semi-compact*
(vi) *for all* $x \in S$, *we have* $\left(\frac{I-A}{B(I-D)^{-1}C}\right)^{-1} B'(I-D)^{-1}C(x) \in \mathscr{P}_{cv}(S)$,
(vii) $\left(\frac{I-A}{B(I-D)^{-1}C}\right)^{-1} B'(I-D)^{-1}C$ *is condensing with respect to* β,

Then, the block operator matrix (3.2.6) *has a fixed point in* $S \times S$.

Proof Define the mapping $T : S \longrightarrow \mathscr{P}_{cv}(S)$ by putting

$$T(x) = \left(\frac{I-A}{B(I-D)^{-1}C}\right)^{-1} B'(I-D)^{-1}C(x). \tag{3.3.4}$$

Based on Lemma 3.3.20, we infer that T has a weakly sequentially closed graph. In fact, consider $(x_n)_n$ as a weakly converging sequence contained in the set S to an element x and take a sequence $y_n \in T(x_n)$ such that $y_n \rightharpoonup y$. By using the definition of the multi-valued mapping T, we get

$$y_n \in A(y_n) + B(I-D)^{-1}C(y_n) \cdot B'(I-D)^{-1}C(x_n).$$

This implies that, there exist three sequences $a_n \in A(y_n)$, $b_n \in B(I-D)^{-1}C(y_n)$ and $b'_n \in B'(I-D)^{-1}C(x_n)$ such that

$$y_n = a_n + b_n \cdot b'_n.$$

Going back to equality (3.1.7), and using Lemma 1.5.9, we infer that

$$\begin{aligned}\beta\left(\{(I-D)^{-1}C(x_n),\ n \in \mathbb{N}\Big\}\right) &\leq \beta(\{C(x_n),\ n \in \mathbb{N}\}) + \beta\left(\Big\{D(I-D)^{-1}C(x_n),\ n \in \mathbb{N}\Big\}\right)\\ &\leq \phi_C(\beta(\{x_n,\ n \in \mathbb{N}\})) + k\beta(\Big\{(I-D)^{-1}C(x_n),\ n \in \mathbb{N}\Big\}).\end{aligned}$$

This implies that

$$\beta(\Big\{(I-D)^{-1}C(x_n),\ n \in \mathbb{N}\Big\}) \leq \varphi(\beta(\{x_n,\ n \in \mathbb{N}\})),$$

where φ is given in (3.1.9). This means that $(x_n)_n$ has a subsequence, say $(x_{n_k})_k$, satisfying

$$(I-D)^{-1}C(x_{n_k}) \rightharpoonup \gamma, \quad \text{for some } \gamma \in S.$$

Thus, on account of equality (3.1.7), we can see that $\gamma = D(\gamma) + C(x)$ and consequently,

$$(I-D)^{-1}C(x_{n_k}) \rightharpoonup (I-D)^{-1}C(x).$$

It follows from assumption (v) and condition $(\mathscr{P})$, that the sequence $(y_n)_n$ has a subsequence, say $(y_{n_k})_k$, such that

$$y_{n_k} = a_{n_k} + b_{n_k} \cdot b'_{n_k} \rightharpoonup a + b \cdot b' \in A(y) + B(I-D)^{-1}C(y) \cdot B'(I-D)^{-1}C(x)$$

for some $a \in Ay, b \in B(I-D)^{-1}C(y)$ and $b' \in B'(I-D)^{-1}C(x)$. Moreover, by using our assumptions, we can show that

$$\begin{aligned} T(S) &\subseteq AT(S) + B(I-D)^{-1}CT(S) \cdot B'(I-D)^{-1}C(S) \\ &\subseteq A(S) + B(I-D)^{-1}C(S) \cdot B'(I-D)^{-1}C(S) \\ &\subseteq A(S) + B(S) \cdot B'(S). \end{aligned}$$

The desired conclusion follows by a direct application of Theorem 1.5.36. Q.E.D.

As an immediate consequence of Theorem 3.3.21, we have the following result.

Corollary 3.3.22 *Let S be a nonempty, closed, and convex subset of a Banach algebra X satisfying the sequential condition $(\mathcal{P})$ and let $A, B, B' : S \longrightarrow \mathcal{P}(X)$ be three multi-valued mappings with weakly sequentially closed graph and $C : S \longrightarrow X$ and $D : X \longrightarrow X$ be two weakly sequentially continuous operators such that*

(i) A and B are $\mathcal{D}$-set-Lipschitzians with $\mathcal{D}$-functions ϕ_A and ϕ_B, respectively,
(ii) A, B and B' are sequentially weakly upper semi-compact,
(iii) $A(S)$, $B(S)$ and $B'(S)$ are bounded,
(iv) C is $\mathcal{D}$-Lipschitzian with $\mathcal{D}$-function ϕ_C,
(v) D is a contraction with a constant k such that $C(S) \subseteq (I-D)(S)$,
(vi) for all $y \in S$, we have $\left(\dfrac{I-A}{B(I-D)^{-1}C}\right)^{-1} B'(I-D)^{-1}C(y) \in \mathcal{P}_{cv}(S)$,
(vii) $\left(\dfrac{I-A}{B(I-D)^{-1}C}\right)^{-1} B'(I-D)^{-1}C$ is condensing with respect to β,

Then, the block operator matrix (3.2.6) *has a fixed point in $S \times S$ provided that*

$$M\phi_B \circ \varphi(r) + \phi_A(r) < r,$$

for all $r > 0$, where φ is given in (3.1.9) *and $M = \|B'(I-D)^{-1}C\|$.*

Theorem 3.3.23 *Let S be a nonempty, closed, and convex subset of a Banach algebra X satisfying condition $(\mathcal{P})$ and let $A, B, B' : S \longrightarrow \mathcal{P}_{cv}(X)$ be three multi-valued mappings with weakly sequentially closed graph, $C : S \longrightarrow X$ and $D : X \longrightarrow X$ be two weakly sequentially continuous operators such that*

(i) A and B are $\mathcal{D}$-set-Lipschitzians with $\mathcal{D}$-functions ϕ_A and ϕ_B, respectively,
(ii) A and B are sequentially weakly upper semi-compact,
(iii) C is $\mathcal{D}$-Lipschitzian with $\mathcal{D}$-function ϕ_C,
(iv) D is a contraction with a constant k,
(v) B' is weakly completely continuous,
(vi) $A(S)$, $B(S)$ and $B'(S)$ are bounded and $C(S) \subseteq (I-D)(S)$,

(vii) $x \in A(x) + B(I - D)^{-1}C(x) \cdot B(I - D)^{-1}C(y)$, *for* $y \in S \Rightarrow x \in S$,

(viii) for all $y \in S$, *the set* $\left(\dfrac{I - A}{B(I - D)^{-1}C}\right)^{-1} B'(I - D)^{-1}C(y)$ *is convex.*

Then, the block operator matrix (3.2.6) *has a fixed point in* $S \times S$ *provided that*

$$M\phi_B \circ \varphi(r) + \phi_A(r) < r, \quad \text{for all } r > 0,$$

where φ *is given in* (3.1.9) *and* $M = \|B'(I - D)^{-1}C\|$.

Proof It follows from Lemma 3.3.20, that the mapping T, where T is given in equality (3.3.4), is well defined on S. In the similar way as in Theorem 3.3.21, we can show that T is condensing with respect to β with a weakly sequentially closed graph. Indeed, take S_1 as a bounded subset contained in the set S with $\beta(S_1) > 0$. We first prove the following inequality

$$\left\| \overline{B'(I - D)^{-1}C(S_1)}^{w} \right\|_{\mathscr{P}} \leq \left\| B'(S) \right\|_{\mathscr{P}}.$$

For an arbitrary element $x \in \overline{B'(I - D)^{-1}C(S_1)}^{w}$, there exists a sequence $(x_n)_n$ contained in the set $B'(I - D)^{-1}C(S_1)$ such that $x_n \rightharpoonup x$. Due to the fact that

$$\|x\| \leq \liminf_{n} \|x_n\|,$$

we infer that

$$\begin{aligned} \|x\| &\leq \left\| B'(I - D)^{-1}C(S_1) \right\|_{\mathscr{P}} \\ &\leq \left\| B'(S) \right\|_{\mathscr{P}}. \end{aligned}$$

Going back to the following equality

$$T = AT + B(I - D)^{-1}CT \cdot B'(I - D)^{-1}C,$$

we can show that

$$\begin{aligned} \beta\left(T(S_1)\right) &\leq \beta\left(B(I - D)^{-1}CT(S_1) \cdot \overline{B'(I - D)^{-1}C(S_1)}^{w}\right) + \beta\left(AT(S_1)\right) \\ &\leq M\phi_B \circ \left(\tfrac{1}{1-k}\phi_C\right)(\beta(TS_1)) + \phi_A(\beta(TS_1)). \end{aligned}$$

This means, in particular, that T is condensing with respect to β, and when we apply Theorem 1.5.36, we reach the desired conclusion. Q.E.D.

As an immediate consequence of Theorem 3.3.23, we get the following.

Corollary 3.3.24 *Let* S *be a nonempty, closed, and convex subset of a Banach algebra* X *satisfying condition* $(\mathscr{P})$ *and let* A, B, $B' : S \longrightarrow \mathscr{P}_{cv}(X)$ *be three multi-valued mappings with weakly sequentially closed graphs and let* $C : S \longrightarrow X$ *and*

$D : X \longrightarrow X$ be two weakly sequentially continuous single-valued mappings such that

(i) A and B are $\mathcal{D}$-set-Lipschitzians with $\mathcal{D}$-functions ϕ_A and ϕ_B, respectively,
(ii) C is $\mathcal{D}$-Lipschitzian with $\mathcal{D}$-function ϕ_C,
(iii) D is a contraction with a constant k and $C(S) \subseteq (I - D)(S)$,
(iv) B' is weakly compact and $A(S)$, $B(S)$ and $B'(S)$ are bounded,
(v) A and B are sequentially weakly upper semi-compact,
(vi) $x \in A(x) + B(I - D)^{-1}C(x) \cdot B'(I - D)^{-1}C(y)$, for $y \in S \Rightarrow x \in S$ and
(vii) the operator $\left(\frac{I - A}{B(I - D)^{-1}C}\right)^{-1} B'(I - D)^{-1}C$ has a convex value on S.

Then, the block operator matrix (3.2.6) *has a fixed point in $S \times S$ provided that*

$$M\phi_B \circ \varphi(r) + \phi_A(r) < r, \quad \text{for all } r > 0,$$

where φ is given in (3.1.9) *and $M = \|B'(I - D)^{-1}C(S)\|$.*

Corollary 3.3.25 *Let S be a nonempty, closed, and convex subset of a Banach algebra X satisfying condition $(\mathcal{P})$ and let A, $C : S \longrightarrow X$ and B, B', $D : X \longrightarrow X$ be five weakly sequentially continuous operators such that*

(i) A, B, and C are $\mathcal{D}$-Lipschitzians with $\mathcal{D}$-functions ϕ_A, ϕ_B and ϕ_C, respectively,
(ii) D is a contraction with a constant k such that $C(S) \subseteq (I - D)(S)$,
(iii) $A(S)$ and $B(S)$ are bounded,
(iv) B' is weakly compact,
(v) $x = A(x) + B(I - D)^{-1}C(x) \cdot B'(I - D)^{-1}C(y)$, for $y \in S \Rightarrow x \in S$.

Then, the block operator matrix (3.2.6) *has a fixed point in $S \times S$ provided that*

$$M\phi_B \circ \varphi(r) + \phi_A(r) < r, \quad \text{for all } r > 0,$$

where φ is given in (3.1.9) *and $M = \|B'(I - D)^{-1}C(S)\|$.*

Remark 3.3.26 Like Corollary 3.3.25, we can easily prove that the weak closure of the set

$$\left(\frac{I - A}{B(I - D)^{-1}C}\right)^{-1} B'(I - D)^{-1}C(M)$$

is weakly compact, for any arbitrary bounded subset M contained in the set S. Then, Corollary 3.3.25 is a generalization of Theorem 3.1.17, since we omit the regularity the mapping $B(I - D)^{-1}C$.

If we put A as a weakly compact and convex subset of X in Corollary 3.3.8, we obtain the following results. We omit their proofs.

Corollary 3.3.27 *Let S be a nonempty, closed, and convex subset of a Banach algebra X satisfying condition $(\mathcal{P})$ and let $B : S \longrightarrow \mathcal{P}_{cv}(X)$, $B' : S \longrightarrow \mathcal{P}(S)$ be two* *multi-valued mappings with weakly sequentially closed graphs and* $C : S \longrightarrow X$ *and* $D : X \longrightarrow X$ *be two weakly sequentially continuous operators satisfying*

(i) *B is* $\mathcal{D}$*-set-Lipschitzian with* $\mathcal{D}$*-function* ϕ_B,
(ii) $B(S)$ *and* $B'(S)$ *are bounded,*
(iii) *C is* $\mathcal{D}$*-Lipschitzian with* $\mathcal{D}$*-function* ϕ_C,
(iv) *D is a contraction with a constant k such that* $C(S) \subseteq (I - D)(S)$,
(v) *B and* B' *are sequentially weakly upper semi-compact,*
(vi) $B'(I-D)^{-1}C(S) \subset \left(\dfrac{I-A}{B(I-D)^{-1}C}\right)(S)$,
(vi) *for all* $y \in S$, *we have*

$$\left(\frac{I-A}{B(I-D)^{-1}C}\right)^{-1} B'(I-D)^{-1}C(y) \in \mathcal{P}_{cv}(S).$$

Then, the block operator matrix (3.2.6) *has a fixed point in* $S \times S$ *provided that*

$$M\phi_B\left(\frac{1}{1-k}\phi_C\right)(r) < r,$$

for all $r > 0$, *where* $M = \|B'(I-D)^{-1}C\|$.

Corollary 3.3.28 *Let S be a nonempty, closed, and convex subset of a Banach algebra X satisfying condition* $(\mathcal{P})$ *and let* $B : S \longrightarrow \mathcal{P}_{cv}(X)$, $B' : S \longrightarrow \mathcal{P}(S)$ *be two multi-valued mappings with weakly sequentially closed graphs and* $C : S \longrightarrow X$ *and* $D : X \longrightarrow X$ *be two weakly sequentially continuous single-valued mappings such that*

(i) *B is* $\mathcal{D}$*-set-Lipschitzian with* $\mathcal{D}$*-function* ϕ_B,
(ii) *B is sequentially weakly upper semi-compact such that* $B(S)$ *is bounded,*
(iii) *C is* $\mathcal{D}$*-Lipschitzian with* $\mathcal{D}$*-function* ϕ_C,
(iv) *D is a contraction with a constant k and* $C(S) \subseteq (I - D)(S)$,
(v) B' *is weakly completely continuous and* $B'(S)$ *is bounded,*
(vi) $B'(I-D)^{-1}C(S) \subset \left(\dfrac{I-A}{B(I-D)^{-1}C}\right)(S)$,
(vii) *for all* $y \in S$, *we have* $\left(\dfrac{I-A}{B(I-D)^{-1}C}\right)^{-1} B'(I-D)^{-1}C(y) \in \mathcal{P}_{cv}(S)$.

Then, the block operator matrix (3.2.6) *has a fixed point in* $S \times S$ *provided that*

$$M\phi_B \circ \varphi(r) < r, \quad \textit{for all } r > 0,$$

where φ *is given in* (3.1.9) *and* $M = \|B'(I-D)^{-1}C(S)\|$.

Corollary 3.3.29 *Let S be a nonempty, closed, and convex subset of a Banach algebra X satisfying the sequential condition* $(\mathcal{P})$ *and let* $B : S \longrightarrow \mathcal{P}_{bd,cl,cv}(X)$ *and* $B' : S \longrightarrow \mathcal{P}(S)$ *be two multi-valued mappings with a weakly sequentially closed* *graphs and* $C,\ D : S \longrightarrow X$ *be two weakly sequentially continuous single-valued mappings such that*

(i) *B and C are* $\mathcal{D}$*-Lipschitzians with* $\mathcal{D}$*-functions* ϕ_B *and* ϕ_C*, respectively,*
(ii) *B and* B' *are sequentially weakly upper semi-compact,*
(iii) $B(S)$ *and* $B'(S)$ *are bounded,*
(iv) *D is a contraction with a constant k and* $C(S) \subseteq (I - D)(S)$,
(v) $\left(\frac{I - A}{B(I - D)^{-1}C}\right)^{-1} B'(I - D)^{-1}C$ *is condensing with respect to* β,
(vi) $B'(I - D)^{-1}C(S) \subset \left(\frac{I - A}{B(I - D)^{-1}C}\right)(S)$,
(vii) *for all* $y \in S$, *we have*

$$\left(\frac{I - A}{B(I - D)^{-1}C}\right)^{-1} B'(I - D)^{-1}C(y) \in \mathcal{P}_{cv}(S).$$

Then, the block operator matrix (3.2.6) *has a fixed point in* $S \times S$ *provided that*

$$M\phi_B \circ \varphi(r) < r, \quad \text{for all } r > 0,$$

where φ *is given in* (3.1.9) *and* $M = \|B'(I - D)^{-1}C(S)\|$.

If we only take $A = 0_X$ in Corollary 3.3.29, we get the following.

Corollary 3.3.30 *Let S be a nonempty, closed, and convex subset of a Banach algebra X satisfying condition* $(\mathcal{P})$ *and let* $B : S \longrightarrow \mathcal{P}_{cv}(X)$, $B' : S \longrightarrow \mathcal{P}(S)$ *be two multi-valued mappings with weakly sequentially closed graphs and* $C,\ D : S \longrightarrow X$ *be two weakly sequentially continuous operators such that*

(i) *B is* $\mathcal{D}$*-set-Lipschitzian with* $\mathcal{D}$*-function* ϕ_B *such that* $B(S)$ *is bounded,*
(ii) *C is* $\mathcal{D}$*-Lipschitzian with* $\mathcal{D}$*-function* ϕ_C,
(iii) *D is a contraction with a constant k,*
(iv) B' *is weakly completely continuous such that* $B'(S)$ *is bounded,*
(v) $C(S) \subseteq (I - D)(S)$ *and* $B'(I - D)^{-1}C(S) \subset \left(\frac{I}{B(I - D)^{-1}C}\right)(S)$,
(vi) *for all* $y \in S$, *the set* $\left(\frac{I}{B(I - D)^{-1}C}\right)^{-1} B'(I - D)^{-1}C(y)$ *is convex in S.*

Then, the block operator matrix (3.2.6) *has a fixed point in* $S \times S$ *provided that*

$$M\phi_B \circ \varphi(r) < r, \quad \text{for all } r > 0,$$

where φ *is given in* (3.1.9) *and* $M = \|B'(I - D)^{-1}C(S)\|$.

Remark 3.3.31 We have the following particular cases that constitute the versions of fixed point theorems involving three operators. Precisely, if we only take $C = \mathbf{1}_X$, $D = \mathbf{0}_X$ and B as a weakly compact and convex subset of X in Theorem 3.3.29, we get the following two Corollaries 3.3.32 and 3.3.33, which extends one of the results obtained in [3].

Corollary 3.3.32 *Let S be a nonempty, closed, and convex subset of a Banach algebra X satisfying the sequential condition $(\mathscr{P})$ and let $A : X \longrightarrow \mathscr{P}_{cv}(X)$ and $B' : S \longrightarrow \mathscr{P}(S)$ be two multi-valued mappings with weakly sequentially closed graphs such that*

(i) A is $\mathscr{D}$-set-contraction with $\mathscr{D}$-function ϕ_A such that $A(S)$ is bounded,
(ii) A and B' are sequentially weakly upper semi-compact,
(iii) the operator $\left(\frac{I-A}{B}\right)^{-1} B'$ is condensing with respect to β
(iv) If $x \in Ax + B \cdot B'y$, for $y \in S$, then $x \in S$,
(v) for all $y \in S$, the set $\left(\frac{I-A}{B}\right)^{-1} B'(y)$ is convex.

Then, the operator inclusion $x \in Ax + B \cdot B'x$ has a solution in S.

Corollary 3.3.33 *Let S be a nonempty, closed, and convex subset of a Banach algebra X satisfying the sequential condition $(\mathscr{P})$ and let $A : X \longrightarrow \mathscr{P}_{cv}(X)$ and $B' : S \longrightarrow \mathscr{P}(S)$ be two multi-valued mappings with weakly sequentially closed graphs such that*

(i) A is $\mathscr{D}$-set-contraction with $\mathscr{D}$-function ϕ_A such that $A(S)$ is bounded,
(ii) A is sequentially weakly upper semi-compact,
(iii) B' is weakly compact,
(iv) If $x \in Ax + B \cdot B'y$, for $y \in S$, then $x \in S$,
(v) for all $y \in S$, the set $\left(\frac{I-A}{B}\right)^{-1} B'(y)$ is convex.

Then, the operator inclusion $x \in Ax + B \cdot B'x$ has a solution in S.

3.3.4 Countably Multi-Valued in Banach Space

We study new results on the existence of fixed points of the following multi-valued block operator matrix

$$\begin{pmatrix} A & B \\ C & D \end{pmatrix}, \tag{3.3.5}$$

where the inputs are, in general, nonlinear multi-valued mappings, by using the concepts of countably condensing multi-valued mapping with respect to β. The obtained results, which can be found in [27], improve and generalize some recent results in [6, 11, 24, 28] and extend several earlier works. We begin this section by establishing some results providing necessary hypotheses for the inputs, so that the block operator matrix (3.3.5) has a fixed point. Throughout this section, X denotes a Banach space and by S as a nonempty, closed, and convex subset of X. Let us denote the set $\Xi^{(A,B,C,D)}(S)$ by

$$\Xi^{(A,B,C,D)}(S) = \left\{(I - A)(x) \bigcap B(I - D)^{-1}C(y), (x, y) \in S \times S\right\}. \quad (3.3.6)$$

Now, we are ready to state our first fixed point theorem.

Theorem 3.3.34 *Let S be a nonempty, closed, and convex subset of a Banach space X and let $A : S \longrightarrow \mathscr{P}(X)$ and $B : X \longrightarrow \mathscr{P}(X)$ be two multi-valued mappings and $C : S \longrightarrow X$, $D : X \longrightarrow X$ be two nonlinear mapping such that*

(i) A is countably $\mathscr{D}$-set-contraction with $\mathscr{D}$-function ϕ_A,
(ii) $C(S) \subset (I - D)(S)$,
(iii) A and $B(I - D)^{-1}C$ have a weakly sequentially closed graph,
(iv) the set $\Xi^{(A,B,C,D)}(S)$ is relatively weakly compact,
(v) for any $x \in S$, the set $(I - A)^{-1}B(I - D)^{-1}C(x)$ is convex in S.

Then, the block operator matrix (3.3.5) *has a fixed point in $S \times S$.*

Proof Define the multi-valued mapping $Q : S \longrightarrow \mathscr{P}_{cl,cv}(S)$ by putting

$$Q(x) = (I - A)^{-1}B(I - D)^{-1}C(x). \quad (3.3.7)$$

Let us assume that $(x_n)_n$ is a sequence contained in the set S, which is weakly convergent to an element x and take a sequence $y_n \in Q(x_n)$, which is weakly convergent to an element y. By the definition of Q, we get

$$(I - A)(y_n) \bigcap B(I - D)^{-1}C(x_n) \neq \emptyset, \ \forall n \in \mathbb{N}. \quad (3.3.8)$$

Then, there exists a sequence $(z_n)_n$ contained in the set S such that

$$\{z_n, \ n \in \mathbb{N}\} \subset \Xi^{(A,B,C,D)}(S),$$

where the set $\Xi^{(A,B,C,D)}(S)$ is given in (3.3.6). Taking into account assumption (iv), together with Theorem 1.3.29, we obtain $(z_n)_n$ possesses a weakly converging subsequence, say $\left(z_{n_k}\right)_k$, to a point z. Going back to the inclusion (3.3.8), we get $z \in B(I - D)^{-1}C(x)$ and

$$z_{n_k} = y_{n_k} - w_{n_k},$$

for some $w_{n_k} \in A(y_{n_k})$. Then, in virtue of assumption (iii), we have $z \in (I - A)(y)$. This implies that

$$(I - A)(y) \bigcap B(I - D)^{-1}C(x) \neq \emptyset.$$

In other words, we have $y \in Q(x)$ and so, G has a weakly sequentially closed graph. We prove that the set $\overline{Q(S)}^w$ is weakly compact. To prove the above-announced fact, let us assume that $(x_n)_n$ is a sequence of elements contained in the set S and take a sequence $y_n \in Q(x_n)$. Thus, there is two sequences $u_n \in B(I-D)^{-1}C(x_n)$ and $v_n \in A(y_n)$ such that

$$y_n = u_n + v_n.$$

This implies that $u_n \in \Xi^{(A,B,C,D)}(S)$. From the above discussion, we get $(u_n)_n$ possesses a weakly convergent subsequence, say $\left(u_{n_k}\right)_k$, to an element u. Taking into account the subadditivity of the De Blasi measure β, and using the following equality

$$y_{n_k} = y_{n_k} - v_{n_k} + v_{n_k},$$

we arrive at the following

$$\begin{aligned}\beta(\{y_{n_k},\ k \in \mathbb{N}\}) &\leq \beta(\{y_{n_k} - v_{n_k},\ k \in \mathbb{N}\}) + \beta(\{A(y_{n_k}),\ k \in \mathbb{N}\})\\ &\leq \phi_A(\beta(\{y_{n_k},\ k \in \mathbb{N}\})),\end{aligned}$$

which is absurd and so, the sequence $(y_{n_k})_k$ has a weakly convergent subsequence. It follows from Theorem 1.3.29, that the set $Q(S)$ is relatively weakly compact. The desired conclusion follows from Theorem 1.4.56. Q.E.D.

Remark 3.3.35 Let S be a nonempty, closed, and convex subset of a Banach space X and let $A : S \longrightarrow \mathscr{P}(X)$ and $B : X \longrightarrow \mathscr{P}(X)$ be two multi-valued mappings and $C : S \longrightarrow X$, $D : X \longrightarrow X$ be two nonlinear mapping. When the graph of the operator A is convex and $Bx \in \mathscr{P}_{cv}(X)$, for all $x \in S$, then the mapping Qx is convex, for all $x \in S$, where Q is given in (3.3.7). In fact, if we take $z_1, z_2 \in Q(x)$, then there exist two points $w_1, w_2 \in B(I-D)^{-1}C(x)$ such that

$$\begin{cases} w_1 = z_1 - a_1, & \text{for some } a_1 \in A(z_1),\\ w_2 = z_2 - a_2, & \text{for some } a_2 \in A(z_2).\end{cases}$$

Consider $t \in (0, 1)$. After simple calculations, we can see that

$$tw_1 + (1-t)w_2 = tz_1 + (1-t)z_2 - (ta_1 + (1-t)a_2),$$

and so,

$$\begin{cases} ta_1 + (1-t)a_2 \in A(tz_1 + (1-t)z_2),\\ tw_1 + (1-t)w_2 \in B(I-D)^{-1}C(x).\end{cases}$$

This yields that

$$(I-A)(tz_1 + (1-t)z_2) \bigcap B(I-D)^{-1}C(x) \neq \emptyset.$$

In other words, we obtain $tz_1 + (1-t)z_2 \in Q(x)$.

Remark 3.3.36 Let S be a nonempty, closed, and convex subset of a Banach space X and let $C : S \longrightarrow X$, $D : X \longrightarrow X$ be two nonlinear mapping. If the set $C(S)$

is relatively weakly compact and D is nonlinear contraction with $\mathscr{D}$-function ϕ_D, then the operator inverse $(I - D)^{-1}$ is well defined and the set $\Xi^{(\theta,I,C,D)}(S)$, as is described in (3.3.6), is relatively weakly compact, where I denote the identity map and θ denote the null map. Again, assumption of stability in Theorem 2.1.1 give

$$(I - D)^{-1}C(x) \in S, \quad \text{for every } x \in S.$$

Due to the fact that S is bounded, and using the equality (3.1.7), we obtain $(I - D)^{-1}C$ is weakly sequentially continuous. Then, Theorem 2.1.1 becomes as a consequence of Theorem 3.3.34 by putting $\phi_C(r) = 0$.

Corollary 3.3.37 *Let S be a nonempty, closed, and convex subset of a Banach space X and let $A : S \longrightarrow \mathscr{P}(X)$ and $B : X \longrightarrow \mathscr{P}(X)$ be two multi-valued mappings with weakly sequentially closed graph, $C : S \longrightarrow X$ and $D : X \longrightarrow X$ be two weakly sequentially continuous single-valued mappings such that*

(i) A is countably $\mathscr{D}$-set-contraction with $\mathscr{D}$-function ϕ_A,
(ii) B is weakly compact,
(iii) $C(S) \subset (I - D)(S)$ and $(I - D)^{-1}C(S)$ is bounded,
(iv) $(I - A)^{-1}B(I - D)^{-1}Cx \in \mathscr{P}_{cv}(S)$, for all $x \in S$.

Then, the block operator matrix (3.3.5) *has a fixed point in $S \times S$.*

Proof After simple calculations, we can see that the operator $B(I - D)^{-1}C$ has a weakly sequentially closed graph. We show that the set $\Xi^{(A,B,C,D)}(S)$, as is described in the equality (3.3.6), is relatively weakly compact. Indeed, consider $(z_n)_n$ as a sequence contained in the set $\Xi^{(A,B,C,D)}(S)$. Since the set $(I - D)^{-1}C(S)$ is bounded, then $\left(B(I - D)^{-1}C(y_n)\right)_n$ possesses a weakly convergent subsequence, and when we apply Theorem 1.3.29, we reach the desired conclusion. Q.E.D.

In the following result, a new version of fixed point theorems for the sum of two multi-valued mappings having weakly sequentially closed graphs is given by putting $C \equiv I$ and $D \equiv \theta$ in Theorem 3.3.34, where I denote the identity map and θ denote the null map.

Corollary 3.3.38 *Let S be a nonempty, closed, and convex subset of a Banach space X and let $A : S \longrightarrow \mathscr{P}(X)$ and $B : X \longrightarrow \mathscr{P}(X)$ be two multi-valued mappings with weakly sequentially closed graphs such that*

(i) A is hemi-weakly compact,
(ii) the set $\Xi^{(A,B,I,\theta)}(S)$, as is described in (3.3.6), *is relatively weakly compact,*
(iii) for every $x \in S$, $(I - A)^{-1}B(x) \in \mathscr{P}_{cv}(S)$.

Then, the operator inclusion $x \in A(x) + B(x)$ has a solution in S.

Remark 3.3.39 Corollary 3.3.38 improves and extends Corollary 3.3.40, which represents a new version of Krasnoselskii's fixed point theorems obtained in [28].

Corollary 3.3.40 *Let S be a nonempty, closed, and convex subset of a Banach space X and let $A : S \longrightarrow \mathscr{P}(X)$ and $B : X \longrightarrow \mathscr{P}(X)$ be two multi-valued mappings with weakly sequentially closed graphs such that*

(i) $A(S)$ is relatively weakly compact,
(ii) B is condensing such that $B(S)$ is bounded,
(iii) for every $x \in S$, we have $(I - A)^{-1}B(x) \in \mathscr{P}_{cv}(S)$.

Then, the operator inclusion $x \in A(x) + B(x)$ has a solution in S.

Theorem 3.3.41 *Let S be a nonempty, closed, and convex subset of a Banach space X and let $A : S \longrightarrow \mathscr{P}_{cl,cv}(X)$ and $B : X \longrightarrow \mathscr{P}_{cl,cv}(X)$ be two multi-valued mappings with weakly sequentially closed graphs, $C : S \longrightarrow X$ and $D : X \longrightarrow X$ be two weakly sequentially continuous single-valued mappings such that*

(i) A is nonlinear $\mathscr{D}$-set-contraction with $\mathscr{D}$-function ϕ_A,
(ii) B is countably q-set-contraction,
(iii) C is countably γ-set-contraction,
(iv) D is nonlinear contraction with $\mathscr{D}$-function ϕ_D,
(v) $\phi_A(r) < (1-q)r$ and $\phi_D(r) < (1-\gamma)r$, for $r > 0$,
(vi) $A(S)$ and $(I - D)^{-1}C(S)$ are bounded,
(vii) for every $x \in S$, we have $(I - A)^{-1}B(I - D)^{-1}C(x) \in \mathscr{P}_{cl,cv}(S)$.

Then, the block operator matrix (3.3.5) *has a fixed point in $S \times S$.*

Proof Define the multi-valued mapping $Q : S \longrightarrow \mathscr{P}_{cl,cv}(S)$ by putting

$$Q(x) = (I - A)^{-1}B(I - D)^{-1}C(x).$$

At first, we show that the operator inverse $(I - D)^{-1}$ is weakly sequentially continuous. Due to the fact that D is nonlinear contraction and Theorem 1.2.26, we can deduce that $(I - D)^{-1}$ exists on X. Let $(x_n)_n$ be a weakly convergent sequence of X to an element $x \in X$. Going back to the following equality

$$(I - D)^{-1}(x_n) = x_n + D(I - D)^{-1}(x_n),\ n \in \mathbb{N}, \tag{3.3.9}$$

and using the subadditivity of the De Blasi measure β, we arrive at the following

$$\begin{aligned}\beta\left(\left\{(I - D)^{-1}(x_n),\ n \in \mathbb{N}\right\}\right) &\leq \beta\left(\{x_n,\ n \in \mathbb{N}\}\right) + \beta\left(\left\{D(I - D)^{-1}(x_n),\ n \in \mathbb{N}\right\}\right)\\ &\leq \beta\left(\left\{D(I - D)^{-1}(x_n),\ n \in \mathbb{N}\right\}\right)\\ &\leq \phi_D\left(\beta\left(\left\{(I - D)^{-1}(x_n),\ n \in \mathbb{N}\right\}\right)\right).\end{aligned}$$

Then, we can extract a subsequence, say $(x_{n_k})_k$, of $(x_n)_n$ such that

$$(I - D)^{-1}(x_{n_k}) \rightharpoonup y.$$

It follows from the equality (3.3.9), that $y = (I - D)^{-1}x$. We prove that

$$y_n = (I - D)^{-1}(x_n) \rightharpoonup y.$$

Suppose that this is not the case, then we may extract a subsequence, say $\left(y_{n_k}\right)_k$, of $(y_n)_n$ and there exists a weak neighborhood V^w of y, that satisfies,

$$y_{n_k} \notin V^w, \text{ for all } k \geq 1.$$

Similarly, there exists a weakly convergent subsequence, say $\left(y_{n_{k_j}}\right)_j$, of the sequence $(y_{n_k})_k$ to y, which is absurd. Let $(u_n)_n$ be a weakly convergent sequence contained in the set S to a point u and take $v_n \in Q(u_n)$ such that $v_n \rightharpoonup v$. Thus, there exists a sequence $(w_n)_n$, that satisfies

$$w_n \in (I - A)(v_n) \bigcap B(I - D)^{-1}C(u_n).$$

Using assumption (ii) to obtain

$$\beta\left(\left\{B(I - D)^{-1}C(u_n),\ n \in \mathbb{N}\right\}\right) \leq q\, \beta\left(\left\{(I - D)^{-1}C(u_n),\ n \in \mathbb{N}\right\}\right). \quad (3.3.10)$$

Due to the fact that the sequence $\left((I - D)^{-1}C(u_n)\right)_n$ is weakly convergent, and using the estimate (3.3.10), we infer that

$$\beta\left(\left\{B(I - D)^{-1}C(u_n),\ n \in \mathbb{N}\right\}\right) = 0.$$

So, the sequence $(w_n)_n$ possesses a weakly convergent subsequence, say $\left(w_{n_k}\right)_k$, to some point w. Taking into account that B has a weakly sequentially closed graph, we obtain $w \in B(I - D)^{-1}C(x)$. This implies that

$$(I - A)(y) \bigcap B(I - D)^{-1}C(x) \neq \emptyset.$$

In other words, we get $y \in Q(x)$. This means, in particular, that the graph of the multi-valued mapping Q is weakly sequentially closed. Our next task is to prove that Q is countably condensing with respect to β. For this purpose, let M be a countably bounded subset contained in the set S with $\beta(M) > 0$. The use of assumption (iii) as well as Lemma 1.5.9 leads to

$$\begin{aligned}\beta((I - D)^{-1}C(M)) &\leq \beta(C(M)) + \beta(D(I - D)^{-1}C(M)) \\ &\leq \gamma\beta(M) + \phi_D\left(\beta((I - D)^{-1}C(M))\right).\end{aligned} \quad (3.3.11)$$

If $\gamma = 0$, then the desired result is valid. Without loss of generality, we assume that $\gamma \neq 0$. By the inequality (3.3.11), we arrive at

$$\beta((I - D)^{-1}C(M)) < \gamma\, \beta(M) + (1 - \gamma)\, \beta((I - D)^{-1}C(M).$$

Then,

$$\beta((I-D)^{-1}C(M)) < \beta(M). \tag{3.3.12}$$

Since the following inclusion

$$Q(M) \subset AQ(M) + B(I-D)^{-1}C(M) \tag{3.3.13}$$

is satisfied, for any countable and bounded subset M of S, we get

$$\beta(Q(M)) \leq \phi_A(\beta(Q(M))) + \beta(B(I-D)^{-1}C(M)). \tag{3.3.14}$$

If we take $q = 0$ or $\gamma = 0$, then the set $Q(M)$ is relatively weakly compact. In the other case, by using (3.3.12) and (3.3.14), we obtain

$$\beta(Q(M)) < (1-q)\,\beta(Q(M)) + q\,\beta(M).$$

This means, in particular, that Q is countably condensing with respect to β. Let K be a weakly compact subset contained in the set S. From the above discussion, we can prove that $B(I-D)^{-1}C(K)$ is relatively weakly sequentially compact and so, it is relatively weakly compact, in view Theorem 1.3.29. Going back to the inclusion (3.3.13), we obtain

$$\beta(Q(K)) \leq \phi_A(\beta(Q(K))).$$

By a contradiction argument, we may prove that $Q(K)$ is relatively weakly compact. The desired conclusion follows by a direct application of Theorem 1.5.34. Q.E.D.

Lemma 3.3.42 *Let S be a nonempty, closed, and convex subset of a Banach space X and let $A : S \longrightarrow \mathscr{P}_{cv}(X)$ and $B : S \longrightarrow \mathscr{P}_{cl,rcp}(X)$ be two multi-valued mappings, where $\mathscr{P}_{cv,rcp}(X)$ denote the family of nonempty, convex and relatively compact subsets of X, $C : S \longrightarrow X$ and $D : X \longrightarrow X$ be two weakly sequentially continuous single-valued mappings such that*

(i) A is nonlinear $\mathscr{D}$-set-contraction with $\mathscr{D}$-function ϕ_A,
(ii) A has a weakly sequentially closed graph,
(iii) for each fixed y, the set

$$\Delta = \{A(x) + B(I-D)^{-1}C(y),\ x \in S\}$$

is bounded.

Then, the multi-valued mapping $x \mapsto (I-A)^{-1}B(I-D)^{-1}C(x)$ is well defined on S.

Proof Let $y \in S$ be fixed and define the mapping $Q_y : S \longrightarrow \mathscr{P}(S)$ by putting

$$Q_y(x) = A(x) + B(I-D)^{-1}C(y).$$

For any bounded subset M in S, we get

$$\begin{aligned}\beta(Q_y(M)) &\leq \beta(A(M)) + \beta(B(I-D)^{-1}C(y)) \\ &\leq \beta(A(M)) \\ &\leq \phi_A(\beta(M)).\end{aligned}$$

This means that Q_y is a nonlinear $\mathscr{D}$-set-contraction mapping with $\mathscr{D}$-function ϕ_A. Based on the convexity of the set $A(x)$, we can deduce that $Q_y(x) \in \mathscr{P}_{cv}(S)$, for all $x \in S$. On the other hand, via simple calculations, one can show that the graph of the mapping Q_y is weakly sequentially closed such that $Q_y(S)$ is bounded. Thus, on account of Theorem 1.5.39, there exists an element $x \in S \bigcap Q_y(x)$. In other words, we have $x = a + b$, for some $a \in A(x)$ and $b \in B(I-D)^{-1}C(y)$. This implies that

$$(I-A)(x) \bigcap B(I-D)^{-1}C(y) \neq \emptyset.$$

This yields that $x \in (I-A)^{-1}B(I-D)^{-1}C(y)$. Q.E.D.

Theorem 3.3.43 *Let S be a nonempty, closed, and convex subset of a Banach space X and let $A : S \longrightarrow \mathscr{P}_{cl,cv}(X)$ and $B : X \longrightarrow \mathscr{P}_{cl,cv}(X)$ be two multi-valued mappings with weakly sequentially closed graphs, $C : S \longrightarrow X$ and $D : X \longrightarrow X$ be two weakly sequentially continuous single-valued mappings such that*

(i) A is countably $\mathscr{D}$-set-contraction with $\mathscr{D}$-function φ_A,
(ii) for each $\lambda \in [0, 1]$, we have $\lambda a \in Ax$, for all $a \in Ax$,
(iii) A is sequentially weakly upper semi-compact ,
(iv) A and B have a w-weakly closed graph,
(v) C is weakly compact and $C(S) \subset (I-D)(S)$,
(vi) $(I-D)^{-1}$ is weakly sequentially continuous on X,
(vii) $A(S) + B(I-D)^{-1}C(S)$ is bounded.

Then, the block operator matrix (3.3.5) *has a fixed point in $S \times S$.*

Proof Define the multi-valued mapping $Q : S \longrightarrow \mathscr{P}_{cl,cv}(S)$ by putting

$$Q(x) = A(x) + B(I-D)^{-1}C(x).$$

Choose a bounded subset M of S with $\beta(M) > 0$. Let $(x_n)_n$ be a sequence of M and take a sequence $z_n \in B(I-D)^{-1}C(x_n)$. Due to the fact that the sequence $(x_n)_n$ is bounded and $C(M)$ is relatively weakly compact, we infer that $(C(x_n))_n$ possesses a weakly convergent subsequence, say $\left(Cx_{n_k}\right)_k$. Since B is sequentially weakly upper semi-compact, then the sequence $(z_{n_k})_k$ possesses a weakly convergent subsequence. Thus, on account of Theorem 1.3.29, we can deduce that $B(I-D)^{-1}C(M)$ is relatively weakly compact. Now, if we choose D as a countable and bounded subset of S with $\beta(D) > 0$, then

$$\begin{aligned}\beta(Q(D)) &\leq \beta(A(D)) + \beta(B(I-D)^{-1}C(D)) \\ &\leq \varphi_A(\beta(D)) \\ &< \beta(D).\end{aligned}$$

This means, in particular, that Q is countably condensing operator with respect to β. Next, we prove that Q has a convex value on S. For this purpose, let us fix an element $x \in S$ and take $z_1, z_2 \in Q(x)$. Then,

$$\begin{cases} z_1 = a_1 + b_1 \\ z_2 = a_2 + b_2, \end{cases}$$

for some $a_1, a_2 \in Ax$ and some $b_1, b_2 \in B(I - D)^{-1}C(x)$. Consider $t \in (0, 1)$, then

$$tz_1 + (1-t)z_2 = ta_1 + (1-t)a_2 + tb_1 + (1-t)b_2.$$

This implies that

$$tz_1 + (1-t)z_2 \in A(x) + B(I - D)^{-1}C(x).$$

In an equivalent way, we have $Q(x) \in \mathscr{P}_{cv}(S)$. However, the operator Q possesses a w-weakly closed graph. Indeed, let us assume that $(u_n)_n$ is a sequence contained in the set S, which is weakly convergent to a point $u \in S$ and take $v_n \in Q(u_n)$ such that $v_n \rightharpoonup v$. Then, there exit two sequences $a_n \in A(u_n)$ and $b_n \in B(I - D)^{-1}C(u_n)$, that satisfies

$$v_n = a_n + b_n.$$

Based on assumption (iii), we can suppose that $a_n \rightharpoonup a \in Ax$. This implies that

$$b_n \rightharpoonup v - a.$$

Due to the fact that $B(I - D)^{-1}C$ possess a w-weakly closed graph, we deduce that

$$B(I - D)^{-1}C(u) \bigcap L(u, y - a) \neq \emptyset,$$

where L is given in the equality (1.4.2). Then, there exists $\lambda \in [0, 1]$ such that

$$(1-\lambda)u + \lambda(v - a) \in B(I - D)^{-1}C(u).$$

On the other hand, via simple calculations, one can show that

$$(1-\lambda)u + \lambda v \in A(u) + B(I - D)^{-1}C(u).$$

Moreover, for an arbitrary weakly compact subset K of S, we can easily prove that

$$Q(K) \subset A(K) + B(I - D)^{-1}C(K)..$$

This implies that

$$\beta(Q(K)) \leq \beta(A(K)).$$

This means that $Q(K)$ is relatively weakly compact. The desired conclusion of the theorem follows by a direct application of Theorem 1.5.37. Q.E.D.

As an consequence of Theorem 3.3.43, we have the following result.

Corollary 3.3.44 *Let S be a nonempty, closed, and convex subset of a Banach space X and let $A : S \longrightarrow \mathscr{P}_{cl,cv}(X)$ and $B : X \longrightarrow \mathscr{P}_{cl,cv}(X)$ be two multi-valued mappings with w-weakly closed graphs, $C : S \longrightarrow X$ and $D : X \longrightarrow X$ be two weakly sequentially continuous single-valued mappings such that*

(i) A is countably $\mathscr{D}$-set-contraction with $\mathscr{D}$-function φ_A,
(ii) for each $\lambda \in [0, 1]$, we have $\lambda a \in Ax$, for all $a \in Ax$,
(iii) A is sequentially weakly upper semi-compact ,
(iv) C is weakly compact and $C(S) \subset (I - D)(S)$,
(v) D is nonlinear contraction with $\mathscr{D}$-function ϕ_D,
(vi) $A(S) + B(I - D)^{-1}C(S)$ is a bounded subset of S.

Then, the block operator matrix (3.3.5) *has a fixed point in $S \times S$.*

Theorem 3.3.45 *Let S be a nonempty, bounded, closed, and convex subset of a Banach space X and let A, $B : S \longrightarrow \mathscr{P}(X)$ be two multi-valued mappings with weakly sequentially closed graphs and C, $D : S \longrightarrow X$ are two nonlinear single-valued mappings such that*

(i) A is countably β-nonexpansive and hemi-weakly compact,
(ii) B is weakly compact,
(iii) $C(S) \subset (I - D)(S)$,
(iv) $(I - D)^{-1}C$ is weakly sequentially continuous,
(v) for each $\lambda \in (0, 1)$ and $x \in S$, we have $\lambda A(x) + B(I - D)^{-1}C(x) \in \mathscr{P}_{cl,cv}(S)$.

Then, the block operator matrix (3.3.5) *has a fixed point in $S \times S$.*

Proof Consider $\lambda \in (0, 1)$ and define the mapping $Q_\lambda : S \longrightarrow \mathscr{P}_{cv}(S)$ by putting

$$Q_\lambda(x) = \lambda A(x) + B(I - D)^{-1}C(x).$$

We show that the graph of the multi-valued mapping Q_λ is weakly sequentially closed graph. To prove this, consider $(x_n)_n$ as a weakly convergent sequence in S to a point x and take $y_n \in Q_\lambda(x_n)$ such that $y_n \rightharpoonup y$. Then,

$$y_n = \lambda u_n + v_n,$$

whenever $u_n \in A(x_n)$ and $v_n \in B(I - D)^{-1}C(x_n)$. Due to the fact that $(I - D)^{-1}C(S)$ is bounded and Theorem 1.3.29, we conclude that the sequence $(v_n)_n$ possesses a weakly convergent subsequence, say $\left(v_{n_k}\right)_k$, to a point $v \in B(I - D)^{-1}C(x)$. After simple calculations, we can see that

$$u_{n_k} = \frac{1}{\lambda}(y_{n_k} - v_{n_k}) \rightharpoonup \frac{1}{\lambda}(y - v).$$

This implies that $y - v \in \lambda A(x)$, or equivalently

$$y \in \lambda A(x) + B(I - D)^{-1}C(x).$$

Our next task is to check that Q_λ is countably condensing with respect to β. To this end, let us fix an arbitrarily countably bounded subset M of S with $\beta(M) > 0$. Then, the set $Q_\lambda(M)$ is bounded and

$$\begin{aligned}\beta(Q_\lambda(M)) &\leq \lambda\, \beta(A(M)) + \beta(B(I - D)^{-1}C(M))\\ &\leq \lambda\beta(M)\\ &< \beta(M).\end{aligned}$$

This means that Q_λ is countably condensing with respect to β. Now, if we take K as a weakly compact subset in S, then

$$\beta(Q_\lambda(K)) \leq \lambda\, \beta(A(K)).$$

It is easy to verify that $A(K)$ is relatively sequentially weakly compact, and so the set $Q_\lambda(K)$ is relatively weakly compact. An application of Theorem 1.5.34 yields that, there exists an element x_λ such that $x_\lambda = Q_\lambda(x_\lambda)$. Choose a sequence $(\lambda_n)_n$ contained in the set $(0, 1)$ such that $\lambda_n \to 1$ and take $(x_n)_n$ as the corresponding sequence. Then,

$$x_n \in \lambda_n A(x_n) + B(I - D)^{-1}C(x_n).$$

In other words, there is two sequence $u_n \in A(x_n)$ and $v_n \in B(I - D)^{-1}C(x_n)$ satisfying

$$x_n = \lambda_n u_n + v_n.$$

It follows from assumption (ii), that $(v_n)_n$ possesses a weakly convergent subsequence, say $\left(v_{n_k}\right)_k$, to a point $v \in B(I - D)^{-1}C(x)$. This implies that

$$x_{n_k} - u_{n_k} = (\lambda_{n_k} - 1)u_{n_k} + v_{n_k} \rightharpoonup v. \tag{3.3.15}$$

Since A is hemi-weakly compact, then $(x_{n_k})_k$ has a weakly convergent subsequence to an element x. Going back to the equality (3.3.15) shows that

$$u_{n_k} \rightharpoonup x - v.$$

So, $x \in v + Ax$, or equivalently, we can write $x \in A(x) + B(I - D)^{-1}C(x)$. Q.E.D.

References

1. Banaś, J., Lecko, M.: Fixed points of the product of operators in Banach algebras. PanAmer. Math. J. **12**(2), 101–109 (2002)
2. Ben Amar, A., Boumiza, M., O'Regan, D.: Hybrid fixed point theorems for multi-valued mappings in Banach algebras under a weak topology setting. J. Fixed Point Theory Appl. **18**, 327–350 (2016)
3. Ben Amar, A., Chouayekh, S., Jeribi, A.: New fixed point theorems in Banach algebras under weak topology features and applications to nonlinear integral equations. J. Funct. Anal. **259**, 2215–2237 (2010)
4. Ben Amar, A., Chouayekh, S., Jeribi, A.: Fixed point theory in a new class of Banach algebras and application. Afr. Mat. **24**, 705–724 (2013)
5. Jeribi, A., Krichen, B.: Nonlinear Functional Analysis in Banach Spaces and Banach Algebras Fixed Point Theory under Weak Topology for Nonlinear Operators and Block Operator Matrices with Applications. Monographs and Research Notes in Mathematics. CRC Press, Boca Raton, FL (2016)
6. Jeribi, A., Krichen, B., Mefteh, B.: Fixed point theory in WC-Banach algebras. Turkish J. Math. **40**, 283–291 (2016)
7. Jeribi, A., Krichen, B., Mefteh, B.: Existence solutions of a two-dimensional boundary value problem for a system of nonlinear equations arising in growing cell populations. J. Biol. Dyn. **7**, 218–232 (2013)
8. Kaddachi, N.: Existence results of ordinary differential inclusions in Banach algebra under weak topology. J. Phys. Math. **9** (2018)
9. Kaddachi, N.: Generalized form of fixed point theorems in Banach algebras under weak topology with an application. Filomat **33**, 4281–4296 (2019)
10. Kaddachi, N., Jeribi, A., Krichen, B.: Fixed point theorems of block operator matrices on Banach algebras and an application to functional integral equations. Math. Methods Appl. Sci. **36**, 659–673 (2013)
11. Khchine, A., Maniar, L., Taoudi, M.A.: Krasnosel'skii-type fixed point theorems for convex-power condensing mappings in locally convex spaces. J. Fixed Point Theory Appl. **19**, 2985–3012 (2017)
12. Dhage, B.C.: On a fixed point theorem in Banach algebras with applications. Appl. Math. Lett. **18**, 273–280 (2005)
13. Dhage, B.C.: A fixed point theorem and applications to nonlinear integral equations. In: Proceeding of International Symposium in Nonlinear Analysis, Applications and Bio-Mathematics, Waltair, India, pp. 53–59 (1987)
14. Dhage, B.C., O'Regan, D.: A fixed point theorem in Banach algebras with applications to functional integral equations. Funct. Differ. Equ. **7**, 259–267 (2000)
15. Burton, T.A.: A fixed point theorem of Krasnoselskii. Appl. Math. Lett. **11**, 85–88 (1998)
16. Jeribi, A., Kaddachi, N., Krichen, B.: Existence results for a system of nonlinear functional integral equations in Banach algebras under weak topology. Fixed Point Theory **18**, 247–268 (2017)
17. Jeribi, A., Kaddachi, N., Laouar, Z.: Fixed point theorems for weakly asymptotically regular mappings in Banach spaces with an application. Numer. Funct. Anal. Optim. **43**(1), 68–87 (2022)
18. Edelstein, M., O'Brien, R.C.: nonexpansive mappings, asymptotic regularity and successive approximations. J. London Math. Soc. **17**, 547–554 (1978)
19. Górnicki, J.: Remarks on asymptotic regularity and fixed points. J. Fixed Point Theory Appl. **21**(1), Paper No. 29, 20 (2019)
20. Ben Amar, A., O'Regan, D., Touati, A.: Fixed point theorems for the sum of (ws)-compact and asymptotically Φ-nonexpansive mappings. J. Fixed Point Theory Appl. **18**, 771–784 (2016)
21. Ben Amar, A., Derbel, S., O'Regan, D., Xiang, T.: Fixed point theory for countably weakly condensing maps and multimaps in non-separable Banach spaces. J. Fixed Point Theory Appl. **21**, 8–25 (2019)

22. Liu, Y., Li, Z.: Schaefer type theorem and periodic solutions of evolution equations. J. Math. Anal. Appl. **316**, 237–255 (2006)
23. Xiang, T., Yuan, R.: Critical type of Krasnosel'skii fixed point theorem. Proc. Amer. Math. Soc. **139**, 1033–1044 (2011)
24. Jeribi, A., Kaddachi, N., Krichen, B.: Fixed point theorems for multi-valued operator matrix under weak topology with an application. Bull. Malays. Math. Sci. Soc. **43**, 1047–1067 (2020)
25. Ben Amar, A., Sikorska-Nowak, A.: On some fixed point theorems for 1-set weakly contractive multi-valued mappings with weakly sequentially closed graph. Adv. Pure Math. **1**, 163–169 (2011)
26. Jeribi, A., Kaddachi, N., Krichen, B.: Existence results for a coupled system of perturbed functional differential inclusions in Banach algebras. Bull. Malays. Math. Sci. Soc. **41**, 893–918 (2018)
27. Ben Amara, K., Kaddachi, N., Jeribi, A.: New fixed point theorems for countably condensing maps with an application to functional integral inclusions. Math. Slovaca **71**, 1487–1510 (2021)
28. Alsheikh, A.A., Ben Amar, A., O'Regan, D.: Fixed point theorems for the sum of two multi-valued mappings and an application to an integral inclusion. Bull. Malays. Math. Sci. Soc. **40**(3), 1307–1320 (2017)

Part II
Applications Arising in Some Physical and Biological Problems

Chapter 4
Nonlinear Integral and Differential Inclusions

This chapter deals with the existence of solution to nonlinear integral inclusions and functional integral equation carried in Banach space and in suitable Banach algebras (Banach algebra, Banach algebra satisfying condition $(\mathscr{P})$, RWC-Banach algebra and generalized Banach algebra). This class of nonlinear inclusions, as special cases, includes some class of nonlinear functional differential inclusions which studied in [1–3]. The studies of nonlinear integral equations and differential equations in Banach algebras have been discussed for a long time in the literature. This study was performed via fixed point techniques (Schauder, Darbo,...), see for example [4–11] and the references therein. The first one deals with some open problems carried from [5, 12–14] concerning the existence of solutions of Volterra integral equation on a bounded interval. More precisely, we discuss the existence of solutions of a nonlinear one-dimensional functional integral inclusions arising in quantic physics on general Banach algebra. The second one, introduced by Perov [15], for solving many problems in diverse areas, such as nonlinear analysis, system of ordinary differential and semi-linear differential equations, see also [16–19]. The third one, introduced by Ben Amar et al. [12] and by Banaś and Taoudi [20], discusses the solvability of an integral equation arising in radiative transfer [21] and the Hammerstein integral equation [21–24]. These three types can be transformed into a fixed point problem.

4.1 Generalization of Volterra Integral Inclusion

Let $J = [0, 1]$, X be a Banach algebra satisfying condition $(\mathscr{P})$ and let $E := C(J, X)$, the space of all continuous functions defined of J into X, endowed with norm

$$\|f\|_\infty = \sup_{t \in J} \|f(t)\|, \quad \text{for all } f \in E.$$

A. Jeribi and N. Kaddachi, *Topological Fixed-Point Theory in Suitable Banach Algebras with Applications*, Infosys Science Foundation Series,
https://doi.org/10.1007/978-981-96-5542-7_4

Let $Q : J \times X \longrightarrow \mathscr{P}(X)$ be a multi-valued mapping. Recall that for an arbitrary fixed element $x \in E$, we associate the integral of Q by putting

$$\int_0^t Q(s, x(s))\, ds = \left\{ \int_0^t v(s)\, ds,\ v \text{ is Pettis-integrable and } v(t) \in Q(t, x(t)) \right\}.$$

The aim of this section is to sketch the existence on E of solutions of the following problem, which is presented in [25], of nonlinear integral inclusion

$$x(t) \in T_1(t, x(t)) \cdot \left(q(t) + \int_0^t k(t, s)\, H(s, x(s))\, ds \right) + T_2(t, x(t)), \quad (4.1.1)$$

where q, k are continuous on J, $T_i : J \times X \longrightarrow X$ and $H : J \times X \longrightarrow \mathscr{P}_{cp,cv}(X)$. Henceforth, we will suppose that the entries of problem (4.1.1) satisfy
(H_1) $k : J \times J \longrightarrow \mathbb{R}$ and $q : J \longrightarrow X$ are continuous.
(H_2) The functions $T_i : J \times X \longrightarrow X$, for $i = 1, 2$, are such that

(i) the partial $x \longrightarrow T_i(t, x)$ is weakly sequentially continuous, for all $t \in J$,

(ii) there is two continuous functions $\vartheta_i : J \longrightarrow [0, \infty)$, $i = 1, 2$, such that

$$\begin{cases} \left\| T_1(t, x(t)) - T_1(t', x(t')) \right\| \leq \vartheta_1(t)\, \|x(t) - x(t')\| + |\vartheta_1(t) - \vartheta_1(t')|\, \|x\|, \text{ and} \\ \left\| T_2(t, x(t)) - T_2(t', x(t')) \right\| \leq |\vartheta_2(t) - \vartheta_2(t')|\, \|x\|_\infty, \end{cases}$$

for all $x \in E$ and $t, t' \in J$,

(iii) there exists a continuous and nondecreasing function $\Psi : [0, +\infty) \longrightarrow [0, +\infty)$ with $\Psi(0) = 0$ such that

$$\beta(T_2(J \times M)) \leq \Psi(\beta(M)),$$

for any nonempty countable and bounded subset M of X.
(H_3) $H : J \times X \longrightarrow \mathscr{P}(X)$ is such that

(i) for each $x \in C(J, X)$, there exists a scalarly measurable function $w : J \longrightarrow X$ satisfying $w(t) \in H(t, x(t))$ a.e. $t \in J$ such that w is Pettis-integrable on J,

(ii) there exists $R > 0$ and there exists a function $h \in L^1(J, \mathbb{R}_+)$ such that

$$\|H(s, x)\|_{\mathscr{P}} \leq h(s)\|x\|, \text{ for all } x \in \overline{\mathscr{B}}_X(R),$$

(iii) there exists a continuous and nondecreasing function $\Phi : [0, +\infty) \longrightarrow [0, +\infty)$ with $\Phi(0) = 0$ such that

$$\beta(H(J \times M(J))) \leq \Phi(\beta(M(J))),$$

for all countably subset M of $\overline{\mathscr{B}}_E(R)$,

(iv) $H(t, \cdot)$ has a weakly sequentially closed graph on $\overline{\mathscr{B}}_X(R)$.

Theorem 4.1.1 *Suppose that the conditions (H_1)-(H_3) are satisfied. Then, problem (4.1.1) has, at least, one solution in $\overline{\mathscr{B}}_E(R)$ provided that*

$$
\begin{cases} \left[\|q\|_\infty + \|k(\cdot,\cdot)\|_\infty \|h\|_{L_1}\ R\right] \|\vartheta_1\|_\infty R + 2\|\vartheta_2\|_\infty R + \|T_2(0, x(0))\| \leq R, \ \ \textit{and} \\ \left(\|q\|_\infty + \|k(\cdot,\cdot)\|_\infty \|h\|_{L^1} R\right) \|\vartheta_1\|_\infty r + 2\, \|\vartheta_1\|_\infty\, R\, \|k(\cdot,\cdot)\|_\infty\, \Phi(r) + \Psi(r) < r, \end{cases} \tag{4.1.2}
$$

for all $0 < r \leq R$.

Proof Put the set S in E by

$$
S = \{x \in E : \ \|x\|_\infty \leq R \text{ and } \|x(t) - x(\tau)\| \leq b(t, \tau), \ t, \tau \in J\},
$$

where

$$
\begin{aligned}
b(t,\tau) = \tfrac{1}{(1-\delta_1\|\vartheta_1\|_\infty)} \Bigg\{ & R\delta_1 |\vartheta_1(t) - \vartheta_1(\tau)| + R\, |\vartheta_2(t) - \vartheta_2(\tau)| \\
& + \delta_2 \Big[\|q(t) - q(\tau)\| + R \sup_{s\in J} |k(t,s) - k(\tau,s))|\ \|h\|_{L^1} \\
& + R \int_t^\tau |k(\tau,s)|\ h(s)\ ds \Big] \Bigg\},
\end{aligned}
$$

with

$$
\begin{cases} \delta_1 = \|q\|_\infty + \|k\|_\infty \|h\|_{L^1} R, \\ \delta_2 = \|\vartheta_1\|_\infty R + \|T_1(0, \theta)\|. \end{cases}
$$

It is obvious that the set S is a nonempty, bounded, convex, closed, and equicontinuous subset in E. Define the operators A, B, and C on S by putting

$$
\begin{cases} (Ax)(t) = T_1(t, x(t)) \\ (Cx)(t) = T_2(t, x(t)) \\ Bx = \left\{ q(t) + \int_0^t k(t,s)\ \xi(s)\ ds, \ \xi \text{ is Pettis-integrable, } \xi(s) \in H(s, x(s)) \right\}. \end{cases}
$$

Problem (4.1.1) can be written in the following way

$$
x(t) \in Ax(t) \cdot Bx(t) + Cx(t).
$$

Thus, on account of Theorem 2.2.9, we need six steps.

Step 1. Let us check that $A, C : S \longrightarrow E$ and $B : S \longrightarrow \mathscr{P}(E)$. It is easy to verify that $Ax, Cx \in C(J, X)$, for all $x \in S$. This is an immediate consequence of assumption (H_2). Taking into account assumption (H_3), and using Proposition 1.4.31, it is shown that, B is well defined. Let x be an arbitrarily fixed element contained in the set S and take an element $y \in Bx$. Then, there exists a Pettis-integrable mapping $\xi : J \longrightarrow X$ with $\xi(s) \in H(s, x(s))$ such that the following equality

$$
y(t) = q(t) + \int_0^t k(t,s)\ \xi(s)\ ds, \quad \text{for all } t \in J,
$$

is satisfied. Choose a sequence $(t_n)_n$ contained in the set $[0, 1]$ such that $t_n \to t$. Then,

$$\|y(t_n) - y(t)\| \le \|q(t_n) - q(t)\| + \left\| \int_0^{t_n} (k(t_n, s) - k(t, s))\, \xi(s)\, ds \right\| \left\| \int_{t_n}^{t} k(t, s)\, \xi(s)\, ds \right\|.$$

It follows from Theorem 1.3.18 that, there exists a function $x^* \in X^*$, that satisfies, $\|x^*\| = 1$ such that

$$\left\| \int_0^{t_n} (k(t_n, s) - k(t, s))\, \xi(s)\, ds \right\| = x^* \left(\int_0^{t_n} (k(t_n, s) - k(t, s))\, \xi(s)\, ds \right)$$

and

$$\left\| \int_{t_n}^{t} k(t, s) \xi(s)\, ds \right\| = x^* \left(\int_{t_n}^{t} k(t, s)\, \xi(s)\, ds \right).$$

Then, in virtue of assumption (H_3), one has

$$\begin{aligned} \|y(t_n) - y(t)\| &\le \|q(t_n) - q(t)\| + \int_0^{t_n} |k(t_n, s) - k(t, s))|\, \|\xi(s)\|\, ds \\ &\quad + \int_{t_n}^{t} |k(t, s)|\, \|\xi(s)\|\, ds \\ &\le \|q(t_n) - q(t)\| + R \left(\int_0^{t_n} |k(t_n, s) - k(t, s))|\, h(s)\, ds \right. \\ &\quad \left. + \int_{t_n}^{t} |k(t, s)|\, h(s)\, ds \right) \\ &\le \|q(t_n) - q(t)\| + \sum_{i=1}^{2} I_i(t_n, t), \end{aligned}$$

where

$$\begin{cases} I_1(t_n, t) = R \displaystyle\int_0^{t_n} |k(t_n, s) - k(t, s))|\, h(s)\, ds, \quad \text{and} \\ I_2(t_n, t) = R \displaystyle\int_{t_n}^{t} |k(t, s)|\, h(s)\, ds. \end{cases}$$

Due to the fact that $h \in L^1(J, X)$, and dominated convergence theorem, we infer that

$$I_1(t_n, t) \to 0 \text{ as } t_n \to t.$$

In the similar way, we can show that

$$\lim_{n \to \infty} I_2(t_n, t) = 0,$$

since $k \in L^{\infty}(J \times J)$. After simple calculations, we can verify that

$$y(t_n) \to y(t).$$

This means that $B(x)$ is continuous.

Step 2. A is countably $\mathcal{D}$-set-Lipschitzian. It is not hard to verify that, A is Lipschitzian with a Lipschitz constant $\|\vartheta_1\|_{\infty}$. Taking into account the weak sequential continuity of the function $T_1(t, \cdot)$ and using Lemma 1.5.9, we get A is countably $\mathcal{D}$-set-Lipschitzian with $\mathcal{D}$-function Φ_A, where

$$\Phi_A(r) = \|\vartheta_1\|_{\infty}\, r.$$

Step 3. B is countably $\mathcal{D}$-set-Lipschitzian and it possess a weakly sequentially closed graph. Let $(x_n)_n$ be a sequence contained in the set S, which is weakly convergent to an element $x \in S$ and take a weakly convergent sequence $y_n \in B(x_n)$ to a point y. Then, there exists a sequence of Pettis-integrable functions $(\xi_n)_n$ with $\xi_n(s) \in H(s, x_n(s))$ such that

$$y_n(t) = q(t) + \int_0^t k(t,s)\, \xi_n(s)\, ds.$$

Thus, on account of Theorem 1.3.50, we have

$$x_n(s) \rightharpoonup x(s), \text{ for any } s \in [0, t].$$

Based on assumption (H_3), we get

$$\begin{aligned}\beta\left(H\left([0,t] \times \{x_n(s),\ n \in \mathbb{N}\}\right)\right) &\leq \beta\left(H\left(J \times \{x_n(s),\ n \in \mathbb{N}\}\right)\right)\\ &\leq \Phi\left(\beta\left(\{x_n(s),\ n \in \mathbb{N}\}\right)\right).\end{aligned}$$

This implies that

$$\beta\left(H\left([0,t] \times \{x_n(s),\ n \in \mathbb{N}\}\right)\right) = 0. \tag{4.1.3}$$

Going back to equality (4.1.3) and using the following inclusion

$$\{\xi_n(s),\ n \in \mathbb{N}\} \subset H\left([0,t] \times \{x_n(s),\ n \in \mathbb{N}\}\right), \text{ for all } s \in [0, t],$$

we get the set $\{\xi_n(s),\ n \in \mathbb{N}\}$ is relatively weakly compact. Then, on account of Theorem 1.3.29, there exists a weakly convergent subsequence, say $(\xi_{n_k}(s))_k$, of $(\xi_n(s))_n$ to a function $\xi(s)$. Due to the fact that, the partial $H(t, \cdot)$ possess a weakly sequentially closed graph, for all $t \in J$, we infer that $\xi(s) \in H(s, x(s))$. Based on assumption (H_3), it is shown that

$$\|\xi_n(s)\| \leq h(s)\, R, \text{ for all } s \in J.$$

This yields that ξ is Pettis-integrable, in view of dominated convergence theorem and, we have

$$y(t) = q(t) + \int_0^t k(t, s)\, \xi(s)\, ds,\ t \in J.$$

Our next task is to show that B is countably $\mathscr{D}$-set-Lipschitzian. In order to prove this fact, let us take a countable subset M of S with $\beta(M) > 0$. Making use of assumption (H_3), together with Lemma 1.4.30, enables us to have that

$$\begin{aligned}
\beta(B(M)(t)) &\leq \beta\left(\bigcup\left\{\int_0^t k(t, s)\, H(s, x(s))\, ds,\quad x \in M\right\}\right)\\
&\leq \beta\left(\int_0^t k(t, s)\, H(s, M(s))\, ds\right)\\
&\leq \|k(\cdot, \cdot)\|_\infty \beta\left(\overline{co}(H(J, M(J)))\right)\\
&\leq \|k(\cdot, \cdot)\|_\infty \Phi(\beta(M(J))).
\end{aligned}$$

In the similar way as in Step 1, we can see that the set $B(S)$ is equi-continuous. Invoking Lemma 1.5.41, and using the equi-continuity of S, we obtain

$$\beta(B(M)) \leq \|k(\cdot, \cdot)\|_\infty \Phi(\beta(M)).$$

This means, in particular, that the operator B is countably $\mathscr{D}$-set-Lipschitzian with $\mathscr{D}$-function Φ_B, where

$$\Phi_B(r) = \|k(\cdot, \cdot)\|_\infty \Phi(r).$$

Step 4. We show that C is countably $\mathscr{D}$-set-contraction and it is weakly sequentially continuous. The claim regarding the weak sequential continuity is immediate, in view of assumption (H_2). Let us take a countable subset M of S with $\beta(M) > 0$. Since Ψ is nondecreasing, it follows that

$$\begin{aligned}
\beta(C(M)(t)) &\leq \beta\left(T_2(t \times M(t))\right)\\
&\leq \beta((T_2(J \times M(t)))\\
&\leq \Psi(\beta(M(t))).
\end{aligned}$$

Then, on account of Theorem 1.4.30, we can prove that

$$\beta(C(M)(t)) \leq \Psi(\beta(M)).$$

This implies that the set $C(S)$ is equi-continuous, in light of assumption (H_2). Making use the fact that the set $C(S)$ is bounded, together with Theorem 1.5.41, enables us to have

$$\beta(C(M)) \leq \Psi(\beta(M)).$$

This show that C is a countably $\mathscr{D}$-set-Lipschitzian mapping with $\mathscr{D}$-function Ψ.
Step 5. For all $x \in S$, we will prove that

$$Ax \cdot Bx + Cx \in \mathscr{P}_{cv}(S).$$

Let us take an element $y \in Ax \cdot Bx + Cx$. Then, there exists a Pettis-integrable mapping $\zeta : J \longrightarrow X$ with $\zeta(s) \in H(s, x(s))$ such that

$$y(t) = T_1(t, x(t)) \left(q(t) + \int_0^t k(t, s)\zeta(s)\, ds \right) + T_2(t, x(t)). \tag{4.1.4}$$

This implies that

$$\begin{aligned}
\|y(t)\| &\leq \|T_1(t, x(t))\| \left\| q(t) + \int_0^t k(t, s)\zeta(s)\, ds \right\| + \|T_2(t, x(t))\| \\
&\leq \|T_1(t, x(t))\| \left(\|q(t)\| + \int_0^t |k(t, s)|\, \|\zeta(s)\|\, ds \right) + \|T_2(t, x(t))\| \\
&\leq \|\vartheta_1\|_\infty R \left(\|q\|_\infty + \|k(t, \cdot)\|_\infty \|h\|_{L_1} R \right) + 2\|\vartheta_2\|_\infty R + \|T_2(0, x(0))\|.
\end{aligned}$$

Going back the inequality (4.1.2), we get $\|y\|_\infty \leq R$. Let x be an arbitrarily fixed element of the set S. Going back to equality (4.1.4), we can prove that

$$\begin{aligned}
\|y(t) - y(\tau)\| &\leq \Bigg\| A(x(t)) \left(q(t) + \int_0^t k(t, s)\zeta(s)\, ds \right) \\
&\qquad - A(x(\tau)) \left(q(\tau) + \int_0^\tau k(\tau, s)\zeta(s)\, ds \right) \Bigg\| \\
&\qquad + \|C(x(t)) - C(x(\tau))\| \\
&\leq \|A(x(t)) - A(x(\tau))\| \left\| \left(q(t) + \int_0^t k(t, s)\zeta(s)\, ds \right) \right\| \\
&\qquad + \|A(x(\tau))\| \Bigg\| \left(q(t) + \int_0^t k(t, s)\zeta(s)\, ds \right) \\
&\qquad - \left(q(\tau) + \int_0^\tau k(\tau, s)\zeta(s)\, ds \right) \Bigg\| \\
&\qquad + \|C(x(t)) - C(x(\tau))\| \\
&\leq \sum_{i=1}^{3} J_i(t, \tau),
\end{aligned}$$

whenever $t, \tau \in J$, where

$$\begin{cases} J_1(t,\tau) = \|A(x(t)) - A(x(\tau))\| \left\| \left(q(t) + \int_0^t k(t,s)\zeta(s)\, ds \right) \right\| \\ J_2(t,\tau) = \|A(x(\tau))\| \left\| \left(q(t) + \int_0^t k(t,s)\zeta(s)\, ds \right) - \left(q(\tau) + \int_0^\tau k(\tau,s)\zeta(s)\, ds \right) \right\| \\ J_3(t,\tau) = \|C(x(t)) - C(x(\tau))\|. \end{cases}$$

After simple calculations, we can see that

$$\begin{aligned} J_1(t,\tau) &= \|A(x(t)) - A(x(\tau))\| \left\| \left(q(t) + \int_0^t k(t,s)\zeta(s)\, ds \right) \right\| \\ &\le \delta_1 \; (\vartheta_1(t)\, \|x(t) - x(\tau)\| + |\vartheta_1(t) - \vartheta_1(\tau)|\, \|x(\tau)\|) \\ &\le \delta_1 \; [\|\vartheta_1\|_\infty\, b(t,\tau) + |\vartheta_1(t) - \vartheta_1(\tau)|\, R]. \end{aligned}$$

In the similar way as in Step 1, we can show that

$$\begin{aligned} J_2(t,\tau) &= \|A(x(\tau))\| \left\| \left(q(t) + \int_0^t k(t,s)\zeta(s)\, ds \right) - \left(q(\tau) + \int_0^\tau k(\tau,s)\zeta(s)\, ds \right) \right\| \\ &\le \delta_2 \left[\|q(t) - q(\tau)\| + R \sup_{s\in J} |k(t,s) - k(\tau,s))|\, \|h\|_{L^1} + R \int_t^\tau |k(\tau,s)|\, h(s)\, ds \right]. \end{aligned}$$

Based on assumption (H_2), we get

$$\begin{aligned} J_3(t,\tau) &= \|T_2(t,x(t)) - T_2(\tau,x(\tau))\| \\ &\le R|\vartheta_2(t) - \vartheta_2(\tau)|. \end{aligned}$$

Then,

$$\begin{aligned} \|y(t) - y(\tau)\| \le \delta_1 \; &[\|\vartheta_1\|_\infty\, b(t,\tau) + |\vartheta_1(t) - \vartheta_1(\tau) \;\; R] + R|\vartheta_2(t) - \vartheta_2(\tau)| \\ &+\delta_2 \left[\|q(t) - q(\tau)\| + R \sup_{s\in J} |k(t,s) - k(\tau,s))|\, \|h\|_{L^1} \right. \\ &\left. + R \int_t^\tau |k(\tau,s)|\, h(s)\, ds \right]. \end{aligned}$$

Going back to the following equality

$$\begin{aligned} (1 - \delta_1 \|\vartheta_1\|_\infty)\, b(t,\tau) = \delta_1 \; &|\vartheta_1(t) - \vartheta_1(\tau)|\, R + R\, |\vartheta_2(t) - \vartheta_2(\tau)| + \\ &\delta_2 \left[\|q(t) - q(\tau)\| + R \sup_{s\in J} |k(t,s) - k(\tau,s))|\, \|h\|_{L^1} \right. \\ &\left. +R \int_t^\tau |k(\tau,s)|\, h(s)\, ds \right], \end{aligned}$$

we obtain

$$\|y(t) - y(\tau)\| \leq b(t, \tau).$$

Thus, $Ax \cdot Bx + Cx \subset S$, for all $x \in S$. Let x be an arbitrarily fixed element of S and take two elements $y_1, y_2 \in Ax \cdot Bx + Cx$. Then, there exist two Pettis-integrable functions $\zeta_1, \zeta_2 : J \longrightarrow X$ with $\zeta_1(s), \zeta_2(s) \in H(s, x(s))$ such that

$$\begin{cases} y_1(t) = T_1(t, x(t)) \left(q(t) + \displaystyle\int_0^t k(t, s)\zeta_1(s)\, ds \right) + T_2(t, x(t)), \text{ and} \\ y_2(t) = T_1(t, x(t)) \left(q(t) + \displaystyle\int_0^t k(t, s)\zeta_2(s)\, ds \right) + T_2(t, x(t)). \end{cases}$$

Then, for all $\lambda \in (0, 1)$, we obtain

$$\begin{aligned} \lambda y_1(t) + (1 - \lambda) y_2(t) &= \lambda T_1(t, x(t)) \left(q(t) + \int_0^t k(t, s)\zeta_1(s)\, ds \right) \\ &\quad + (1 - \lambda) T_1(t, x(t)) \left(q(t) + \int_0^t k(t, s)\zeta_2(s)\, ds \right) + T_2(t, x(t)) \\ &= T_1(t, x(t))\, q(t) + T_1(t, x(t)) \int_0^t k(t, s)(\lambda\zeta_1(s) + (1 - \lambda)\zeta_2)\, ds \\ &\quad + T_2(t, x(t)). \end{aligned}$$

This implies that

$$\lambda\zeta_1(s) + (1 - \lambda)\zeta_2(s) \in H(s, x(s)).$$

It is easily seen that $\lambda\zeta_1(\cdot) + (1 - \lambda)\zeta_2(\cdot)$ is Pettis-integrable, and so

$$\lambda y_1 + (1 - \lambda) y_2 \in Ax \cdot Bx + Cx.$$

Let $x \in S$ be fixed. We show that

$$Ax \cdot Bx + Cx \in \mathscr{P}_{cl}(S).$$

Indeed, consider $(y_n)_n$ as a convergent sequence contained in the set $Ax \cdot Bx + Cx$ to a point $y \in E$. Let us denote

$$y_n(t) = T_1(t, x(t)) \left(q(t) + \int_0^t k(t, s)\xi_n(s)\, ds \right) + T_2(t, x(t)),$$

for some sequence of Pettis-integrable mapping $\xi_n : J \longrightarrow X$ with $\xi_n(s) \in H(s, x(s))$. Using assumption (H_3), we get $Bx \in \mathscr{P}_{rcp}(S)$, for all $x \in S$. Due to the fact that

$$\{\xi_n(s),\ n \in \mathbb{N}\} \subset H(s, x(s))$$

and using Theorem 1.3.29, we infer that there exists a convergent subsequence, say $(\xi_{n_k}(s))_k$, to a point $\xi(s) \in H(s, x(s))$. This yields, in view of the dominated convergence theorem, that ξ is Pettis-integrable and

$$\int_0^t k(t,s)\xi_n(s)\,ds \to \int_0^t k(t,s)\xi(s)\,ds.$$

This imply that

$$y_n(t) \to T_1(t, x(t)) \left(q(t) + \int_0^t k(t,s)\xi(s)\,ds \right) + T_2(t, x(t)).$$

By uniqueness of limit, we deduce that

$$y \in Ax \cdot Bx + Cx, \quad \text{for all } x \in S.$$

In other words, $Ax \cdot Bx + Cx \in \mathscr{P}_{cl}(X)$.

Step 6. For any $0 < r < \|S\|$, we prove that

$$\|B(S)\|\ \Phi_A(r) + \|A(S)\|\ \Phi_B(r) + \Phi_A(r)\ \Phi_B(r) + \Phi_C(r) < r.$$

At first, let us observe that $A(S)$ and $B(S)$ are two bounded subsets with bounds $\|\vartheta_1\|_\infty\ \|S\|$ and $\|q\|_\infty + \|k(\cdot,\cdot)\|_\infty\ \|h\|_{L^1}\|S\|$, respectively. From the above discussion, we can choose

$$\begin{cases} \Phi_A(r) = \|\vartheta_1\|_\infty\ r \\ \Phi_B(r) = \|k(\cdot,\cdot)\|_\infty\ \Phi(r) \\ \Phi_C(r) = \Psi(r). \end{cases}$$

Going back to estimate (4.1.2) yields that

$$\begin{aligned} &[\|q\|_\infty + \|k(\cdot,\cdot)\|_\infty\ \|h\|_{L^1}\|S\|]\ \|\ \vartheta_1\|_\infty\ r + \|\vartheta_1\|_\infty\ \|S\|\ \|k(\cdot,\cdot)\|_\infty\ \Phi(r) \\ &\quad + \|\vartheta_1\|_\infty\ \|k(\cdot,\cdot)\|_\infty\ r\ \Phi(r) + \Psi(r) < r, \end{aligned}$$

whenever $0 < r \le \|S\|$. Q.E.D.

Besides these progresses and as application, we study the existence results for the following functional differential problem (FDI)

$$\left(\frac{x(t) - T_2(t, x(t))}{T_1(t, x(t))} \right)' \in k(t)\ H(t, x(t)),\ t \in [0, 1], \tag{4.1.5}$$

satisfying the initial condition

$$x(0) = x_0 \in X. \tag{4.1.6}$$

By a solution of the operators equations (4.1.5)–(4.1.6), we understand a absolutely continuous function on J into X such that the above inclusions (4.1.5)–(4.1.6) are satisfied on J.

Theorem 4.1.2 *Under the assumptions* (H_2)–(H_3), *the (FDI) (4.1.5)–(4.1.6) has a solution x in* $\overline{\mathscr{B}_E}(R)$ *provided that, there exists* $R > 0$ *such that the estimates (4.1.2) is satisfied with*

$$\|q\|_\infty = \frac{\|x_0 - T_2(0, x_0)\|}{\|T_1(0, x_0)\|}.$$

Proof Observe that problem (4.1.5)–(4.1.6) can be written as the following fixed point

$$\begin{cases} x(t) \in T_1(t, x(t)) \cdot \left[\dfrac{x_0 - T_2(0, x_0)}{T_1(0, x_0)} + \displaystyle\int_0^t k(s)\, H(s, x(s))\, ds \right] + T_2(t, x(t)), \\ x(0) = x_0. \end{cases}$$

Let us notice that when we apply Theorem 4.1.1 by putting

$$q(t) = \frac{x_0 - T_2(0, x_0)}{T_1(0, x_0)}$$

and $k(t, s) = k(s)$, for all $t, s \in J = [0, 1]$, we reach the desired conclusion. Q.E.D.

Remark 4.1.3 Observe that problem (4.1.5)–(4.1.6) contains many special types of functional integral inclusions. For example, if we take $T_1(t, x) = 1_X$ and $q(t) = 0$, then Theorem 4.1.2 reduces to the existence of solutions of the following perturbed Volterra integral inclusion

$$x(t) \in \int_0^t k(t, s)\, H(s, x(s))\, ds + T(t, x(t)),\ t \in [0, 1].$$

4.2 Existence Theory in Generalized Banach Algebra

Assume that X is a reflexive Banach space and take $E = C(J, X)$, where $J = [0, \mathscr{T}]$, $\mathscr{T} \in \mathbb{R}_+^*$. The aim of this section is to apply Theorem 2.3.5, in order to discuss on $E \times E$, the existence results for the following two-dimensional problem, which is presented in [26], of quadratic integral equations of fractional order

$$\begin{aligned} x_i(t) = {} & f_i(t, x_1(t), x_2(t)) \cdot \int_0^t \frac{(t-s)^{\gamma_i - 1}}{\Gamma(\gamma_i)} g_i(s, x_1(s), x_2(s))\, ds \\ & + \sum_{k=1}^m I^{\sigma_i^k} h_i^k(t, x_1(t), x_2(t)) \end{aligned} \tag{4.2.1}$$

for $i = 1, 2$ and for all $t, \gamma_i \in (0, 1)$, where $I^{\sigma_i^k}$ denote the fractional Pettis-integral of order $\sigma_i^k > 0$ and f_i, g_i, h_i^k are weakly sequentially continuous mappings on $J \times X \times X$ with respect to the second variable. When dealing with this problem, some technical difficulties arise. So, we need the following assumptions.

(H_4) For $i = 1, 2$, the mapping $f_i : J \times X \times X \longrightarrow X$ is such that
 (i) the partial $x \longrightarrow f_1(t, x, y)$ is regular on X,
 (ii) the partial $y \longrightarrow f_2(t, x, y)$ is regular on X,
 (iii) the partial $t \longrightarrow f_i(t, x, y)$ is continuous,
 (iv) the partial $(x, y) \longrightarrow f_i(t, x, y)$ is weakly sequentially continuous,
 (v) there exist $a_{i1}, a_{i2} \in \mathbb{R}_+^*$, for $i = 1, 2$ such that

$$\| f_i(t, x, y) - f_i(t, \tilde{x}, \tilde{y})\| \leq a_{i1} \|x - \tilde{x}\| + a_{i2}\|y - \tilde{y}\|,$$

for all $(x, y), (\tilde{x}, \tilde{y}) \in X \times X$.

(H_5) The mapping $g_i : J \times X \times X \longrightarrow X, i = 1, 2$ is such that
 (i) the partial $t \longrightarrow g_i(t, x, y)$ is Riemann integrable,
 (ii) the partial $(x, y) \longrightarrow g_i(t, x, y)$ is weakly sequentially continuous.

(H_6) For $i = 1, 2$, the mapping $h_i^k : J \times X \times X \longrightarrow X$, $k = 1, \ldots, m$, is such that
 (i) the partial $t \longrightarrow h_i^k(t, x, y)$ is Riemann integrable,
 (ii) the partial $(x, y) \longrightarrow h_i^k(t, x, y)$ is weakly sequentially continuous,
 (iii) there exists nonnegative real numbers b_{i1}^k and b_{i2}^k such that

$$\|h_i^k(t, x, y) - h_i^k(t, \tilde{x}, \tilde{y})\| \leq b_{i1}^k \|x - \tilde{x}\| + b_{i2}^k\|y - \tilde{y}\|,$$

for all $(x, y), (\tilde{x}, \tilde{y}) \in X \times X$.

Theorem 4.2.1 *Under the assumptions* (H_4)-(H_6), *the coupled system (4.2.1) has, at least, one solution provided that there exist two numbers* r_0, $P \in \mathbb{R}_+^*$ *such that the following estimates*

$$\begin{cases} \|g_i(t, s, x, y)\| \leq P \quad \text{with } \|x\| \leq r_0 \text{ and } \|y\| \leq r_0 \\ \rho \left[P \left(\frac{\mathcal{T}^{\gamma_1}}{\Gamma(\gamma_1+1)} + \frac{\mathcal{T}^{\gamma_2}}{\Gamma(\gamma_2+1)} \right) + \sum_{k=1}^{m} \left(\frac{\mathcal{T}^{\sigma_1^k}}{\Gamma(\sigma_1^k + 1)} + \frac{\mathcal{T}^{\sigma_2^k}}{\Gamma(\sigma_2^k + 1)} \right) \right] < 1, \\ M_A \text{ and } M_C \text{ are two matrices converging to zero,} \end{cases} \quad (4.2.2)$$

are satisfied, where

$$\rho = \max\{a_{11}, a_{12}, a_{21}, a_{22}, b_{11}^k, b_{12}^k, b_{21}^K, b_{21}^k, b_{22}^k, \quad k = 1, \ldots, m\},$$

$$M_A = \begin{pmatrix} a_{11} & a_{12} \\ a_{21} & a_{22} \end{pmatrix} \quad \text{and} \quad M_C = \begin{pmatrix} \sum_{k=1}^{m} \frac{\mathcal{T}^{\sigma_1^k}}{\Gamma(\sigma_1^k + 1)} b_{11}^k & \sum_{k=1}^{m} \frac{\mathcal{T}^{\sigma_1^k}}{\Gamma(\sigma_1^k + 1)} b_{12}^k \\ \sum_{k=1}^{m} \frac{\mathcal{T}^{\sigma_2^k}}{\Gamma(\sigma_2^k + 1)} b_{21}^k & \sum_{k=1}^{m} \frac{\mathcal{T}^{\sigma_2^k}}{\Gamma(\sigma_2^k + 1)} b_{22}^k \end{pmatrix}.$$

Proof Let us consider the mapping A_i, B_i and C_i, $i = 1, 2$, on the space $E \times E$ defined by

$$\begin{cases} A_i(x, y)(t) = f_i(t, x(t), y(t)), \\ B_i(x, y)(t) = \displaystyle\int_0^t \frac{(t-s)^{\gamma_1 - 1}}{\Gamma(\gamma_1)} g_i(s, x(s), y(s))\, ds, \\ C_i(x, y)(t) = \displaystyle\sum_{k=1}^{m} I^{\sigma_1^i} h_i^k(t, x(t), y(t)). \end{cases}$$

Then, problem (4.2.1) is equivalent to the following equation

$$(x, y) = (A_1(x, y), A_2(x, y))(B_1(x, y), B_2(x, y)) + (C_1(x, y), C_2(x, y)).$$

We will prove that, the operators A_i, B_i and C_i, $i = 1, 2$ satisfy all the conditions of Theorem 2.3.5. For this purpose, define the set S of $E \times E$ by

$$S = \left\{ (x, y) \in E \times E,\ \ \|(x, y)\| = \begin{pmatrix} \|x\|_\infty \\ \|y\|_\infty \end{pmatrix} \leq \begin{pmatrix} r_0 \\ r_0 \end{pmatrix} \right\},$$

where

$$r_0 \geq \frac{F_0 P\left(\frac{T^{\gamma_1}}{\Gamma(\gamma_1+1)} + \frac{\mathcal{T}^{\gamma_2}}{\Gamma(\gamma_2+1)}\right) + R_0 \displaystyle\sum_{k=1}^{m} \left(\frac{\mathcal{T}^{\sigma_1^k}}{\Gamma(\sigma_1^k + 1)} + \frac{\mathcal{T}^{\sigma_2^k}}{\Gamma(\sigma_2^k + 1)} \right)}{1 - \rho\left[P\left(\frac{\mathcal{T}^{\gamma_1}}{\Gamma(\gamma_1+1)} + \frac{\mathcal{T}^{\gamma_2}}{\Gamma(\gamma_2+1)}\right) + \displaystyle\sum_{k=1}^{m} \left(\frac{\mathcal{T}^{\sigma_1^k}}{\Gamma(\sigma_1^k + 1)} + \frac{\mathcal{T}^{\sigma_2^k}}{\Gamma(\sigma_2^k + 1)} \right) \right]},$$

here

$$\begin{cases} F_0 = \max\ \{\|f_1(\cdot, 0, 0)\|, \|f_2(\cdot, 0, 0)\|\} \\ R_0 = \max\ \left\{\|h_1^k(\cdot, 0, 0)\|, \|h_2^k(\cdot, 0, 0)\|,\ \ k = 1, \ldots, m\right\}. \end{cases}$$

It is obvious that, the set S is nonempty, bounded, convex and closed of $E \times E$. We start by proving the continuity of the mapping $A_i(x, y)$, for $i = 1, 2$, for all $(x, y) \in E \times E$. Consider $(t_n)_n$ as a converging sequence contained in the set J to an element t. Based on assumption (H_4), we get

$$\begin{aligned} \|A_i(x, y)(t_n) - A_i(x, y)(t)\| &= \|f_i(t_n, x(t_n), y(t_n)) - f_i(t, x(t), y(t))\| \\ &\leq a_{i1}\|x(t_n) - x(t)\| + a_{i2}\|y(t_n) - y(t)\| \\ &\quad + \|f_i(t_n, x(t), y(t) - f_i(t, x(t), y(t)\|. \end{aligned}$$

Since the partial $t \longrightarrow f_i(t, x, y)$ is continuous, for all $x,\ y \in X$, it follows that

$$A_i(x, y) \in C(J, X), \quad \text{for each } x,\ y \in E.$$

Let us assume that $((x_n, y_n))_n$ is a sequence contained in the set $E \times E$, which is weakly convergent to a point (x, y). It is easy to verify that, the sequence $((x_n, y_n))_n$ is bounded. Then, on account of Theorem 1.3.50, we get

$$(x_n(t), y_n(t)) \rightharpoonup (x(t), y(t)).$$

Thus, in virtue of assumption (H_4), we obtain

$$A_i(x_n, y_n)(t) \rightharpoonup A_i(x, y)(t).$$

As a result, A is weakly sequentially continuous. Moreover, taking into account assumptions on maps f_i and h_i and using Theorem 1.3.50, one obtains the weak sequential continuity of maps B and C. Now, we show that the mappings A and C are contractive. The claim regarding A is clear, by using assumption (H_4). Let $(x, y), (\tilde{x}, \tilde{y})$ be two arbitrarily fixed element of the set $E \times E$ and take $t \in [0, 1]$. Then, for each $i = 1, 2$, we have

$$\begin{aligned}\|C_i(x, y)(t) - C_i(\tilde{x}, \tilde{y})(t)\| &= \left\| \sum_{k=1}^{m} I^{\sigma_i^k} h_i^k(t, x(t), y(t)) - \sum_{k=1}^{m} I^{\sigma_i^k} h_i^k(t, \tilde{x}(t), \tilde{y}(t)) \right\| \\ &\leq \sum_{k=1}^{m} I^{\sigma_i^k} \|h_i^k(t, x(t), y(t)) - h_i^k(t, \tilde{x}(t), \tilde{y}(t))\| \\ &\leq \sum_{k=1}^{m} I^{\sigma_i^k} \left(b_{i1}^k \|x(t) - \tilde{x}(t)\| + b_{i1}^k \|y(t) - \tilde{y}(t)\| \right) \\ &\leq \sum_{k=1}^{m} \frac{\mathcal{T}^{\sigma_i^k}}{\Gamma(\sigma_i^k + 1)} \left(b_{i1}^k \|x(t) - \tilde{x}(t)\| + b_{i1}^k \|y(t) - \tilde{y}(t)\| \right).\end{aligned}$$

This implies that

$$\|C(x, y) - C(\tilde{x}, \tilde{y})\| \leq M_C \|(x, y) - (\tilde{x}, \tilde{y})\|,$$

where

$$M_C = \begin{pmatrix} \sum_{k=1}^{m} \frac{\mathcal{T}^{\sigma_1^k}}{\Gamma(\sigma_1^k + 1)} b_{11}^k & \sum_{k=1}^{m} \frac{\mathcal{T}^{\sigma_1^k}}{\Gamma(\sigma_1^k + 1)} b_{12}^k \\ \sum_{k=1}^{m} \frac{\mathcal{T}^{\sigma_2^k}}{\Gamma(\sigma_2^k + 1)} b_{21}^k & \sum_{k=1}^{m} \frac{\mathcal{T}^{\sigma_2^k}}{\Gamma(\sigma_2^k + 1)} b_{22}^k \end{pmatrix}.$$

Our next task is to show that, the set $B(S)$ is relatively weakly compact. Let us fix an element $\varepsilon > 0$ and take $t, t' \in J$ with $|t' - t| < \varepsilon$. It follows from Theorem 1.3.18, that there exists a linear function $\phi \in X^*$ with $\|\phi\| = 1$ and

$$\left\| B_i(x, y)(t') - B_i(x, y)(t) \right\| = \phi\left(B_i(x, y)(t') - B_i(x, y)(t) \right), \quad \text{for } i = 1, 2.$$

Based on the first inequality of (4.2.2), it is shown that

$$\begin{aligned}\|B_i(x,y)(t') - B_i(x,y)(t)\| &\leq \int_0^t \frac{|(t'-s)^{\gamma_i-1} - (t-s)^{\gamma_i-1}|}{\Gamma(\gamma_i)} \phi(g_i(s,x(s),y(s)))\,ds \\ &\quad + \int_t^{t'} \frac{(t'-s)^{\gamma_i-1}}{\Gamma(\gamma_i)} \phi(g_i(s,x(s),y(s)))\,ds \\ &\leq P \int_0^t \frac{|(t'-s)^{\gamma_i-1} - (t-s)^{\gamma_i-1}|}{\Gamma(\gamma_i)} ds + P \int_t^{t'} \frac{(t'-s)^{\gamma_i-1}}{\Gamma(\gamma_i)} ds.\end{aligned}$$

This means, in particular, that the set $B(S)$ is weakly equi-continuous. Take a sequence $((x_n, y_n))_n$ contained in the set S. Going back to the inequality (4.2.2), we obtain

$$\begin{aligned}\|B_i(t, x_n(t), y_n(t))\| &\leq \int_0^t \frac{(t-s)^{\gamma_i-1}}{\Gamma(\gamma_i)} \|g_i(s, x_n(s), y_n(s))\|\,ds \\ &\leq \frac{P\mathcal{T}^{\gamma_i}}{\Gamma(\gamma_i+1)}.\end{aligned}$$

Then, the set $\{B_i(x_n, y_n),\ n \in \mathbb{N}\}$ is uniformly bounded in $B(S)$. In other words, we get $B(S)(t)$ is a sequentially relatively weakly compact subset of $B(S)(t)$. Thus, on account of Theorem 1.3.42, we get $B(S)$ is sequentially relatively weakly compact, and when we apply Theorem 1.3.29, we reach $B(S)$ is relatively weakly compact. Finally, we check that there are satisfied assumption (iv) imposed in Theorem 2.3.5. Indeed, let $(x, y) \in E \times E$ and $(u, v) \in S$ such that

$$(x, y) = A(x, y) \cdot B(u, v) + C(x, y).$$

After simple calculations, we can see that

$$\begin{aligned}\|x(t)\| &\leq \|f_1(t, x(t), y(t))\| \int_0^t \frac{(t-s)^{\gamma_1-1}}{\Gamma(\gamma_1)} \|g_1(s, u(s), v(s))\|\,ds \\ &\quad + \sum_{i=1}^m I^{\sigma_1^i} \|h_1^i(t, x(t), y(t))\| \\ &\leq [\rho(\|x\|_\infty + \|y\|_\infty) + F_0] \frac{P\mathcal{T}^{\gamma_1}}{\Gamma(\gamma_1+1)} \\ &\quad + \sum_{k=1}^m \frac{\mathcal{T}^{\sigma_1^k}}{\Gamma(\sigma_1^k+1)} [\rho(\|x\|_\infty + \|y\|_\infty) + R_0],\end{aligned}$$

whenever $t \in [0, 1]$. This yields that

$$\begin{aligned}\|x(t)\| + \|y(t)\| &\leq [\rho(\|x\|_\infty + \|y\|_\infty) + F_0]\, P \left(\frac{\mathcal{T}^{\gamma_1}}{\Gamma(\gamma_1+1)} + \frac{\mathcal{T}^{\gamma_2}}{\Gamma(\gamma_2+1)} \right) \\ &\quad + \sum_{k=1}^m \left(\frac{\mathcal{T}^{\sigma_1^k}}{\Gamma(\sigma_1^k+1)} + \frac{\mathcal{T}^{\sigma_2^k}}{\Gamma(\sigma_2^k+1)} \right) [\rho(\|x\|_\infty + \|y\|_\infty) + R_0].\end{aligned}$$

This implies that $\|x\|_\infty \le r_0$ and $\|y\|_\infty \le r_0$, and when we apply Theorem 2.3.5, we deduce that the coupled system (4.2.1) has a solution in $E \times E$. Q.E.D.

Example 4.2.1 Let us consider the following system on $C(J,\mathbb{R}) \times C(J,\mathbb{R})$ of nonlinear functional differential equations

$$\begin{cases} D^{\frac{1}{2}}\left[\dfrac{x(t) - \sum_{k=1}^{2} I^{\sigma_1^k} h_1^k(t, x(t), y(t))}{f_1(t, x(t), y(t))}\right] = \dfrac{3}{35(13-t^2)}(7|x(t)| + 15|y(t)|), \\ D^{\frac{1}{2}}\left[\dfrac{y(t) - \sum_{k=1}^{2} I^{\sigma_2^k} h_2^k(t, x(t), y(t))}{f_2(t, x(t), y(t))}\right] = \dfrac{3}{35(13-t^2)}(7|x(t)| + 15|y(t)|), \\ x(0) = y(0) = 0. \end{cases} \tag{4.2.3}$$

It should be noticed that the coupled system (4.2.3) can be transformed into problem (4.2.1), whenever

$$\sum_{k=1}^{2} I^{\sigma_1^k} h_1^k(t, x(t), y(t)) = I^{1/3}\frac{2te^{-3t}}{15(3+t)}\left(\frac{x^2(t) + 9|x(t)|}{|x(t)| + 5} + \frac{12e^{3t}}{5}\right) + I^{10/3}\frac{2t\sin\pi t}{14+t^2}\left(\frac{x^2(t) + 5|x(t)|}{|x(t)| + 8} + \frac{1}{3}\right),$$

$$\sum_{k=1}^{2} I^{\sigma_2^k} h_2^k(t, x(t), y(t)) = I^{7/4}\frac{t\sin t}{7(4+e^t)}\left(\frac{y^2(t) + 4|y(t)|}{|y(t)| + 3} + \cos t\right) + I^{29/6}\frac{3t\cos t}{10(4-t^2)}\left(\frac{y^2(t)+5|y(t)|}{|y(t)|+4} + \frac{t}{t+2}\right),$$

$$\begin{cases} f_1(t, x(t), y(t)) = \dfrac{3\cos\pi t + 2t}{5(2+10t^2)(|x(t)|+3)}, \\ f_2(t, x(t), y(t)) = \dfrac{4\cos\pi t + 3t}{7(3+8t^2)(|y(t)|+6)}, \\ g_1(t, x(t), y(t)) = g_2(t, x(t), y(t)) = \dfrac{3}{35(13-t^2)}(7|x(t)| + 15|y(t)|), \end{cases}$$

here $\gamma_1 = \gamma_2 = \frac{1}{2}$, $T = 1, m = 2, \sigma_1^1 = \frac{1}{3}, \sigma_1^2 = \frac{10}{3}, \sigma_2^1 = \frac{7}{4}$ and $\sigma_2^2 = \frac{29}{6}$. On the other hand, via simple calculations, one can prove that

$$\begin{cases} |f_1(t,x,y)-f_1(t,\tilde{x},\tilde{y})| \le \dfrac{3+2t}{5(2+10t^2)}|x-\tilde{x}|, \\ |f_2(t,x,y)-f_2(t,\tilde{x},\tilde{y})| \le \dfrac{4+3t}{7(1+5t^2)}|y-\tilde{y}|, \end{cases}$$

and

$$\begin{cases} |h_1^1(t,x,y)-h_1^1(t,\tilde{x},\tilde{y})| \le \dfrac{18t}{75(3+t)}|x-\tilde{x}|, \\ |h_1^2(t,x,y)-h_1^2(t,\tilde{x},\tilde{y})| \le \dfrac{10t}{8(14+t^2)}|x-\tilde{x}|, \\ |h_2^1(t,x,y)-h_2^1(t,\tilde{x},\tilde{y})| \le \dfrac{4t}{21(4+e^t)}|y-\tilde{y}|, \\ |h_2^2(t,x,y)-h_2^2(t,x,\tilde{y})| \le \dfrac{3t}{8(4-t^2)}|y-\tilde{y}|. \end{cases}$$

Then,

$$\begin{cases} a_{11} = \dfrac{1}{12},\ a_{21} = 0,\ a_{12} = 0,\ a_{22} = \dfrac{1}{6}, \\ b_{11}^1 = \dfrac{3}{50},\ b_{12}^1 = 0,\ b_{21}^1 = 0,\ b_{22}^1 = \dfrac{4}{21(4+e)}, \\ b_{11}^2 = \dfrac{1}{12},\ b_{12}^2 = 0,\ b_{21}^2 = 0,\ b_{22}^2 = \dfrac{1}{8}. \end{cases}$$

Observe that

$$|g_1(t,x,y)| = |g_2(t,x,y)| \le \frac{1}{4}.$$

It is easily seen that $\rho = \dfrac{1}{6}$, $F_0 = \dfrac{1}{36}$, $H_0 = \dfrac{2}{25}$. Besides, the assumption (H_4) is followed with $r_0 = 2$. Moreover,

$$\begin{cases} \dfrac{PT^{\gamma_1}}{\Gamma(\gamma_1+1)}a_{11} + \displaystyle\sum_{k=1}^{m} \dfrac{T^{\sigma_1^k}}{\Gamma(\sigma_1^k+1)}b_{11}^k \simeq 0.0990 \\ \dfrac{PT^{\gamma_2}}{\Gamma(\gamma_2+1)}a_{12} + \displaystyle\sum_{k=1}^{m} \dfrac{T^{\sigma_2^k}}{\Gamma(\sigma_2^k+1)}b_{12}^k = \dfrac{PT^{\gamma_1}}{\Gamma(\gamma_1+1)}a_{21} + \displaystyle\sum_{k=1}^{m} \dfrac{T^{\sigma_1^k}}{\Gamma(\sigma_1^k+1)}b_{21}^k = 0 \\ \dfrac{PT^{\gamma_2}}{\Gamma(\gamma_2+1)}a_{22} + \displaystyle\sum_{k=1}^{m} \dfrac{T^{\sigma_2^k}}{\Gamma(\sigma_2^k+1)}b_{22}^k \simeq 0.0653. \end{cases}$$

Accordingly, an application of Theorem 4.2.1 yields that, the system (4.2.3) has, at least, one solution in $C(J,\mathbb{R}) \times C(J,\mathbb{R})$.

4.3 Existence Theory in RWC-Banach Algebra

Let X be a RWC-Banach algebra. The aim of this section is to apply Corollary 2.4.12, in order to prove the existence of solutions on $E = C([0, 1], X)$, for the following nonlinear functional integral equation (NFIE)

$$x(t) = \left[\left(q(t) + \int_0^t k(t, s)\, \mathscr{G}(s, x(s))\, ds\right) v\right] \cdot \mathscr{F}(t, x(t)), \qquad (4.3.1)$$

for all $t \in J = [0, 1]$, where the functions $q : J \longrightarrow X$ and $k : J \times X \longrightarrow X$ are continuous. Here, $v \in X\backslash\{0\}$ and $E = C([0, 1], X)$ is endowed with the norm $\| \cdot \|_\infty$ and with the pointwise multiplication as is given in the equality (2.4.2). Now, let us introduce the following hypotheses.

(H_7) The function $\mathscr{F} : J \times X \longrightarrow X$ is such that

(i) the partial $x \longrightarrow \mathscr{F}(t, x)$ is weakly sequentially continuous,

(ii) there exists $r_0 > 1$ such that, $\mathscr{F}$ is continuous with respect to the first variable uniformly on $\overline{\mathscr{B}}_X(r_0)$,

(iii) there exists a continuous real function $\sigma : J \to \mathbb{R}_+$ such that

$$\|\mathscr{F}(t, x) - \mathscr{F}(t, y)\| \leq \sigma(t)\|x - y\|, \quad \text{for } t \in J \text{ and } x,\ y \in \overline{\mathscr{B}}_X(r_0).$$

(H_8) The function $\mathscr{G} : J \times X \longrightarrow \mathbb{R}$ is such that

(i) $x \longrightarrow \mathscr{G}(t, x)$ is weakly sequentially continuous,

(ii) $\mathscr{G}(\cdot, x) \in L^1(J)$, for all $x \in \overline{\mathscr{B}}_X(r_0)$,

(iii) there exists an integrable function γ such that

$$|\mathscr{G}(t, x) - \mathscr{G}(t, y)| \leq \gamma(t)\|x - y\|,$$

for all $t \in J$ and $x,\ y \in X$.

Theorem 4.3.1 *Suppose that the hypotheses* (H_7)*-*(H_8) *are satisfied. If*

$$\begin{cases} \left(\|q\|_\infty + \|k\|_\infty a_{r_0}\right)\left(\|\sigma\|_\infty r_0 + \|\mathscr{F}(\cdot, \theta)\|_\infty\right)\|v\| \leq r_0, \\ \max\left(r_0, 2\left(\|\sigma\|_\infty r_0 + \|\mathscr{F}(\cdot, \theta)\|_{L^1}\right)\right)\|k\|_\infty \|\gamma\|_{L^1}\|v\| + \left(\|q\|_\infty + \|k\|_\infty \|v\| a_{r_0}\right)\|\sigma\|_\infty < 1, \end{cases}$$

where $a_{r_0} = \|\gamma\|_{L^1} r_0 + \|\mathscr{G}(\cdot, \theta)\|_{L^1}$, *then the problem (NFIE)* (4.3.1) *has, at least, one solution in* $\overline{\mathscr{B}}_E(r_0)$.

Proof Let us introduce on E two operators $\mathscr{T}$ and $\mathscr{S}$ by putting

$$\begin{cases} (\mathscr{T}x)(t) = \left(q(t) + \displaystyle\int_0^t k(t, s)\mathscr{G}(s, x(s))\, ds\right) \cdot v \\ (\mathscr{S}x)(t) = \mathscr{F}(t, x(t)). \end{cases}$$

It is clear that Eq. (4.3.1) can be transformed into the following problem

$$x(t) = (\mathscr{T}x)(t) \cdot (\mathscr{S}x)(t).$$

Define the set S on E by

$$\Omega = \{x \in E, \ \|x\|_\infty \leq r_0\}.$$

First, let us prove that $\mathscr{T} : E \longrightarrow E$ and $\mathscr{S} : \Omega \longrightarrow E$. The claim regarding $\mathscr{S}$ is clear, by assumption (H_7). Let x be an arbitrarily fixed element of the set Ω and take $t, t' \in J$. By hypothesis (H_8)-(iii), we have

$$\begin{aligned}\int_0^t |k(t,s) - k(t',s)|\, |\mathscr{G}(s,x(s))|\, ds &\leq \int_0^t |k(t,s) - k(t',s)|\, |\mathscr{G}(s,x(s)) - \mathscr{G}(s,\theta)|\, ds \\ &\quad + \int_0^t |k(t,s) - k(t',s)|\, |\mathscr{G}(s,\theta)|\, ds \\ &\leq r_0 \left[\int_0^t |k(t,s) - k(t',s)| \left(\gamma(s) + \sup_{s \in J} |\mathscr{G}(s,\theta)| \right) ds \right].\end{aligned}$$

This implies that

$$\begin{aligned}\|\mathscr{T}x(t) - \mathscr{T}x(t')\| &\leq |q(t) - q(t')| \\ &\quad + r_0 \left(\int_0^t |k(t,s) - k(t',s)| \left(\gamma(s) + \sup_{s \in J} |\mathscr{G}(s,\theta)| \right) ds \right) \|v\|.\end{aligned}$$

Hence, by using our hypotheses, together with the dominated convergence theorem, we obtain the continuity of $\mathscr{T}x$, which achieves the claim. Next, we will show that $\mathscr{T}$ and $\mathscr{S}$ satisfy all hypothesis of Corollary 2.4.12.

Claim 1. Let us begin by showing that $\mathscr{T}$ is strongly continuous. Indeed, suppose that $(x_n)_n$ is a weakly convergent sequence contained in the set Ω to a point $x \in \Omega$. By using Theorem 1.3.50, we infer that

$$x_n(t) \rightharpoonup x(t), \ \text{ for each } t \in J.$$

For any $t \in J$, we get

$$\|\mathscr{T}x_n(t) - \mathscr{T}x(t)\| \leq \|k\|_\infty \|v\| \int_0^1 |\mathscr{G}(s,x_n(s)) - \mathscr{G}(s,x(s))|\, ds.$$

Since $\mathscr{G}$ is weakly sequentially continuous with respect to the second variable, then

$$\mathscr{G}(s,x_n(s)) \longrightarrow \mathscr{G}(s,x(s)), \ \text{ for all } s \in J.$$

Using the dominated convergence theorem, we get

$$\mathscr{T}x_n \longrightarrow \mathscr{T}x.$$

The assumption (H_8) guaranties that $\mathscr{T}$ is $\mathscr{D}$-Lipschitzian with $\mathscr{D}$-function Φ, where

$$\Phi(t) = \|k\|_\infty \|\gamma\|_{L^1} \|v\| \, t.$$

Consequently, $\mathscr{T}(\Omega)$ is bounded with bound

$$M_{\mathscr{T}} = \|q\|_\infty + \|k\|_\infty \|v\| \left(\|\gamma\|_{L^1} r_0 + \|\mathscr{G}(\cdot, \theta)\|_{L^1}\right).$$

Claim 2. $\mathscr{F}$ is weakly sequentially continuous and countably δ-set-Lipschitzian on Ω such that $\mathscr{F}(\Omega)$ is bounded. At first, in virtue of hypothesis (H_7), and via simple calculations, one can show that

$$\|\mathscr{F}(\Omega)\| \leq \|\sigma\|_\infty r_0 + \|\mathscr{F}(\cdot, \theta)\|_\infty.$$

Next, we shall verify the weak sequential continuity of the operator $\mathscr{F}$. Indeed, assume that $(x_n)_n$ is a weakly convergent sequence contained in the set Ω to an element $x \in \Omega$. Then, on account of Theorem 1.3.50, we have

$$x_n(t) \rightharpoonup x(t), \quad \text{for each } t \in J.$$

Since $x \mapsto \mathscr{F}(t, x)$ is weakly sequentially continuous, for all $t \in J$, it follows that

$$\mathscr{F}x_n(t) \rightharpoonup \mathscr{F}x(t), \quad \text{for each } t \in J.$$

From Theorem 1.3.50, we get $\mathscr{F}x_n \rightharpoonup \mathscr{F}x$, since the set $\{\mathscr{F}x_n, \ n \in \mathbb{N}\}$ is bounded. The use of hypothesis (H_7) and Lemma 1.5.9 leads to $\mathscr{F}$ is countably δ-set-Lipschitzian, where $\delta = \|\sigma\|_\infty$.

Claim 3. Let y be an arbitrarily fixed element of the set Ω and take $x \in E$ with

$$x = \mathscr{T}x \cdot \mathscr{S}y.$$

Then

$$x(t) = \left[\left(q(t) + \int_0^t k(t,s)\, \mathscr{G}(s, x(s))\, ds\right) v\right] \cdot \mathscr{F}(t, y(t)), \quad \text{for all } t \in J.$$

This implies that

$$\begin{aligned}\|x(t)\| &= \left\| \left[\left(q(t) + \int_0^t k(t,s)\, \mathscr{G}(s, x(s))\, ds \right) v \right] \cdot \mathscr{F}(t, y(t)) \right\| \\ &\leq \left(\|q\|_\infty + \left\| \int_0^t k(t,s)\, \mathscr{G}(s, x(s))\, ds \right\| \right) \|v\| \; \|\mathscr{F}(t, y(t))\| .\end{aligned}$$

According to assumption (H_8), we obtain

$$\begin{aligned}\|x(t)\| &\leq \left(\|q\|_\infty + \int_0^t |k(t,s)| \; (\gamma(s) r_0 + |\mathscr{G}(s,\theta)|)\; ds \right) (\|\sigma\|_\infty r_0 + \|\mathscr{F}(\cdot,\theta)\|_\infty) \|v\| \\ &\leq \left(\|q\|_\infty + \|k\|_\infty \left[\|\gamma\|_{L^1} r_0 + \|\mathscr{G}(\cdot,\theta)\|_{L^1} \right] \right) (\|\sigma\|_\infty r_0 + \|\mathscr{F}(\cdot,\theta)\|_\infty) \|v\|.\end{aligned}$$

Thus, $x \in \Omega$. Therefore, all the conditions of Corollary 2.4.12 are satisfied and so, Equation (4.3.1) has a solution in Ω. Q.E.D.

References

1. Dhage, B.C.: Some nonlinear alternatives in Banach algebras with applications I. Nonlinear Stud. **12**, 271–283 (2005)
2. Dhage, B.C.: Some nonlinear alternatives in Banach algebras with applications II. Kyungpook Math. J. **45**, 281–292 (2005)
3. Dhage, B.C.: multi-valued mappings and fixed points II. Tamkang J. Math. **37**, 27–46 (2006)
4. Banaś, J., Lecko, M.: Fixed points of the product of operators in Banach algebras. PanAmer. Math. J. **12**(2), 101–109 (2002)
5. Caballero, J., Lopez, B., Sadarangani, K.: Existence of nondecreasing and continuous solutions of an integral equation with linear modification of the argument. Acta Math. Sin. (Engl. Ser.) **23**, 1719–1728 (2007)
6. Chaker, W., Ghribi, A., Jeribi, A., Krichen, B.: Fixed point theorem for quasi-contractive mapping in partial cone metric space. Afr. Mat. **26**(5–6), 1153–1159 (2015)
7. Chaker, W., Ghribi, A., Jeribi, A., Krichen, B.: Fixed point theorems for (p, q)-quasi-contraction mappings in cone metric spaces. Chin. Ann. Math. Ser. B **37**(2), 211–220 (2016)
8. Chaker, W., Jeribi, A.: Common fixed point of single-valued and multivalued mappings. Afr. Mat. **25**(3), 617–622 (2014)
9. Dhage, B.C.: On some nonlinear alternatives of Leray-Schauder type and functional integral equations. Arch. Math. (Brno) **42**(1), 11–23 (2006)
10. Dhage, B.C.: On a fixed point theorem in Banach algebras with applications. Appl. Math. Lett. **18**, 273–280 (2005)
11. Smart, D.R.: Fixed Point Theorems. Cambridge University Press, Cambridge (1980)
12. Ben Amar, A., Chouayekh, S., Jeribi, A.: New fixed point theorems in Banach algebras under weak topology features and applications to nonlinear integral equations. J. Funct. Anal. **259**, 2215–2237 (2010)
13. Burton, T.A.: Volterra Integral and Differential Equations. Academic, New York (1983)
14. Dhage, B.C., O'Regan, D.: A fixed point theorem in Banach algebras with applications to functional integral equations. Funct. Differ. Equ. **7**, 259–267 (2000)
15. Perov, A.I.: On the Cauchy problem for a system of ordinary differential equation (Russian). Pribliz. Metod. Resen. Differ. Uravn. **2**, 115–134 (1964)

16. Graef, J.R., Henderson, J., Ouahab, A.: Topological methods for differential equations and inclusions. Monographs and Research Notes in Mathematics. CRC Press (2019)
17. Nieto, J.J., Ouahab, A., Rodríguez-López, R.: Fixed point theorems in generalized Banach algebras and applications. Fixed Point Theory. **19**, 707–732 (2018)
18. Ouahab, A.: Some Perov's and Krasnosel'skii type fixed point results and application. Comm. Appl. Nonlinear Anal. 623–642 (2015)
19. Petre, I.R.: A multi-valued version of Krasnosel'skii's theorem in generalized Banach spaces. An. Ştiinţ. Univ. "Ovidius" Constanţa Ser. Mat. **22**, 177–192 (2014)
20. Banaś, J., Taoudi, M.A.: Fixed points and solutions of operator equations for the weak topology in Banach algebras. Taiwanese J. Math. **18**(3), 871–893 (2014)
21. Cichon, M., Kubiaczyk, I., Sikorska, A.: The Henstock-Kurzweil-Pettis integrals and existence theorems for the Cauchy problem. Czechoslovak. Math. J. **54**, 279–289 (2004)
22. Chandrasekhar, S.: Radiative Transfer. Dover Publications, New York (1960)
23. El-Sayed, A.M.A., Hashem, H.H.G.: Integrable and continuous solutions of nonlinear quadratic integral equation. Electr. J. Qual. Theory Diff. Equ. **25**, 1–10 (2008)
24. El-Sayed, A.M.A., Hashem, H.H.G.: Monotonic positive solution of a nonlinear quadratic functional integral equation. Appl. Math. Comput. **216**, 2576–2580 (2010)
25. Ben Amara, K., Kaddachi, N., Jeribi, A.: Existence results for nonexpansive multi-valued operators and nonlinear integral inclusions. Afr. Mat. **34**(3), Paper No. 46, 25 (2023)
26. Jeribi, A., Kaddachi, N., Laouar, Z.: Generalized form of fixed point theorems in generalized Banach algebra relative to the weak topology with an application. Filomat **36**(18), 6253–6268 (2022)

Chapter 5
Two-Dimensional Boundary Value Problems

The objective of this chapter is to study several coupled systems of nonlinear functional integral and differential equations arising in some physical and biological problems in the Banach algebra $C([0, 1], X)$, where X is a Banach algebra satisfying the condition $(\mathscr{P})$. We should mention that all the involved operator equations are generated by block matrices with the dimensions 2×2 with nonlinear inputs. We will use the ideas developed in Chap. 3. The results of this chapter can be found in [1–3]. Certain assumptions on its entries involving the measure of weak noncompactness to ensure the existence of solutions are therefore given. Precisely, we extend several earlier works which used the condition $(\mathscr{P})$.

5.1 Functional Integral Equation in Banach Algebra

This section is dedicated to show the applicability of Theorem 3.1.5 to examine a result on the existence of solutions of the following coupled system

$$\begin{cases} x(t) = a(t)x(t) + y(t)\left[r(t) + \int_0^{\sigma(t)} h\left(t, y(\eta(s))\right)\, ds\right] \\ y(t) = \dfrac{1}{1 + |x(\theta(t))|} - f\left(t, x(\theta(t))\right) + g(t, y(t)) \end{cases} \tag{5.1.1}$$

for all $t \in J$, where $J = [0, 1]$ and x and y are an unknown functions in the Banach algebra $C(J, \mathbb{R})$. Let us specify the assumptions under which the system (5.1.1) will be investigated. These assumptions are as follows

A. Jeribi and N. Kaddachi, *Topological Fixed-Point Theory in Suitable Banach Algebras with Applications*, Infosys Science Foundation Series,
https://doi.org/10.1007/978-981-96-5542-7_5

(H_9) The mapping $f : J \times \mathbb{R} \longrightarrow \mathbb{R}$ is Lipschitzian with a Lipschitz constant $q > 1$.
(H_{10}) The mapping $g : J \times \mathbb{R} \longrightarrow \mathbb{R}$ is such that
  (i) the partial $x \longrightarrow g(t, x)$ is Lipschitzian with a Lipschitz constant $k \in]\frac{2}{1+q}, 1[$,
  (ii) $|g(t, x(t))| \leq (1 - k)|x(t)|$, for all $x \in C(J, \mathbb{R})$,
  (iii) $\|f(\cdot, x(\cdot))\|_\infty \leq 1$, for all $x \in C(J, \mathbb{R})$ with $\|x\|_\infty \leq 1 + k$.
(H_{11}) $a : J \longrightarrow \mathbb{R}$ is continuous such that $\|a\|_\infty < \frac{1}{2}$.
(H_{12}) $\theta, \sigma, \eta : J \longrightarrow J$ are continuous and nondecreasing with $\sigma(t) \leq t$.
(H_{13}) $r : J \longrightarrow \mathbb{R}$ is continuous with $\|r\|_\infty < \frac{1-\|a\|_\infty}{2(1+q)}(1 - k)$.
(H_{14}) The mapping $h : J \times \mathbb{R} \longrightarrow \mathbb{R}$ is such that
  (i) h is continuous,
  (ii) $\|h(\cdot, x(\cdot))\|_\infty \leq \frac{1-\|a\|_\infty}{2q(1+q)}(1 - k) - \|r\|_\infty$, with $\|x\| \leq 1 + q$.

Theorem 5.1.1 *Under the assumptions* (H_9)-(H_{14}), *the coupled system* (5.1.1) *has a solution in* $C(J, \mathbb{R}) \times C(J, \mathbb{R})$.

Proof Let us define, on the space E, five mappings A, B, C, D, and B' by putting

$$\begin{cases} (Ax)(t) = a(t)x(t), \\ (Bx)(t) = x(t), \\ (Cx)(t) = \dfrac{1}{1 + |x(\theta(t))|} - f\left(t, x(\theta(t))\right), \\ (B'x)(t) = r(t) + \displaystyle\int_0^{\sigma(t)} h(t, x(\eta(s)))\, ds, \\ (Dx)(t) = g\left(t, x(t)\right). \end{cases}$$

Observe that the system (5.1.1) can be written in the following form

$$\begin{cases} x(t) = Ax(t) + By(t) \cdot B'y(t) \\ y(t) = Cx(t) + Dy(t). \end{cases}$$

For the following, define on $C(J, \mathbb{R})$ the sets S and S' by

$$S := \{y \in C(J, \mathbb{R}), \|y\| \leq 1 + k\} \text{ and } S' := \{y \in C(J, \mathbb{R}), \|y\| \leq 1 + q\}.$$

The assumption (H_9) guaranties that

$$\begin{aligned} |Cx(t) - Cy(t)| &= \left| \frac{1}{1 + |x(\theta(t))|} - f\left(t, x(\theta(t))\right) - \frac{1}{1 + |y(\theta(t))|)} - f\left(t, y(\theta(t))\right) \right| \\ &\leq \left| \frac{|x(\theta(t))| - |y(\theta(t))|}{(1 + |x(\theta(t))|)(1 + |y(\theta(t))|)} \right| + |f\left(t, y(\theta(t))\right) - f\left(t, x(\theta(t))\right)| \\ &\leq (1 + q)\|x - y\|. \end{aligned}$$

whenever $x, y \in S$. This implies that

$$\|Cx - Cy\| \leq (1 + q)\|x - y\|.$$

Similarly, we can show that A and B are Lipschitzians with constants $\|a\|_\infty$ and 1, respectively and, the operator D is a contraction with constant k, in view of assumption (H_{10}). Now, we prove that the set $C(S)$ is relatively compact. For this purpose, consider $(x_n)_n$ as a sequence contained in the set S and take a sequence $y_n = C(x_n)$. Based on assumption (H_9), we get

$$|y_n(t)| = \left| \frac{1}{1+|x_n(\theta(t))|} - f(t, x(\theta(t))) \right| \leq 2.$$

Applying the Bolzano–Weierstrass theorem, we can see that $(y_n(t))_n$ possesses a convergent subsequence, say $(y_{n_k}(t))_k$, to a function $y(t)$ and so, the set $C(S)$ is a sequentially relatively compact subset in X. The result follows from Theorem 1.3.29. Next, we show the inclusion

$$C(S) \subseteq (I - D)(S').$$

Let $x \in S$ be fixed element and define the mapping $\varphi_x : C(J, \mathbb{R}) \longrightarrow C(J, \mathbb{R})$ by putting

$$\varphi_x(y)(t) = \frac{1}{1+|x(\theta(t))|} - f(t, x(\theta(t))) + g(t, y(t)).$$

Then, in virtue of hypothesis (H_{10}), we can demonstrate that φ_x is a contraction with a constant k, and when we apply Theorem 1.2.4, we reach that, there exists exactly one point $y \in C(J, \mathbb{R})$ satisfying

$$Cx + Dy = y.$$

This implies that

$$C(S) \subseteq (I - D)(C(J, \mathbb{R})).$$

Due to the fact that the function y is continuous on J, we conclude that there exists a point $t^* \in J$ such that

$$\begin{aligned} \|y\|_\infty &= |y(t^*)| \\ &= |Cx(t^*) + Dy(t^*)| \\ &\leq \frac{1}{1+|x(\theta(t^*))|} + 1 + (1-k)|y(t^*)| \\ &\leq 2 + (1-k)|y(t^*)|. \end{aligned}$$

This means, in particular, that $\|y\|_\infty < 1 + q$ and consequently,

$$C(S) \subseteq (I - D)(S').$$

Besides, for all $x \in S'$, we have $B'x \in S'$. To see that, consider $(t_n)_n$ as a convergent sequence of J to a point $t \in J$. Then,

$$
\begin{aligned}
\left|B'x(t_n) - B'x(t)\right| &= \left|r(t_n) + \int_0^{\sigma(t_n)} h(t_n, x(\eta(s)))\,ds - r(t) - \int_0^{\sigma(t)} h(t, x(\eta(s)))\,ds\right| \\
&\le |r(t_n) - r(t)| + \left|\int_0^{\sigma(t_n)} h(t_n, x(\eta(s)))\,ds - \int_0^{\sigma(t)} h(t, x(\eta(s)))\,ds\right| \\
&\le |r(t_n) - r(t)| + \left|\int_0^{\sigma(t_n)} |(h(t_n, x(\eta(s))) - h(t, x(\eta(s))))|\,ds\right| \\
&\quad + \left|\int_{\sigma(t)}^{\sigma(t_n)} h\,(t, x(\eta(s)))\,ds\right| \\
&\le |r(t_n) - r(t)| + \int_0^1 |(h(t_n, x(\eta(s))) - h(t, x(\eta(s))))|\,ds \\
&\quad + (1 - \|r\|_\infty)\,|\sigma(t_n) - \sigma(t)|\,.
\end{aligned}
$$

Using the assumption (H_{14}), we obtain

$$
h(t_n, x(\eta(s))) \to h(t,\ x(\eta(s))) \text{ in } \mathbb{R}.
$$

and

$$
|h(t_n,\ x(\eta(s))) - h(t,\ x(\eta(s)))| \le 2 - \|r\|_\infty
$$

for all $t,\ s \in J$. Define the mapping $\varphi : J \longrightarrow \mathbb{R}$ by putting

$$
\varphi(s) = 2 - \|r\|_\infty.
$$

We can easily verify that $\varphi \in L^1(J)$. Then, on account of dominated convergence theorem, we have

$$
\lim_{n\to+\infty} \left|B'x(t_n) - B'x(t)\right| = 0.
$$

This means that $B'x$ is a continuous mapping on J. Using a similar argument, we can deduce that the operator B' is continuous on $(I-D)^{-1}C(S)$. Based on assumption (H_{14}), we infer that

$$
\begin{aligned}
M &= \sup_{x\in S} \|B'(I-D)^{-1}C(x)\| \\
&\le \sup\left\{\sup_{t\in J}\left\{|r(t)| + \int_0^{\sigma(t)} |h(t,\ x_1(\eta(s)))|\,ds\right\},\ \ x_1 \in S'\right\} \\
&\le \sup\left\{\sup_{t\in J}|r(t)| + \sup_{t\in J}\int_0^{\sigma(t)} |h(t,\ x_1(\eta(s)))|\,ds,\ x_1 \in S'\right\} \\
&\le \sup\left\{\sup_{t\in J}|r(t)| + \sup_{t\in J}\int_0^{\sigma(t)} \sup_{t\in J}|h(t,\ x_1(\eta(s)))|\,ds,\ x_1 \in S'\right\} \\
&\le \sup\left\{\sup_{t\in J}|r(t)| + \sup_{t\in J}\int_0^{t} \sup_{s\in J}|h(t,\ x_1(\eta(s)))|\,ds,\ x_1 \in S'\right\} \\
&< \frac{(1-\|a\|_\infty)(1-k)}{2(1+q)},
\end{aligned}
$$

where $x_1 = (I - D)^{-1}Cx \in S'$. After simple calculations, we can see that

$$\frac{1}{1-k}M + \|a\|_\infty < 1.$$

Our next task is to show that

$$A(x) + B(I - D)^{-1}C(x) \cdot B'(I - D)^{-1}C(y) \in S, \quad \text{for all } x, \ y \in S.$$

In order to achieve this, let x and z be two arbitrarily fixed elements of the set S, then

$$\begin{aligned}\left|Ax(t) + B(I - D)^{-1}Cx(t) \cdot B'(I - D)^{-1}Cz(t)\right| &\leq \|a\|_\infty \|x\| + |x_1(t)||B'(z_1(t))| \\ &\leq \tfrac{1+k}{2} + (1+q)\left(\tfrac{(1-\|a\|_\infty)(1-k)}{2(1+q)}\right) \\ &\leq \tfrac{1+k}{2} + \tfrac{1+k}{2} \\ &= 1+k,\end{aligned}$$

whenever $x_1 = (I - D)^{-1}Cx$ and $z_1 = (I - D)^{-1}Cz$. The desired result follows from Theorem 3.1.5. Q.E.D.

5.2 Weak Sequential Continuity and Weak Compactness in FIEs

To elaborate, our results constructed in Chap. 3, here we provide two examples. Precisely, we examine in this section the existence results of two nonlinear integral equations. For the following, assume that X is a Banach algebra satisfying the sequential condition $(\mathcal{P})$.

Example 5.2.1 Let us consider the following system of nonlinear functional integral equations occurring in some quantic physical problems

$$\begin{cases} x(t) = f(t, x(t)) + [a(t)y(t)] \cdot \left[\left(\int_0^{\sigma_1(t)} k(t,s) f_1(s, y(\eta(s)))\, ds\right) u\right] \\ y(t) = \left(q(t) + \int_0^{\sigma_2(t)} p\,(t, s, x(s), x(\lambda s))\, ds\right) v + g(t, y(t)), \end{cases} \tag{5.2.1}$$

whenever $u, v \in X\backslash\{0\}$, where $\lambda \in [0, 1]$, $a, q, \sigma_1, \sigma_2, \eta$ are continuous real functions on $J = [0, T]$, for $0 < T < \infty$.

For the sake of convenience, we impose the following hypotheses
(H_{15}) The functions a and k are such that

(i) $a : J \longrightarrow X$ is continuous,

(ii) $k : J \times J \to \mathbb{R}$ is nonnegative and continuous function.

(H_{16}) $\sigma_1, \sigma_2, \eta : J \longrightarrow J$ are continuous.

(H_{17}) $q : J \longrightarrow \mathbb{R}$ is continuous.
(H_{18}) p is weakly sequentially continuous on $J \times J \times X \times X$.
(H_{19}) $f : J \times X \longrightarrow X$ is such that
 (i) f is continuous with respect to the first variable,
 (ii) f is weakly sequentially continuous,
 (iii) f is a contraction with a constant k' with respect to the second variable.
(H_{20}) $f_1 : J \times X \longrightarrow \mathbb{R}$ is such that
 (i) the partial $x \longrightarrow f_1(t, x)$ is weakly sequentially continuous,
 (ii) $\|f_1(\cdot, x(\cdot))\| \leq \lambda r$, if $\|x\|_\infty \leq r$, for $r > 0$.
(H_{21}) $g : J \times X \longrightarrow X$ is such that
 (i) $x \longrightarrow g(t, x)$ is weakly sequentially continuous,
 (ii) $x \longrightarrow g(t, x)$ is a -nonlinear contraction with $\mathscr{D}$-function Φ such that

$$\Phi(r) < (1 - \lambda)r.$$

Theorem 5.2.1 *Suppose that the assumptions (H_{15})-(H_{21}) are satisfied. If there exists $r_0 > 0$ such that*

$$\begin{cases} |p(t, s, x(s), x(\lambda s))| \leq r_0, \text{ for } x \in C(J, X) \text{ with } \|x\|_\infty \leq r_0, \\ \|f(t, x(t))\| \leq k'\|x(t)\|, \text{ for } t \in J \text{ and } x \in C(J, X) \text{ with } \|x\|_\infty \leq r_0, \\ \|g(\cdot, x(\cdot))\| \leq \lambda\|x\|_\infty, \text{ for } x \in C(J, X) \text{ with } \|x\|_\infty \leq r_0, \\ \Phi(r) < (1 - \lambda)r, \\ \|a\|_\infty \leq \dfrac{(1 - k')r_0}{\delta^2 K T \lambda \|u\|_\infty}, \end{cases} \tag{5.2.2}$$

where $K = \sup\limits_{t,\, s \in J} \{|k(t, s)|\}$ and $\lambda\delta = (\|q\|_\infty + T r_0)\, \|v\|_\infty + r_0$, then the nonlinear system (5.2.1) *has a solution in $C(J, X) \times C(J, X)$.*

Proof Put the set S on $C(J, X)$ by

$$S := \{x \in C(J, \mathbb{R}) \text{ such that } \|x\| \leq r_0\},$$

where r_0 satisfies the inequalities (5.2.2). Define on S five nonlinear operators A, B, C, D, and B' by putting

$$\begin{cases} (Ax)(t) = f(t, x(t)), \\ (Bx)(t) = a(t)x(t), \\ (Cx)(t) = \left(q(t) + \displaystyle\int_0^{\sigma_2(t)} p(t, s, x(s), x(\lambda s))\, ds\right) v, \ 0 < \lambda < 1. \\ (Dx)(t) = g(t, x(t)), \\ (B'x)(t) = \left(\displaystyle\int_0^{\sigma_1(t)} k(t, s) f_1(s, x(\eta(s)))\, ds\right) u, \end{cases} \tag{5.2.3}$$

for all $t \in J$. Then, the coupled system (5.2.1) can be written in the following

$$\begin{cases} x(t) = Ax(t) + By(t) \cdot B'y(t) \\ y(t) = Cx(t) + Dy(t). \end{cases}$$

In order to prove our theorem, we will apply the result contained in Theorem 3.1.11. We start with the assumption $(I - D)^{-1}C(S)$ is bounded. Due to the fact that D is a nonlinear contraction with $\mathscr{D}$-function Φ and Theorem 1.2.26, we get the inverse operator $(I - D)^{-1}$ exists on $C(J, X)$. Take a point $(x, y) \in S \times C(J, X)$ such that $y = (I - D)^{-1}Cx$. Then,

$$y(t) = \left(q(t) + \int_0^{\sigma_2(t)} p\,(t, s, x(s), x(\lambda s))\; ds\right) v + g\,(t, y(t))\,, \text{ for all } t \in J.$$

Using the continuity of the function y, we obtain there exists a point $t^* \in J$ such that

$$\begin{aligned}
\|y\|_\infty &= \|y(t^*)\| \\
&\leq \left|q(t^*) + \int_0^{\sigma_2(t^*)} p\left(t^*, s, x(s), x(\lambda s)\right)\; ds\right| \|v\| \\
&\quad + \|g(t^*, y(t^*)) - g(t^*, x(t^*))\| + \|g(t^*, x(t^*))\| \\
&\leq (\|q\|_\infty + Tr_0)\; \|v\| + \Phi\; (\|x(t^*) - y(t^*)\|) + \|g(t^*, x(t^*))\| \\
&< (\|q\|_\infty + Tr_0)\; \|v\| + (1 - \lambda)\|y(t^*)\| + \|x\|_\infty \\
&\leq (\|q\|_\infty + Tr_0)\; \|v\| + r_0 + (1 - \lambda)\|y\|_\infty .
\end{aligned}$$

This yields that

$$\|y\|_\infty < \frac{1}{\lambda}\left[(\|q\|_\infty + Tr_0)\; \|v\| + r_0\right].$$

This means that, the set $(I - D)^{-1}C(S)$ is bounded with a bound δ, where

$$\delta = \frac{1}{\lambda}\left[(\|q\|_\infty + Tr_0)\; \|v\| + r_0\right].$$

It should be noted that the operators defined in (5.2.3) are well defined. The claim regarding Ax, By and Dy is immediate, for all $(x, y) \in S \times C(J, X)$. We only prove the continuity of the operators Cx and $B'y$ on J, for any $(x, y) \in S \times (I - D)^{-1}C(S)$. To see this, consider $(t_n)_n$ as a convergent sequence of J to an element t. Then,

$$\begin{aligned}
\|(B'y)(t_n) - (B'y)(t)\| &\leq \left|\int_0^{\sigma_1(t_n)} k(t_n, s) f_1(s, y(\eta(s)))\; ds - \int_0^{\sigma_1(t)} k(t, s) f_1(s, y(\eta(s)))\; ds\right| \|u\| \\
&\leq \left[\int_0^{\sigma_1(t_n)} |k(t_n, s) - k(t, s)|\; |f_1(s, y(\eta(s)))|\; ds\right] \|u\| \\
&\quad + \left|\int_{\sigma_1(t_n)}^{\sigma_1(t)} k(t, s) f_1(s, y(\eta(s)))\; ds\right| \|u\|.
\end{aligned}$$

Moreover, taking into account that $(I-D)^{-1}C(S)$ is bounded with a bound δ, and using assumption (H_{20}), we get

$$\begin{aligned}\|(B'y)(t_n)-(B'y)(t)\| &\leq \left[\int_0^T |k(t_n,s)-k(t,s)|\,\lambda\delta ds\right]\|u\| + \left|\int_{\sigma_1(t_n)}^{\sigma_1(t)} K\lambda\delta ds\right|\|u\| \\ &\leq \left[\int_0^T |k(t_n,s)-k(t,s)|\,\lambda\delta ds\right]\|u\| + K\lambda\delta\,|\sigma_1(t_n)-\sigma_1(t)|\,\|u\|.\end{aligned}$$

Due to the fact that k and σ_1 are continuous on J, we infer that $B'y \in C(J,\mathbb{R})$. The use of the first inequality in (5.2.2) shows that, the operator C is well defined. Now, we show that B and B' are weakly sequentially continuous on $(I-D)^{-1}C(S)$ and A, C and D are weakly sequentially continuous on S. The claim regarding the operators A and D is clear, by using Theorem 1.3.50. Let us assume that $(x_n)_n$ is a weakly convergent sequence contained in the set $(I-D)^{-1}C(S)$ to a point x. Then, on account of Theorem 1.3.50, we get

$$x_n(t) \rightharpoonup x(t) \text{ in } X.$$

The condition $(\mathscr{P})$ guaranties that

$$(B(x_n))(t) \rightharpoonup (Bx)(t) \text{ in } X.$$

Due to the fact that $\|B(x_n)\| \leq \|a\|_\infty \delta$ and Theorem 1.3.50, we infer that

$$B(x_n) \rightharpoonup Bx \text{ in } E.$$

Taking into account assumption (H_{20}) and using Theorem 1.3.50, we obtain

$$f_1(t,x_n(t)) \rightharpoonup f_1(t,x(t)) \text{ in } \mathbb{R}.$$

By using the dominated convergence theorem, we can verify that

$$\lim_{n\to+\infty}\int_0^{\sigma_1(t)} k(t,s)f_1(s,x_n(\eta(s)))\,ds = \int_0^{\sigma_1(t)} k(t,s)f_1(s,x(\eta(s)))\,ds.$$

This prove the weak sequential continuity of the operator B' on $(I-D)^{-1}C(S)$. Consider $(x_n)_n$ as a weakly convergent sequence contained in the set S to an element x. Then, in virtue of Theorem 1.3.50, we can see that $x_n(t) \rightharpoonup x(t)$. Thus, in virtue of assumption (H_{18}), we obtain

$$(C(x_n))(t) \to (Cx)(t) \text{ in } X.$$

Thus, $C(x_n) \rightharpoonup Cx$. Let us assume that $(x_n)_n$ is a sequence contained in the set S, then $(C(x_n))(t) = r_n(t)\cdot v$, where

$$r_n(t) = q(t) + \int_0^{\sigma_2(t)} p\,(t, s, x_n(s), x_n(\lambda s))\, ds.$$

Since $|r_n(t)| \leq \|q\|_\infty + Tr_0$, then the sequence $(r_n(t))_n$ possesses a renamed convergent subsequence to a function $r(t)$. This implies that

$$(C(x_n))(t) \rightharpoonup (Cx)(t) \text{ in } X.$$

This prove the weak sequential compactness of the set $\overline{C(S)(t)}^w$. Next, we will show that $C(S)$ is a weakly equi-continuous set. In fact, for a fixed element $\varepsilon > 0$, for an arbitrary $x \in S$, $x^* \in X^*$ and $t, t' \in J$ such that $|t' - t| \leq \varepsilon$, we have

$$\begin{aligned}\left|x^*\left((Cx)(t) - (Cx)(t')\right)\right| &\leq \left[\int_0^{\sigma_2(t)} \left|p\,(t, s, x(s), x(\lambda s)) - p\left(t', s, x(s), x(\lambda s)\right)\right| ds\right] \|x^*(v)\| \\ &\quad + |q(t) - q(t')| \|x^*(v)\| + \left[\int_{\sigma_2(t)}^{\sigma_2(t')} \left|p\left(t', s, x(s), x(\lambda s)\right)\right| ds\right] \|x^*(v)\| \\ &\leq [w(q, \varepsilon) + w(p, \varepsilon) + r_0 w(\sigma, \varepsilon)]\, \|x^*(v)\|,\end{aligned}$$

where

$$\begin{cases} w(q, \varepsilon) = \sup\left\{|q(t) - q(t')|,\ t,\ t' \in J,\ |t - t'| \leq \varepsilon\right\}, \\ w(p, \varepsilon) = \sup\left\{|p(t, s, x, y) - p(t', s, x, y)|,\ t,\ t', s \in J,\ |t - t'| \leq \varepsilon,\ x,\ y \in S\right\}, \text{ and} \\ w(\sigma, \varepsilon) = \sup\left\{|\sigma(t) - \sigma(t')|,\ t,\ t' \in J,\ |t - t'| \leq \varepsilon\right\}. \end{cases}$$

Taking into account assumption (H_{18}) and using the uniform continuity of the operators p, q and σ, one obtains

$$\lim_{\varepsilon \to 0} w(q, \varepsilon) = \lim_{\varepsilon \to 0} w(p, \varepsilon) = \lim_{\varepsilon \to 0} w(\sigma, \varepsilon) = 0.$$

An application of Theorem 1.3.42, we can show that $C(S)$ is sequentially relatively weakly compact, and when we apply Theorem 1.3.29, we reach the desired conclusion. On the other hand, by using Lemma 1.5.9, one can prove that A is condensing with respect to β. In our next step, we show that there is a satisfied assumption (iv) of Theorem 3.1.11. To this end, observe firstly, that for all $x \in (I - D)^{-1}C(S)$, there exists exactly one $z \in C(J, X)$ such that $z = x$, with $\|z\| \leq \delta$. Let us fix an arbitrarily element $x \in S$ and take $y \in C(J, X)$ such that the following equality

$$y = A(x) + B(I - D)^{-1}C(x) \cdot B'(I - D)^{-1}C(x)$$

is satisfied. Then, for all $t \in J$, we have

$$\begin{aligned}\|y(t)\| &\leq \|f(t, x(t))\| + \|a(t)z(t)\| \left\| \left(\int_0^{\sigma_1(t)} k(t, s) f_1(s, z(\eta(s)))\, ds\right) u \right\| \\ &< k'\|x(t)\| + \|a\|_\infty K \|z(t)\| \|u\|_\infty \\ &\leq k' r_0 + K\lambda\delta^2 \|u\|_\infty,\end{aligned}$$

where $K = \sup_{t,\, s \in J} k(t, s)$. Since $y \in C(J, X)$, there exists $t^* \in J$ with

$$\|y\|_\infty = \|y(t^*)\|.$$

Thus, $\|y\|_\infty \leq r_0$, in light of the last inequality in (5.2.2). This implies that $y \in S$. The proof is complete. Q.E.D.

Example 5.2.2 We will focus our attention on the existence of solutions of the following system of nonlinear integral equations

$$\begin{cases} x(t) = \displaystyle\int_0^{\sigma_1(t)} k_1(t, s) f_1(s, x(\eta_1(s)))\, ds + y(t) \cdot \left[\int_0^{\sigma_2(t)} k_2(t, s) f_2(s, y(\eta_2(s)))\, ds \right], \\ y(t) = \dfrac{1}{1 + b(t)|x(t)|} - g\left(t, \dfrac{1}{1 + b(t)|x(t)|}\right) + g(t, y(t)), \end{cases} \quad (5.2.4)$$

for all $t \in [0, 1]$, where σ_i, η_i are continuous on $[0, 1]$ and f_i are a generalized Lipschitzian mapping on $[0, 1] \times \mathbb{R}$, for $i = 1, 2$. The following definition concerning the superposition operators, which will play a crucial role below.

Definition 5.2.2 A mapping $f : [0, 1] \times \mathbb{R} \longrightarrow \mathbb{R}$ is said to satisfy L^1-Carathéodory condition or simply is called to be a L^1-Carathéodory, provided that the following conditions are satisfied

(i) the function f is measurable with respect to the first variable,
(ii) the function f is almost everywhere continuous with respect to the second variable,
(iii) for an arbitrary $r > 0$, there exists a function $h_r \in L^1([0, 1], \mathbb{R})$ such that

$$|f(t, x)| \leq h_r(t), \quad \text{for each } t \in J \text{ and } |x| \leq r.$$

Let us introduce the following assumptions, in order to study the existence of solutions of the nonlinear problem (5.2.4).
(H_{22}) $\sigma_i, \eta_i : [0, 1] \longrightarrow [0, 1]$ are continuous, for $i = 1, 2$.
(H_{23}) $b : [0, 1] \longrightarrow \mathbb{R}$ is a continuous and positive real function.
(H_{24}) $k_i : [0, 1] \times [0, 1] \longrightarrow \mathbb{R}$ are continuous and nonnegative, for $i = 1, 2$.
(H_{25}) $f_1 : [0, 1] \times \mathbb{R} \longrightarrow \mathbb{R}$ is generalized Lipschitzian with function l_1, a.e., there exists a continuous function $l_1 : [0, 1] \longrightarrow \mathbb{R}$ such that

$$|f_1(t, x) - f_1(t, y)| \leq |l_1(t)||x - y|,$$

for all $x, y \in \mathbb{R}$.
(H_{26}) The function $f_2 : [0, 1] \times \mathbb{R} \longrightarrow \mathbb{R}$ is such that
(i) f_2 is $L^1_{\mathbb{R}}$-Carathéodory,
(ii) f_2 is generalized Lipschitzian with function l_2.
(H_{27}) $g : [0, 1] \times \mathbb{R} \longrightarrow \mathbb{R}$ is such that
(i) $x \longrightarrow g(t, x)$ is a contraction with a constant k,

(ii) $t \longrightarrow g(t, x)$ is a continuous mapping on J, for all $x \in C([0, 1], \mathbb{R})$.

(H_{28}) there exists a real number $r > 2$ such that

(i) $|f_1(t, x(t))| \leq |x(t)|$, for $x \in C([0, 1], \mathbb{R})$ with $\|x\|_\infty \leq r$,

(ii) $K_2\|h_r\|_{L^1} \leq (1 - K_1)r$, where $K_1 = \sup_{t,s\in[0,1]} k_1(t, s) < 1$ and $K_2 = \sup_{t,s\in[0,1]} k_2(t, s)$,

(iii) $\dfrac{\|b\|_\infty(1 + k)}{1 - k} K_2\|h_r\|_{L^1} + K_1\|l_1\|_{L^1} < k$,

(iv) $0 \leq \dfrac{rK_2\|l_2\|_{L^1}}{(1 - k)^2} < 1$.

Theorem 5.2.3 *Under assumptions* (H_{22})*-*(H_{28})*, problem* FIE (5.2.4) *has, at least, one solution in* $C([0, 1], \mathbb{R}) \times C([0, 1], \mathbb{R})$.

Proof Let $J = [0, 1]$ and consider on the space $C(J, \mathbb{R})$ five operators A, B, C, D, and B' by putting

$$\begin{cases} (Ax)(t) = \displaystyle\int_0^{\sigma_1(t)} k_1(t, s) f_1(s, x(\eta_1(s)))\, ds, \\ (Bx)(t) = x(t), \\ (Cx)(t) = \dfrac{1}{1 + b(t)|x(t)|} - g\left(t, \dfrac{1}{1 + b(t)|x(t)|}\right), \\ (Dx)(t) = g\,(t, x(t))\,, \\ (B'x)(t) = \displaystyle\int_0^{\sigma_2(t)} k_2(t, s) f_2(s, x(\eta_2(s)))\, ds. \end{cases} \tag{5.2.5}$$

Then, the system (5.2.4) can be written in the following form

$$\begin{cases} x(t) = Ax(t) + By(t) \cdot B'y(t) \\ y(t) = Cx(t) + Dy(t). \end{cases}$$

We will show that these operators satisfy all conditions of Theorem 3.1.17. Since the maps f_2 and k_2 are continuous such that f_2 is L^1-Carathéodory, it follows from the dominated convergence theorem that B' is continuous on S, where $S = \overline{\mathscr{B}}_{C(J,\mathbb{R})}(r)$. Therefore, it is easy to see that, the operators defined in (5.2.5) are well defined.

Claim 1. A, B, C and B' are Lipschitzians. To see this, for all x, $y \in S$, we have

$$\begin{aligned} \|Ax - Ay\| &= \sup_{t\in J}\left|\int_0^{\sigma_1(t)} k_1(t, s) f_1(s, x(\eta_1(s)))\, ds - \int_0^{\sigma_1(t)} k_1(t, s) f_1(s, y(\eta_1(s)))\, ds\right| \\ &\leq \sup_{t\in J}\int_0^{\sigma_1(t)} k_1(t, s)\, |f_1(s, x(\eta_1(s))) - f_1(s, y(\eta_1(s)))|\, ds \\ &\leq \sup_{t\in J}\int_0^{\sigma_1(t)} K_1 l_1(s)\, |x(\eta_1(s)) - y(\eta_1(s))|\, ds \\ &\leq K_1\|x - y\|\int_0^1 l_1(s)\, ds. \end{aligned}$$

This shows that A is Lipschitzian with a Lipschitz constant $K_1\|l_1\|_{L^1}$. On the other hand, via simple calculations, one can show that C and B' are Lipschitzians with constants $(1+k)\|b\|_\infty$ and $K_2\|l_2\|_{L^1}$, respectively.

Claim 2. Let $x \in S$ and $t \in J$, then one can easily verify that

$$(Cx)(t) = (I-D)\left(\frac{1}{1+b|x|}\right)(t). \tag{5.2.6}$$

Hence, there exists an element $y \in S$ satisfying $(Cx)(t) = (I-D)(y)(t)$. Consequently, $C(S) \subseteq (I-D)(S)$ and from our assumptions, we infer that the operator $(I-D)^{-1}C$ is Lipschitzian with a Lipschitz constant $\frac{1+k}{1-k}\|b\|_\infty$.

Claim 3. C is a strongly continuous mapping on S. Indeed, let $(x_n)_n$ be a weakly convergent sequence contained in the set S to a point $x \in S$, since S is weakly closed in $C(J,\mathbb{R})$. By using Theorem 1.3.50, we get

$$x_n(t) \rightharpoonup x(t) \quad \text{for all } t \in J.$$

Since C is Lipschitzian, then C is continuous and so,

$$C(x_n)(t) \to Cx(t).$$

This shows that C is strongly continuous.

Claim 4. Let $x \in S$. It follows from the equality (5.2.6), that the operator $B(I-D)^{-1}C$ is regular on $B'(I-D)^{-1}C(S)$. Therefore, the operator inverse $\left(\frac{I}{B(I-D)^{-1}C}\right)^{-1}$ exists on $B'(S)$. Indeed, let $x, y \in B'(S)$ be with relation

$$x(t)(1+b(t)|x(t)|) = y(t).$$

Then,

$$|x(t)|(1+b(t)|x(t)|) = |y(t)|.$$

After simple calculations, we can verify that

$$\left(\sqrt{b(t)}|x(t)| + \frac{1}{2\sqrt{b(t)}}\right)^2 = \frac{1}{4b(t)} + |y(t)|,$$

which further implies the following equation

$$\sqrt{b(t)}|x(t)| = \frac{-1}{2\sqrt{b(t)}} + \sqrt{\frac{1}{4b(t)} + |y(t)|}.$$

Hence,

$$b(t)|x(t)| = \frac{-1}{2} + \sqrt{\frac{1}{4} + b(t)|y(t)|}.$$

This means that,

$$x(t) = \frac{y(t)}{1 + b(t)|x(t)|} = \frac{y(t)}{\frac{1}{2} + \sqrt{\frac{1}{4} + b(t)|y(t)|}}.$$

Define the mapping $G : C(J, \mathbb{R}) \longrightarrow C(J, \mathbb{R})$ by putting

$$G(x) = \frac{x}{\frac{1}{2} + \sqrt{\frac{1}{4} + b|x|}}.$$

After simple calculations, we can prove that

$$\left(\frac{I}{B(I-D)^{-1}C}\right) G(x) = G\left(\frac{I}{B(I-D)^{-1}C}(x)\right) = x.$$

This yields that

$$\left(\frac{I}{B(I-D)^{-1}C}\right)^{-1} x = \frac{x}{\frac{1}{2} + \sqrt{\frac{1}{4} + b|x|}}.$$

Due to the fact that $K_1\|l_1\|_{L^1} < 1$ and Theorem 1.2.4, we deduce that $(I - A)^{-1}$ exists on $(I - A)(S)$. So, the operator $\left(\frac{I - A}{B(I-D)^{-1}C}\right)^{-1}$ exists on $B'(I - D)^{-1}C(S)$.

Claim 5. By using the assumption (H_{26}), we have

$$\begin{aligned} M_1 &= \sup_{x \in S} \|B'(I - D)^{-1}C(x)\| \\ &\leq \sup_{x \in S} \|B'x\| \\ &\leq \sup_{x \in S} \left\{ \sup_{t \in J} \left| \int_0^{\sigma_2(t)} k_2(t, s) f_2(s, x(\eta_2(s)))\, ds \right| \right\} \\ &\leq K_2\|h_r\|_{L^1}. \end{aligned}$$

Consequently, in view of (H_{28}), we have

$$\frac{(1 + k)\|b\|_\infty}{1 - k} M_1 + K_1\|l_1\|_{L^1} < k.$$

Since $M_2 = \sup_{x \in S} \|B(I - D)^{-1}C(x)\| \leq \sup_{x \in S} \|Bx\| \leq r$, we obtain

$$0 \leq \frac{M_2 K_2\|l_2\|_{L^1}}{(1 - k)^2} < 1.$$

Let us fix an arbitrary $x \in C(J, \mathbb{R})$ and take $y \in S$ such that

$$x = A(x) + B(I - D)^{-1}C(x) \cdot B'(I - D)^{-1}C(y).$$

After simple calculations, we can prove that

$$\begin{aligned}
|x(t)| &\leq |Ax(t)| + |B(I - D)^{-1}Cx(t)||B'(I - D)^{-1}Cy(t)| \\
&\leq \int_0^{\sigma_1(t)} |k_1(t, s) f_1(s, x(\eta_1(s)))| \, ds \quad + \\
&\quad \frac{1}{1 + b(t)|x(t)|} \left| \int_0^{\sigma_2(t)} k_2(t, s) f_2 \left(s, \frac{1}{1 + b(\eta_2(s))|y(\eta_2(s))|}\right) ds \right| \\
&\leq \int_0^{\sigma_1(t)} k_1(t, s) \, |f_1(s, x(\eta_1(s)))| \, ds \quad + \\
&\quad \frac{1}{1 + b(t)|x(t)|} \int_0^1 k_2(t, s) \left| f_2 \left(s, \frac{1}{1 + b(\eta_2(s))|y(\eta_2(s))|}\right) \right| ds \\
&\leq \int_0^1 K_1 \, |x(\eta_1(s))| \, ds + \int_0^1 K_2 h_r(s) \, ds \\
&\leq K_1 \|x\|_\infty + K_2 \|h_r\|_{L^1} \\
&\leq r.
\end{aligned}$$

The desired conclusion follows from a direct application of Theorem 3.1.17. Q.E.D.

5.3 Quadratic Equations of Chandrasekhar Type

The purpose of this section is to study the existence of solutions of the following system, presented in [1], of two-dimensional of quadratic integral equations of Chandrasekhar type

$$\begin{cases} x(t) = f_1(t, x(t)) + g(t, y(t)) \left(\int_0^t \frac{t}{t+s} u_1(s, y(s)) \, ds \right) \cdot u, \\ y(t) = f_2(t, y(t)) + \left(\int_0^t \frac{t}{t+s} u_2(s, x(s)) \, ds \right) \cdot v, \end{cases} \tag{5.3.1}$$

for all $t \in J = [0, b]$, $0 < b < \infty$, where $u, v \in X \backslash \{0\}$. Let us specify the assumptions under which the system (5.3.1) will be investigated. These assumptions are as follows.

(H_{29}) $f_i : J \times X \to X$, for $i = 1, 2$, is such that

(i) the partial $x \longrightarrow f_i(t, x)$ is a contraction with a constant k_i,
(ii) the partial $x \longrightarrow f_i(t, x)$ is weakly sequentially continuous,
(iii) $M_i = \sup_{(t,x) \in J \times X} |f_i(t, x)|$.

(H_{30}) $u_i : J \times X \to X$, for $i = 1, 2$ is such that
(i) u_i is measurable with respect to the first variable,
(ii) u_i is weakly sequentially continuous with respect to the second variable,
(iii) the partial $x \longrightarrow u_i(t, x)$ is a contraction with a constant l_i,
(iv) $\|u_i(t, x)\| \leq m_i(t) \in L^1(J)$, for all $(t, x) \in J \times X$ and,
(v) $\lambda_i = \sup_{t \in J} \int_0^b \frac{1}{t+s} m_i(s)\, ds < \infty$.

(H_{31}) $g : J \times X \longrightarrow X$ is such that
(i) the partial $x \longrightarrow g(t, x)$ is weakly sequentially continuous,
(ii) there exist $L > 0$, $K > 0$ such that

$$0 < g(t, x(t)) - g(t, y(t)) \leq \frac{L\|x - y\|}{K + \|x - y\|}$$

for all $x, y \in X$. Moreover, $L \leq K$,

(iii) $\|g\| = N < \infty$.

Theorem 5.3.1 *Assume that the assumptions (H_{29})-(H_{31}) are satisfied. Furthermore, if*

$$\begin{cases} L\|u\|\|m_1\| \leq K(1 - k), \\ k + bl_2\|v\| \leq 1, \end{cases}$$

then the system of the quadratic integral equations (5.3.1) has, at least, one solution.

Proof Define the set S by

$$S = \{x \in C(J, X),\ \|x\| \leq \inf\{\lambda_1 N\|v\| + M_1, \lambda_2\|v\| + M_2\}\}.$$

Let us define on the space $C(J, X)$ five operators A, B, C, D, and B' on S by putting

$$\begin{cases} (Ax)(t) = f_1(t, x(t)), \\ (By)(t) = g(t, y(t)), \\ (Cx)(t) = \left(\int_0^t \frac{t}{t+s} u_2(s, x(s))\, ds\right) \cdot v, \\ (Dy)(t) = f_2(t, y(t)), \\ (B'y)(t) = \left(\int_0^t \frac{t}{t+s} u_1(s, y(s))\, ds\right) \cdot u. \end{cases}$$

Then, the system (5.3.1) can be written in the following form

$$\begin{cases} x(t) = (Ax)(t) + (By)(t) \cdot (B'y)(t), \\ y(t) = (Cx)(t) + (Dy)(t). \end{cases}$$

We prove that these operators satisfy the assumptions of Theorem 3.1.15.

Claim 1. We show that A, B, and C are $\mathscr{D}$-Lipschitzians. To see this, for an arbitrary $x, y \in S$, we get

$$\begin{aligned}\|Ax(t) - Ay(t)\| &= \|f_1(t, x(t)) - f_1(t, y(t))\| \\ &\leq k_1|x(t) - y(t)|.\end{aligned}$$

Taking the supremum over t, we obtain

$$\|Ax - Ay\| \leq k_1(\|x - y\|).$$

This means that A is $\mathscr{D}$-Lipschitzian with $\mathscr{D}$-function $\psi_A(r) = k_1 r$. Furthermore, by using assumption (H_{31}), we conclude that

$$\begin{aligned}\|Bx(t) - By(t)\| &= \|g(t, x(t)) - g(t, y(t))\| \\ &\leq \frac{L\|x - y\|}{K + \|x - y\|}.\end{aligned}$$

This implies that

$$\|Bx - By\| \leq \frac{L\|x - y\|}{K + \|x - y\|}.$$

Consequently, the mapping B is $\mathscr{D}$-Lipschitzian with $\mathscr{D}$-function ψ_B given by

$$\psi_B(r) = \frac{Lr}{K + r}.$$

On the other hand, via simple calculations, one can show that the mapping C is $\mathscr{D}$-Lipschitzian with $\mathscr{D}$-function ψ_C, where

$$\psi_C(r) = bl_2\|v\|_\infty r.$$

In fact, for arbitrary $x, y \in X$, we get

$$\begin{aligned}\|Cx(t) - Cy(t)\| &= \left\|\left(\int_0^t \frac{t}{t+s} u_2(s, x(s))\, ds\right) \cdot v - \left(\int_0^t \frac{t}{t+s} u_2(s, y(s))\, ds\right) \cdot v\right\| \\ &\leq \|v\| \left\|\int_0^t \frac{t}{t+s} (u_2(s, x(t)) - u_2(s, y(s)))\, ds\right\| \\ &\leq \|v\| \int_0^t \frac{t}{t+s} \|(u_2(s, x(t)) - u_2(s, y(s))\|\, ds \\ &\leq \|v\| \int_0^t \frac{t}{t+s} l_2\|x - y\|\, ds \\ &\leq bl_2\|v\|\|x - y\|.\end{aligned}$$

This yields

$$\|Cx - Cy\| \leq bl_2\|v\|\|x - y\|.$$

Claim 2. Consider $(x_n)_n$ as a sequence contained in the set S. Then,

$$(B'x_n)(t) = r_n(t) \cdot u,$$

where

$$r_n(t) = \int_0^t \frac{t}{t+s} u_1(s, y(s))\, ds.$$

Since $\|r_n(t)\| \leq \lambda_1$, then the sequence $(r_n(t))_n$ possesses a renamed weakly convergent subsequence to a function $r(t)$ and so,

$$(B'x_n)(t) \rightharpoonup (B'x)(t) \ \text{ in } X, \quad \text{for all } t \in [0, b].$$

Thus, the set $B'(S)(t)$ is sequentially relatively weakly compact. Next, we will show that $B'(S)$ is an equi-continuous subset with respect to the weak topology. Let us fix $x^* \in X^*$ and take $t_1, t_2 \in J$ with $t_2 > t_1$. Then,

$$\begin{aligned}
|x^*(B'x(t_2) - B'x(t_1))| &\leq \left| \int_0^{t_2} \frac{t_2}{t_2+s} u_1(s, x(s))\, ds - \int_0^{t_1} \frac{t_1}{t_1+s} u_1(s, x(s))\, ds \right| \|x^*(u)\| \\
&\leq \left| \int_0^{t_2} \frac{t_2}{t_2+s} u_1(s, x(s))\, ds - \int_0^{t_2} \frac{t_1}{t_1+s} u_1(s, x(s))\, ds \right. \\
&\quad \left. + \int_0^{t_2} \frac{t_1}{t_1+s} u_1(s, x(s))\, ds - \int_0^{t_1} \frac{t_1}{t_1+s} u_1(s, x(s))\, ds \right| \|x^*(u)\| \\
&\leq \left(\int_0^{t_2} \frac{t_2 - t_1}{(t_1+s)(t_2+s)} s\, |u_1(s, x(s))|\, ds + \int_{t_1}^{t_2} \frac{t_1}{t_1+s} |u_1(s, x(s))|\, ds \right) \|x^*(u)\| \\
&\leq \left(\int_0^{t_2} \left(\frac{t_2}{t_1} - 1 \right) |u_1(s, x(s))|\, ds + \int_{t_1}^{t_2} \frac{t_1}{t_1+s} |u_1(s, x(s))|\, ds \right) \|x^*(u)\|.
\end{aligned}$$

Due to the fact that $t_2 \to t_1$ and Arzela-Ascoli's theorem, we infer that the set $B'(S)$ is sequentially relatively weakly compact. The desired conclusion follows from a direct application of Theorem 1.3.29.

Claim 3. Let $x \in S$ be fixed. Define the mapping $\phi_x : C(J, X) \longrightarrow C(J, X)$ by putting

$$\phi_x(y) = Cx + Dy.$$

Notice that ϕ_x is a contraction with a constant k_2, in light of assumption (H_{29}). Then, on account of Theorem 1.2.4, there exists an unique element $y \in C(J, X)$ such that

$$y = Cx + Dy.$$

Consequently,

$$C(S) \subseteq (I - D)(C(J, X)).$$

Since $y \in C(J, X)$, then there exists $t^* \in J$ such that

$$\begin{aligned}\|y\|_\infty &= |Cx(t^*) + Dy(t^*)| \\ &\leq \left| \left(\int_0^{t^*} \frac{t^*}{t^* + s} u_2(s, x(s))\, ds \right) v \right| + |f_2(t^*, x(t^*)| \\ &\leq \lambda_2 \|v\| + M_2.\end{aligned}$$

This means, in particular, that $y \in S$ and so,

$$C(S) \subseteq (I - D)(S). \tag{5.3.2}$$

Claim 4. The use of (5.3.2) as well as assumption (H_{30}) leads to

$$\begin{aligned}M &= \|B'(I - D)^{-1} C(S)\| \\ &\leq \sup_{x \in S} \left\{ \sup_{t \in J} \left| \left(\int_0^{t} \frac{t}{t + s} u_1(s, x(s))\, ds \right) u \right| \right\} \\ &\leq \lambda_1 \|u\|.\end{aligned}$$

Therefore,

$$M \psi_B \left(\frac{1}{1 - k_2} \psi_C \right)(r) + \psi_A(r) < r.$$

Next, let $y \in C(J, X)$ be fixed and take $x \in S$ such that

$$y(t) = Ax(t) + B(I - D)^{-1} Cx(t) \cdot B'(I - D)^{-1} Cx(t).$$

Then,

$$\begin{aligned}\|y(t)\| &= \|Ax(t) + B(I - D)^{-1} Cx(t) \cdot B'(I - D)^{-1} Cx(t)\| \\ &\leq \|Ax(t)\| + \|B(I - D)^{-1} Cx(t)\| \|B'(I - D)^{-1} Cx(t)\| \\ &\leq M_1 + N \|u\| \lambda_1.\end{aligned}$$

Due to the fact $y \in C(J, X)$, we infer that $\|y\|_\infty = \|y(t^*)\|$, for some $t^* \in J$ and, then

$$\|y\|_\infty \leq M_1 + N \|u\| \lambda_1.$$

As a result $y \in S$. Hence, an application of Theorem 3.1.15, shows that the system (5.3.1) has, at least, one solution in $S \times S$. Q.E.D.

References

1. Fahem, O., Jeribi, A., Kaddachi, N.: Existence of solutions for a system of Chandrasekhar's equations in Banach algebras under weak topology. Filomat **33**, 5949–5957 (2019)
2. Jeribi, A., Kaddachi, N., Krichen, B.: Existence results for a system of nonlinear functional integral equations in Banach algebras under weak topology. Fixed Point Theory **18**, 247–268 (2017)
3. Kaddachi, N., Jeribi, A., Krichen, B.: Fixed point theorems of block operator matrices on Banach algebras and an application to functional integral equations. Math. Methods Appl. Sci. **36**, 659–673 (2013)

Chapter 6
Existence Theory of Critical-Type of Fixed Point

The aim of this chapter is to sketch applications of the critical-type of fixed point theory for the block operator matrix (3.1.12) in the context of weak topology. The essential case, in which $I - A$ is not invertible, is covered in the next section. This poses an unique challenge within the confines of traditional fixed point theory, as $(I - A)^{-1}$ does not always correspond to a single-valued operator. The results of this chapter can be found in [1, 2].

6.1 A System of Functional Integral Equation

Let X be a Banach algebra. In this section, we deal with the existence, on the space $E = C([0, 1], X)$, of solutions of the following problem (in short FIE)

$$\begin{cases} x(t) = k(t, x(t)) + \left[q(t) + \displaystyle\int_0^{\sigma(t)} g(t, s, y(s))\, ds \right] \cdot u \\ y(t) = \phi \left(t, \displaystyle\int_0^t \frac{t}{t+s} w(s, x(s))\, ds \right) \cdot v + b(t) y(t), \end{cases} \tag{6.1.1}$$

where b and σ are continuous on $J = [0, 1]$ and k, w, ϕ are weakly sequentially continuous on $[0, 1] \times X$ with respect to the second variable. The problem (6.1.1) may be transformed into problem

$$\begin{pmatrix} x(t) \\ y(t) \end{pmatrix} = \begin{pmatrix} k(t, \cdot) & G(t, \cdot) \\ \varphi(t, \cdot) & b(t) \end{pmatrix} \begin{pmatrix} x(t) \\ y(t) \end{pmatrix},$$

A. Jeribi and N. Kaddachi, *Topological Fixed-Point Theory in Suitable Banach Algebras with Applications*, Infosys Science Foundation Series,
https://doi.org/10.1007/978-981-96-5542-7_6

where

$$\begin{cases} G(t, y(t)) = \left[q(t) + \int_0^{\sigma(t)} g(t, s, y(s))\, ds\right] \cdot u, \; and \\ \varphi(t, x(t)) = \phi\left(t, \int_0^t \frac{t}{t+s} w(s, x(s))\, ds\right) \cdot v. \end{cases}$$

Let us consider the following hypotheses needed in the sequel.
(H_{32}) $\sigma : [0, 1] \longrightarrow [0, 1]$ is continuous and nondecreasing.
(H_{33}) $q : [0, 1] \longrightarrow \mathbb{R}$ is continuous.
(H_{34}) $b : [0, 1] \longrightarrow X$ is continuous such that $\frac{1}{2} \leq \|b\|_\infty < 1$.
(H_{35}) $k : [0, 1] \times X \longrightarrow X$ is such that

- (i) k is affine with respect to the second variable,
- (ii) k is weakly sequentially continuous with respect to the second variable,
- (iii) $x \longrightarrow k(t, x)$ is a contraction with constant $\|b\|_\infty$.

(H_{36}) The mapping $g : [0, 1] \times [0, 1] \times X \longrightarrow \mathbb{R}$ is such that

- (i) g is continuous,
- (ii) $t \longrightarrow g(t, x)$ is continuous uniformly, for all $x \in X$,
- (iii) $x \longrightarrow g(t, x)$ is Lipschitzian with a Lipschitz constant $1 - \|b\|_\infty$.

(H_{37}) $w : [0, 1] \times X \longrightarrow \mathbb{R}$ is such that

- (i) w is measurable with respect to the first variable,
- (ii) w is weakly sequentially continuous with respect to the second variable,
- (iii) There exists $a > 0$ and there exists a function $m \in L^1(J, X)$ satisfying

$$|w(t, x)| \leq m(t) \;\text{ and }\; \int_0^1 \frac{1}{t+s} m(s)\, ds \leq a.$$

(H_{38}) $\phi : [0, 1] \times X \longrightarrow \mathbb{R}$ is such that

- (i) $x \longrightarrow \phi(t, x)$ is weakly sequentially continuous,
- (ii) there exists $L > 0$ such that

$$|\phi(t, x) - \phi(t, y)| \leq L\|x - y\|, \;\text{ for all } t \in [0, 1] \text{ and } x \in X.$$

Theorem 6.1.1 *Under the hypotheses* (H_{32})-(H_{38})*, the system* (6.1.1) *has a solution in* $C(J, X) \times C(J, X)$ *provided that there exists* $r_0 > 1$ *such that*

$$\begin{cases} |g(t, s, x(s))| \leq r_0 - \|q\|_\infty \\ \sup\limits_{t\in[0,1]} \|k(t, 0)\| \leq (1 - \|b\|_\infty - \|u\|_\infty) r_0. \end{cases} \tag{6.1.2}$$

Proof In order to demonstrate Theorem 6.1.1, we will apply the result contained in Theorem 3.2.5. Let us define on the space $E = C([0, 1], X)$, four operators A, B, C, and D by putting

$$\begin{cases} (Ax)(t) = k(t, x(t)), \\ (Bx)(t) = \left[q(t) + \displaystyle\int_0^{\sigma(t)} g(t, s, x(s))\, ds \right] \cdot u, \\ (Cx)(t) = \phi\left(t, \displaystyle\int_0^t \frac{t}{t+s} w(s, x(s))\, ds\right) \cdot v, \\ (Dx)(t) = b(t)x(t). \end{cases}$$

Put $S = \overline{\mathscr{B}}_X(r_0)$, where r_0 is given in (6.1.2). We recall that, problem (6.1.1) can be written in the following form

$$\begin{cases} x(t) = Ax(t) + By(t) \\ y(t) = Cx(t) + Dy(t). \end{cases}$$

We shall now verify the following claims.
Claim 1. Based on assumption (H_{35}) and Lemma 1.5.9, we can easily prove that A is condensing with respect to β with a weakly sequentially closed graph.
Claim 2. Consider $(x_n)_n$ as an arbitrary sequence contained in the set S. Then, in virtue of assumption (H_{38}), we have

$$\begin{aligned} \|C(x_n)(t)\| &\leq \left[\left| \phi\left(t, \int_0^t \frac{t}{t+s} w(s, x_n(s))\, ds\right) - \varphi(t, 0) \right| + |\varphi(t, 0)| \right] \|v\|_\infty \\ &\leq \left[L \left| \int_0^t \frac{t}{t+s} w(s, x_n(s))\, ds \right| + \sup_{t\in[0,1]} |w(t, 0)| \right] \|v\|_\infty \\ &\leq \left[L\|m\|_{L^1} + \sup_{t\in[0,1]} |w(t, 0)| \right] \|v\|_\infty, \end{aligned}$$

for all $t \in [0, 1]$. This prove that, the sequence $(C(x_n)(t))_n$ is uniformly bounded, for all $t \in [0, 1]$ and so, the set $C(S)(t)$ is sequentially relatively weakly compact. In order to demonstrate that the set $C(S)$ is relatively weakly compact, we will apply Theorem 1.3.42. Then, we only have to show that $C(S)(t)$ is weakly equicontinuous. Let us fix $\varepsilon > 0$. Let $x \in S$ be fixed and choose $0 \leq t < t' \leq 1$ (assume that $|t - t'| < \varepsilon$). It is easy to see that

$$\left| x^*\left((C(x_n))(t) - (C(x_n))(t')\right) \right| \leq [L|M(\varepsilon)| + |\Phi(\varepsilon)|]\, \|x^*(v)\|,$$

where

$$\begin{cases} M(\varepsilon) = \displaystyle\int_0^t \frac{t}{t+s} w(s, x_n(s))\, ds - \int_0^{t'} \frac{t'}{t'+s} w(s, x_n(s))\, ds, \\ \Phi(\varepsilon) = \phi\left(t, \displaystyle\int_0^{t'} \frac{t'}{t'+s} w(s, x(s))\, ds\right) - \phi\left(t', \displaystyle\int_0^{t'} \frac{t'}{t'+s} w(s, x(s))\, ds\right). \end{cases}$$

Based on assumption (H_{37}), it follows that

$$\begin{aligned}|M(\varepsilon)| &\leq |t-t'|\int_0^1 \frac{s}{(t+s)(t'+s)}w(s,x_n(s))\,ds + \int_{t'}^t \frac{t}{t+s}w(s,x_n(s))\,ds\\ &\leq |t-t'|\int_0^1 \frac{1}{t+s}w(s,x_n(s))\,ds + \int_{t'}^t \frac{t}{t+s}w(s,x_n(s))\,ds\\ &\leq a|t-t'| + \int_{t'}^t m(s)\,ds.\end{aligned}$$

Taking into account the uniform continuity of ϕ on [0, 1], we get

$$\lim_{\varepsilon\to 0} M(\varepsilon) = \lim_{\varepsilon\to 0} \Phi(\varepsilon) = 0.$$

This yields that

$$\lim_{t\to t'} \left|x^*\left((C(x_n))(t) - (C(x_n))(t')\right)\right| = 0.$$

Then, $C(S)$ is sequentially relatively weakly compact, in view of Theorem 1.3.42. The desired conclusion follows from Theorem 1.3.29.

Claim 3. Let $(t_n)_n$ be a convergent sequence contained in the set [0, 1] to a point t. Then,

$$\begin{aligned}\|(Bx)(t_n) - (Bx)(t)\| &\leq \left|\int_0^{\sigma(t_n)} g(t_n,s,x(s))\,ds - \int_0^{\sigma(t)} g(t,s,x(s))\,ds\right| \|u\| +\\ &\quad |q(t_n) - q(t)|\,\|u\|\\ &\leq \left[\int_0^{\sigma(t_n)} |g(t_n,s,x(s)) - g(t,s,x(s))|\,ds\right]\|u\| +\\ &\quad \left[\left|\int_{\sigma(t_n)}^{\sigma(t)} g(t,s,x(s))\,ds\right| + |q(t_n)-q(t)|\right]\|u\|\\ &\leq \left[\int_0^1 |g(t_n,s,x(s)) - g(t,s,x(s))|\,ds\right]\|u\| +\\ &\quad [r_0\,|\sigma(t_n)-\sigma(t)| + |q(t_n)-q(t)|]\,\|u\|,\end{aligned}$$

whenever $x \in E$. Based on hypothesis (H_{36}), we get

$$\lim_{n\to+\infty} |g(t_n,s,x(s)) - g(t,s,x(s))| = 0.$$

This implies that

$$\lim_{n\to+\infty} \int_0^1 g(t_n,s,x(s))\,ds = \int_0^1 g(t,s,x(s))\,ds.$$

Since q and σ are continuous on [0, 1], it follows that $Bx \in E$. In view of Theorem 3.2.5, it is sufficient to show the weak sequential continuity on $(I-D)^{-1}C(S)$ of the operator B and on S of the operator C. Consider $(y_n)_n$ as a weakly convergent sequence of $(I-D)^{-1}C(S)$ to a point y. After simple calculations, we can see that

$$\|y_n(t)\| \leq \|C(x_n)(t)\| + \|b\|_\infty \|y_n(t)\| + \|D(I-D)^{-1}C(0)(t)\|.$$

Since $\|b\|_\infty < 1$, then

$$\begin{aligned}\|y_n(t)\| &\leq \frac{1}{1-\|b\|_\infty}\left[\left(L\|m\|_{L^1} + \sup_{t\in[0,1]}|w(t,0)|\right)\|v\|_\infty + \|D(I-D)^{-1}C(0)(t)\|\right] \\ &\leq \frac{\|b\|_\infty\|v\|_\infty}{1-\|b\|_\infty}\left[L\|m\|_{L^1} + \sup_{t\in[0,1]}|w(t,0)| + \sup_{t\in[0,1]}\left|\phi\left(t, \sup_{t\in[0,1]}|w(t,0)|\right)\right|\right] \\ &\leq \left[L\|m\|_{L^1} + \sup_{t\in[0,1]}|w(t,0)| + \sup_{t\in[0,1]}\left|\phi\left(t, \sup_{t\in[0,1]}|w(t,0)|\right)\right|\right]\|v\|_\infty.\end{aligned}$$

This means that the set $(I-D)^{-1}C(S)$ is bounded with bound ϱ, where

$$\varrho = \left[L\|m\|_{L^1} + \sup_{t\in[0,1]}|w(t,0)| + \sup_{t\in[0,1]}\left|\phi\left(t, \sup_{t\in[0,1]}|w(t,0)|\right)\right|\right]\|v\|_\infty.$$

An application of Theorem 1.3.50 shows that,

$$y_n(t) \rightharpoonup y(t).$$

Besides, for all $t \in [0,1]$, we can write $B(y_n)(t) = r_n(t)\cdot u$, where

$$r_n(t) = q(t) + \int_0^{\sigma(t_n)} g(t_n, s, y_n(s))\,ds.$$

Due to the fact that $(r_n(t))_n$ is a bounded sequence, we infer that there exists a renamed convergent subsequence to a function $r(t)$. This implies that

$$B(x_n) \rightharpoonup Bx.$$

Our next task is to prove the weak sequential continuity of maps C and D. The claim regarding C follows immediately from the Lebesgue dominated convergence theorem and assumption (H_{38}). Consider $(\xi_n)_n$ as a weakly converging sequence contained in the set S to an element ξ. Then, on account of Theorem 1.3.50, we can see that $\xi_n(t) \rightharpoonup \xi(t)$, and so

$$(D\xi_n)(t) \rightharpoonup (D\xi)(t).$$

Due to the fact that the set $\{D\xi_n,\ n\in\mathbb{N}\}$ is bounded with bound $\|b\|_\infty r_0$, we can conclude the properties on D. Now, in the similar way as in Theorem 3.2.1, we can show that $(I-D)^{-1}C$ is weakly sequentially continuous.

Claim 4. The set $\{x \in X,\ (I-A)(x) = y\}$ is convex, for each $y \in B(I-D)^{-1}C(S)$. In fact, let $x \in S$ be fixed and take $y, z \in (I-A)^{-1}B(I-D)^{-1}C(x)$. Then,

$$\begin{cases} y = A(y) + B(I - D)^{-1}C(x), \\ z = Az + B(I - D)^{-1}C(x). \end{cases}$$

Due to the fact that k is affine and $B(I - D)^{-1}C$ is constant, we infer that

$$\begin{aligned} \lambda y + (1 - \lambda)z &= \lambda(Ay)(t) + (1 - \lambda)(Az) + B(I - D)^{-1}C(x) \\ &= (A(\lambda y + (1 - \lambda)z)) + B(I - D)^{-1}C(x), \end{aligned}$$

for each $\lambda \in [0, 1]$. This implies that

$$\lambda y + (1 - \lambda)z \in (I - A)^{-1}B(I - D)^{-1}C(x),$$

for any $x \in S$.

Claim 5. Let us fix $x \in S$. Define the mapping $\varphi_x : C([0, 1], X) \longrightarrow C([0, 1], X)$ by putting

$$\varphi_x(y) = B(I - D)^{-1}C(x) + A(y).$$

Then,

$$\varphi_x(y)(t) = \left[q(t) + \int_0^{\sigma(t)} g(t, s, z(s))\, ds\right] \cdot u + k(t, y(t)),$$

where $z \in C([0, 1], X)$ is unique and verifying

$$\|z\|_\infty \leq \frac{L\|m\|_{L^1} + \sup_{t\in[0,1]} |w(t, 0)|}{1 - \|b\|_\infty}.$$

Notice that the mapping φ_x is a contraction with a constant $\|\sigma\|_\infty$ and so, there exists an unique element $y \in C([0, 1], X)$ such that $\varphi_x(y) = y$. This implies that

$$B(I - D)^{-1}C(S) \subset (I - A)(C([0, 1], X)).$$

Since $y \in C([0, 1], X)$, then $\|y\|_\infty = \|y(t^*)\|$, for some $t^* \in [0, 1]$. Thus,

$$\begin{aligned} \|y\|_\infty &\leq \left|q(t^*) + \int_0^{\sigma(t^*)} g(t^*, s, z(s))\, ds\right| \|u\|_\infty + \|k(t^*, y(t^*))\|_\infty \\ &\leq r_0\|u\|_\infty + \|b\|_\infty\|y(t^*)\| + \|k(t^*, 0)\|. \end{aligned}$$

Then,

$$\|y\|_\infty \leq \frac{r_0\|u\|_\infty + \sup_{t\in[0,1]} \|k(t, 0)\|}{1 - \|b\|_\infty}.$$

This means that $B(I - D)^{-1}C(S) \subset (I - A)(S)$. The desired conclusion follows by a direct application of Theorem 3.2.5.

Remark 6.1.2 Problem (6.1.1) is new in functional analysis. Observe that Theorem 6.1.1 contains many special types.

(i) If we take $\varphi(t, x(t)) = 0$, $\sigma(t) = t$ and $b(t) = 1$, for $t \in [0, 1]$ in system (6.1.1), then Theorem 6.1.1 reduces to existence results, established in [3], via of the following nonlinear integral equation

$$x(t) = q(t) + k(t, x(t), x(\lambda(t))) + \int_0^t g(t, x(s), x(\gamma(t)))\, ds.$$

(ii) If we take $k(t, x(t)) = 0$, $\varphi(t, x(t)) = 0$ and $b(t) = 1$ in system (6.1.1), then Theorem 6.1.1 reduces to existence solutions to the following integral equation

$$x(t) = q(t) + \int_0^t g(t, x(s))\, ds.$$

6.2 Existence of Solutions for (FIE) in Banach Algebras Satisfying Condition ($\mathscr{P}$)

Our main interest is dealing with the existence, in the space $C([0, 1], X)$, of solutions for the following two-dimensional, which is presented in [4], of nonlinear functional integral equations (FIE)

$$\begin{cases} x(t) = \psi\left(t, \int_0^t \frac{t}{t+s} h(s, x(s))\, ds\right) u + Ty(t) \cdot \left(\int_0^{\sigma_1(t)} g(s, y(s))\, ds\right) v, \\ y(t) = \left(\int_0^{\sigma_2(t)} p\,(s, x(s))\, ds\right) w + a(t)y(t), \\ (x(0), y(0)) = (x_0, y_0) \in \mathbb{R}^2, \end{cases} \tag{6.2.1}$$

where X is a Banach algebra satisfying condition ($\mathscr{P}$) and $u, v, w \in X \backslash \{0\}$. Let us denote

$$\begin{cases} \varphi(t, x(t)) = \psi\left(t, \int_0^t \frac{t}{t+s} h(s, x(s))\, ds\right) u, \\ G(t, x(t)) = \left(\int_0^{\sigma_1(t)} g(s, x(s))\, ds\right) v, \quad \text{and} \\ K(t, x(t)) = \left(\int_0^{\sigma_2(t)} p\,(s, x(s))\, ds\right) w. \end{cases}$$

The system (6.2.1) will be considered under the following assumptions.

(H_{39}) The mappings $\sigma_1, \sigma_2 : [0, 1] \longrightarrow [0, 1]$ are continuous and nondecreasing with $\|\sigma_2\|_\infty > 1$.

(H_{40}) The mapping $a : [0, 1] \longrightarrow \mathbb{R}$ is continuous such that $\|a\|_\infty < 1$.

(H_{41}) The mapping $g : [0, 1] \times X \longrightarrow X$ is such that
(i) g is continuous uniformly with respect to the first variable,
(ii) g is a contraction with a constant $\|a\|_\infty$ with respect to the second variable,
(iii) g is weakly sequentially continuous with respect to the second variable.

(H_{42}) The mapping $p : [0, 1] \times X \longrightarrow \mathbb{R}$ is such that
(i) p is affine with respect to the second variable,
(ii) p is weakly sequentially continuous with respect to the second variable,
(iii) p is contraction with constant $\frac{1-\|a\|_\infty}{\|\sigma_2\|_\infty}$ with respect to the second variable.

(H_{43}) The mapping $h : [0, 1] \times X \longrightarrow \mathbb{R}$ is such that
(i) h is affine with respect to the second variable,
(ii) h is measurable with respect to the first variable,
(iii) h is weakly sequentially continuous with respect to the second variable,
(iv) there exists a continuous operator $\gamma : J \longrightarrow \mathbb{R}_+$ satisfying

$$|h(t, x) - h(t, y)| \leq \gamma(t)\|x - y\|, \text{ for all } t \in [0, 1] \text{ and } x,\ y \in X,$$

(v) there exists $b > 0$ and there exists a function $m \in L^1(J, X)$ that satisfies

$$|h(t, x)| \leq m(t) \text{ and } \int_0^1 \frac{1}{t+s} m(s)\, ds \leq b.$$

(H_{44}) The mapping $\psi : [0, 1] \times X \longrightarrow \mathbb{R}$ is such that
(i) ψ is affine with respect to the second variable,
(ii) ψ is weakly sequentially continuous with respect to the second variable,
(iii) there exists $L > 0$ such that

$$|\psi(t, x) - \psi(t, y)| \leq L\|x - y\|, \text{ for all } t \in [0, 1] \text{ and } x,\ y \in X.$$

(H_{45}) The mapping $T : C([0, 1], X) \longrightarrow C([0, 1], X)$ is such that
(i) T is affine and regular,
(ii) T is weakly sequentially continuous,
(iii) there exists a function $\rho : [0, 1] \longrightarrow \mathbb{R}_+$ satisfying

$$|Tx(t) - Ty(t)| \leq \rho(t)\|x(t) - y(t)\|,$$

for all $t \in [0, 1]$ and $x,\ y \in C([0, 1], X)$.

(H_{46}) there exists a function $M_1, M_2 \in \mathbb{R}_+^*$ such that
(i) for all $r > 0$, we have $|g(t, x)| \leq M_1$ and $|p(t, x)| \leq M_2$, for all $x \in X$ with $\|x\| \leq r$,

(ii) $\lambda L\|\gamma\|_\infty\|u\|_\infty \ln(2) + \|\rho\|_\infty\|\sigma_1\|_\infty\|v\|_\infty M < \frac{\lambda}{2}$, for each $\lambda \in [0, 1]$,
(iii) $\sup_{t\in[0,1]} \|k(t, x)\| \leq (1 - \|b\|_\infty - \|u\|_\infty)r$, $r > 0$, for $\|x\| \leq r$.

Theorem 6.2.1 *Suppose that the assumptions* (H_{39})-(H_{46}) *are satisfied. Then, problem* (6.2.1) *has, at least, one solution in* $C([0, 1], X) \times C([0, 1], X)$.

Proof Let $S = \overline{\mathscr{B}}_{C([0,1],X)}(r_0)$, where

$$r_0 = \frac{b\|u\|_\infty}{1 - (\lambda L\|\gamma\|_\infty \ln(2)\|u\|_\infty + \|\rho\|_\infty\|\sigma_1\|_\infty\|v\|_\infty M)}.$$

Let us define, on the space $C([0, 1], X)$, five operators A, B, C, D, and B' by putting

$$\begin{cases} (Ax)(t) = \psi\left(t, \displaystyle\int_0^t \frac{t}{t+s} h(s, x(s))\, ds\right) u, \\ (Bx)(t) = (Tx)(t), \\ (B'x)(t) = \left(\displaystyle\int_0^{\sigma_1(t)} g(t, s, x(s))\, ds\right) v, \\ (Cx)(t) = \left(\displaystyle\int_0^{\sigma_2(t)} p\,(s, x(s))\, ds\right) w, \\ (Dx)(t) = a(t)x(t). \end{cases}$$

We show that these operators satisfy the assumptions of Theorem 3.2.16.
Claim 1. Let $(x_n)_n$ be an arbitrary sequence contained in the set S. Based on assumption (H_{42}), we get

$$\begin{aligned} \|y_n(t)\| &= \|(I - D)^{-1}C(x_n(t))\| \\ &\leq \|C(x_n)(t)\| + \|D(I - D)^{-1}C(x_n)(t)\| \\ &\leq \|C(x_n)(t)\| + \|a\|_\infty\|(I - D)^{-1}C(x_n)(t)\| + \|D(I - D)^{-1}C(0)(t)\|. \end{aligned}$$

It is easy to see that, there exists $t^* \in [0, 1]$ such that

$$\begin{aligned} \|y_n\|_\infty &= \|y_n(t^*)\| \\ &\leq \frac{\|w\|_\infty}{1 - \|a\|_\infty}\left[\int_0^{\sigma_2(t^*)} p\,(s, x_n(s))\, ds + \sup_{t\in[0,1]} |p(t, 0)|\right] \\ &\leq \frac{r_0\|w\|_\infty}{1 - \|a\|_\infty}\left(\|\sigma_2\|_\infty M_2 + 1\right). \end{aligned}$$

Then, the sequence $\left((I - D)^{-1}C(x_n)(t)\right)_n$ is uniformly bounded with bound ϱ, where

$$\varrho = \frac{r_0\|w\|_\infty}{1 - \|a\|_\infty}\left(\|\sigma_2\|_\infty M_2 + 1\right).$$

Consequently, the set $(I - D)^{-1}C(S)(t)$ is sequentially relatively weakly compact. We proceed to prove the weak equi-continuity of the set $(I - D)^{-1}C(S)$. Let us fix $\varepsilon > 0$ and take $x \in S$, $x^* \in X^*$ and $t, t' \in [0, 1]$ (assume that $|t - t'| \leq \varepsilon$). Then,

$$\begin{aligned}
&\left\|x^*\left(((I-D)^{-1}C(x_n))(t')-((I-D)^{-1}C(x_n))(t)\right)\right\| \\
&\leq \frac{1}{1-\|a\|_\infty}\left|\int_{\sigma_2(t)}^{\sigma_2(t')} p\left(s, x(s)\right) ds\right| \|x^*(w)\| \\
&\leq \frac{M_2 r_0 \|x^*(w)\|}{1-\|a\|_\infty} \sup_{|t-t'|\leq\varepsilon} \left\{\left|\sigma_2(t')-\sigma_2(t)\right|\right\}.
\end{aligned}$$

Since σ_2 is continuous on $[0, 1]$, it follows that

$$\lim_{t\to t'} \left|x^*\left(((I-D)^{-1}C(x_n))(t)-((I-D)^{-1}C(x_n))(t')\right)\right| = 0.$$

The use of Theorem 1.3.42 as well as Theorem 1.3.29 leads to $(I-D)^{-1}C(S)$ is relatively weakly compact.

Claim 2. Let $x \in C([0, 1], X)$ be fixed and take a sequence $(t_n)_n$ contained in the set $[0, 1]$ such that $t_n \to t$. Then,

$$\begin{aligned}
\|(Ax)(t_n)-(Ax)(t)\| &\leq L\left|\int_0^{t_n} \frac{t_n}{t_n+s} h(s, x(s))\, ds - \int_0^t \frac{t}{t+s} h(s, x(s))\, ds\right| \|u\|_\infty \\
&\leq \left|\int_0^{t_n} \frac{s(t_n-t)}{(t_n+s)(t+s)} h(s, x(s))\, ds\right| \|u\|_\infty + \left|\int_{t_n}^t \frac{t}{t+s} h(s, x(s))\, ds\right| \\
&\leq \left(b|t_n-t| + \int_{t_n}^t m(s)\, ds\right) \|u\|_\infty.
\end{aligned}$$

Hence, in view on assumption (H_{43}), we get $Ax \in C([0, 1], X)$. Similarly, we can show Cx, $B'x \in C([0, 1], X)$. Consider $(\xi_n)_n$ as a weakly convergent sequence contained in the set S to an element ξ. Then, on account of Theorem 1.3.50, we get

$$\xi_n(t) \rightharpoonup \xi(t) \quad \text{in } X, \quad \text{for all } t \in [0, 1].$$

Using assumption (H_{43}), we obtain

$$\frac{t}{t+s} h\left(s, \xi_n(s)\right) \rightharpoonup \frac{t}{t+s} h\left(s, \xi(s)\right) \text{ in } \mathbb{R}.$$

The use of assumption (H_{44}) and dominated convergence theorem, we infer that

$$(A(\xi_n))(t) \rightharpoonup (A\xi)(t) \text{ in } X.$$

Due to the fact that $(A(\xi_n))_n$ is bounded with a bound η, where

$$\eta = [L\|m\|_{L^1} + \|A(0)\|_\infty]\, \|u\|_\infty,$$

and Theorem 1.3.50, we get

$$A(\xi_n) \rightharpoonup A\xi.$$

By using assumption (H_{42}) and assumption (H_{46}), together with the dominated convergence theorem, we can prove that

$$\lim_{n\to\infty}\int_0^{\sigma_2(t)} p\,(s, \xi_n(s))\; ds = \int_0^{\sigma_2(t)} p\,(s, \xi(s))\; ds.$$

Taking into account that $(C(\xi_n))_n$ is bounded with a bound $\|\sigma_2\|_\infty M_2\|w\|_\infty$, and using the sequential condition ($\mathscr{P}$), we get

$$C(\xi_n) \to C\xi.$$

Similarly, we can show that D is weakly sequentially continuous on S. Our next task is to show the weak sequential continuity of the operator B' on the set $(I - D)^{-1}C(S)$. In fact, let $(y_n)_n$ be a weakly convergent sequence contained in the set $(I - D)^{-1}C(S)$ to an element y. Observe that the set $\{y_n,\ n \in \mathbb{N}\}$ is bounded and so, by Theorem 1.3.50, we obtain

$$y_n(t) \rightharpoonup y(t) \quad \text{in } X.$$

For any $t \in [0, 1]$, we can write $B'y_n(t) = r_n(t)v$, where

$$r_n(t) = \int_0^{\sigma_1(t)} g\,(s, y_n(s))\; ds.$$

Due to the fact that $(r_n(t))_n$ is a bounded sequence with bound $M_1\|\sigma_1\|_\infty$, we infer that there exists a renamed subsequence such that $r_n(t) \to r(t)$. The sequential condition ($\mathscr{P}$) guaranties that $B'y_n \rightharpoonup B'y$.

Claim 3. For all $x, y \in C([0, 1], X)$, we have

$$\begin{aligned}
\|(Ax)(t) - (Ay)(t)\| &\le \left|\psi\left(t, \int_0^t \frac{t}{t+s}h(s, x(s))\, ds\right) - \psi\left(t, \int_0^t \frac{t}{t+s}h(s, y(s))\, ds\right)\right| \|u\|_\infty\\
&\le L\left|\int_0^t \frac{t}{t+s}h(s, x(s))\, ds - \int_0^t \frac{t}{t+s}h(s, y(s))\, ds\right| \|u\|_\infty\\
&\le L\int_0^t \frac{t}{t+s}\,|h(s, x(s)) - h(s, y(s))|\; ds\|u\|_\infty\\
&\le L\int_0^t \frac{1}{t+s}\gamma(s)\|x(s) - y(s)\|\; ds\|u\|_\infty\\
&\le L\|\gamma\|_\infty \ln(2)\|x - y\|_\infty\|u\|_\infty.
\end{aligned}$$

Then, on account of Lemma 1.5.9, we arrive at A is 1-set-weakly contractive.

Claim 4. We show that there is satisfied hypothesis (iv) of Theorem 3.2.5. Let $x \in S$ be fixed and take $y, z \in C([0, 1], X)$ such that

$$\begin{cases} y(t) = \lambda Ay(t) + B(I - D)^{-1}Cy(t)\cdot B'(I - D)^{-1}Cx(t)\\ z(t) = \lambda Az(t) + B(I - D)^{-1}Cz(t)\cdot B'(I - D)^{-1}Cx(t).\end{cases}$$

Then, for all $\tau \in [0, 1]$, we have

$$\begin{aligned}
[\tau y + (1-\tau)z](t) &= \lambda\psi\left(t, \int_0^t \frac{t}{t+s}\tau h(s, y(s))\, ds + \int_0^t \frac{t}{t+s}(1-\tau)h(s, z(s))\, ds\right) u \\
&\quad + [T(\tau y_1 + (1-\tau)z_1)](t) \cdot \left(\int_0^{\sigma_1(t)} g(t, s, x_1(s))\, ds\right) v \\
&= \lambda\psi\left(t, \int_0^t \frac{t}{t+s} h\left(s, \tau y(s) + (1-\tau)z(s)\right)(s)\, ds\right) u \\
&\quad + B\left(\tau y_1 + (1-\tau)z_1\right)(t) \cdot B'\left(x_1\right)(t) \\
&= \lambda A\left(\tau y_1 + (1-\tau)z_1\right)(t) + B\left(\tau y_1 + (1-\tau)z_1\right)(t) \cdot B'(x_1)(t).
\end{aligned}$$

where

$$\begin{cases}
x_1(t) = (I-D)^{-1}Cx(t) = (1-a(t))^{-1} \cdot \left(\int_0^{\sigma_2(t)} p(s, x(s))\, ds\right) w \\
y_1(t) = (I-D)^{-1}Cy(t) = (1-a(t))^{-1} \cdot \left(\int_0^{\sigma_2(t)} p(s, y(s))\, ds\right) w \\
z_1(t) = (I-D)^{-1}Cz(t) = (1-a(t))^{-1} \cdot \left(\int_0^{\sigma_2(t)} p(s, z(s))\, ds\right) w.
\end{cases}$$

Thus, in view of hypothesis (H_{42}), we have

$$[\tau y + (1-\tau)z](t) \in \left(\frac{I - \lambda A}{B(I-D)^{-1}C}\right)^{-1} B'(I-D)^{-1}Cx(t).$$

Claim 5. Let $x \in S$ be fixed. Define the mapping $\varphi_x : C([0,1], X) \longrightarrow C([0,1], X)$ by putting

$$\varphi_x(y)(t) = \lambda\psi\left(t, \int_0^t \frac{t}{t+s} h(s, y(s))\, ds\right) u + Ty_1(t) \cdot \left(\int_0^{\sigma_1(t)} g(s, x_1(s))\, ds\right) v,$$

where $x_1, y_1 \in C([0,1], X)$ satisfying

$$\|x_1\|_\infty \leq \frac{r_0\|w\|_\infty}{1 - \|a\|_\infty}$$

and

$$\|y_1\|_\infty \leq \frac{\|w\|_\infty}{1 - \|a\|_\infty}\|y\|_\infty.$$

Let $y, z \in C([0,1], X)$ be fixed. Then,

$$\begin{aligned}\|\varphi_x(y)-\varphi_x(z)\| &\leq \|\lambda Ay(t)-\lambda Az(t)\| + \|B(I-D)^{-1}Cy(t)-B(I-D)^{-1}Cz(t)\|\|B'(I-D)^{-1}Cx(t)\|\\ &\leq \left\|\lambda\psi\left(t,\int_0^t \frac{t}{t+s}h(s,y(s))\,ds\right)u-\lambda\psi\left(t,\int_0^t \frac{t}{t+s}h(s,z(s))\,ds\right)u\right\|\\ &\quad+\|\rho\|_\infty\left\|(I-D)^{-1}Cy-(I-D)^{-1}Cz\right\|\left\|\left(\int_0^{\sigma_1(t)} g(s,(I-D)^{-1}Cx(s))\,ds\right)v\right\|\\ &\leq \left[\lambda L\|\gamma\|_\infty \ln(2)\|u\|_\infty+\|\rho\|_\infty\|\sigma_1\|_\infty\|v\|_\infty M_1\right]\|y-z\|.\end{aligned}$$

It follows from hypothesis (H_{46}), that φ_x is a contraction with a Lipschitz constant

$$\Delta=\lambda L\|\gamma\|_\infty\ln(2)\|u\|_\infty+\|\rho\|_\infty\|\sigma_1\|_\infty\|v\|_\infty M_1.$$

From Theorem 1.2.4, we get there exists an unique function $y\in C([0,1],X)$ such that $\varphi_x(y)=y$. Thus,

$$B'(I-D)^{-1}C(S)\subset\left(\frac{I-\lambda A}{B(I-D)^{-1}C}\right)(C([0,1],X)).$$

It is obvious that, there exists $t^*\in[0,1]$ such that $\|y\|_\infty=\|y(t^*)\|$ and so,

$$\begin{aligned}\|y\|_\infty &\leq \|\lambda Ay(t^*)\|+\|B(I-D)^{-1}Cy(t^*)\|\|B'(I-D)^{-1}Cx(t^*)\|\\ &\leq \lambda L\left[\int_0^{t^*}\frac{t^*}{t^*+s}|h(s,y(s))-h(s,0)|\,ds+\int_0^{t^*}\frac{t^*}{t^*+s}|h(s,0)|\,ds\right]\|u\|_\infty\\ &\quad+\|\rho\|_\infty\left\|(I-D)^{-1}Cy(t^*)-(I-D)^{-1}C(0)(t^*)\right\|\left\|\left(\int_0^{\sigma_1(t^*)} g(s,(I-D)^{-1}Cx(s))\,ds\right)v\right\|\\ &\leq \lambda L\|\gamma\|_\infty\ln(2)\|u\|_\infty\|y\|_\infty+b\|u\|_\infty+\|\rho\|_\infty\|\sigma_1\|_\infty\|v\|_\infty M_1\|y\|_\infty.\end{aligned}$$

This yields

$$\|y\|_\infty\leq\frac{b\|u\|_\infty}{1-\Delta}.$$

The desired conclusion follows by a direct application of Theorem 3.2.16. Q.E.D.

Remark 6.2.2 Problem (6.2.1) is new in the literature, so the obtained results of this section are new to the theory of differential inclusions and contain many special cases. If we take $a(t)=1$, $\varphi(t,x(t))=c(t)$, $\sigma_2(t)=0$ and

$$G(t,x(t))=\left[q(t)+\int_0^{\sigma(t)}p(t,s,x(\zeta(s)),x(\eta(s)))\,ds\right]u,$$

then Theorem 6.2.1 reduces to the existence results, proved in [5], for the following nonlinear equation

$$\begin{cases}x(t)=c(t)+(Tx)(t)\cdot G(t,x(t))\\ x(0)=x_0\in\mathbb{R},\end{cases}$$

where c, q, ζ, η and σ are continuous on $[0,1]$, T and p are nonlinear functions and u is a non vanishing vector of a Banach algebra X satisfying certain topological conditions of sequential nature. If we take $a(t) = 1$, $\varphi(t, x(t)) = c(t)$, $\sigma_2(t) = 0$ and $(Tx)(t) = f(t, x(t))$ in problem (6.2.1), then Theorem 6.2.1 reduces to the existence solutions, proved in [6], of the following quadratic integral equation

$$\begin{cases} x(t) = c(t) + g(t, x(t)) \cdot \int_0^t f(s, x(s))\, ds \\ x(0) = x_0 \in \mathbb{R}. \end{cases}$$

References

1. Jeribi, A., Kaddachi, N., Krichen, B.: Fixed point theorems for multi-valued operator matrix under weak topology with an application. Bull. Malays. Math. Sci. Soc. **43**, 1047–1067 (2020)
2. Kaddachi, N.: Existence results of ordinary differential inclusions in Banach algebra under weak topology. J. Phys. Math. **9** (2018)
3. Dhage, B.C.: Global attractivity results for nonlinear functional integral equations via a Krasnoselskii type fixed point theorem. Nonlinear Anal. **70**, 2485–2493 (2009)
4. Kaddachi, N.: Generalized form of fixed point theorems in Banach algebras under weak topology with an application. Filomat **33**, 4281–4296 (2019)
5. Ben Amar, A., Chouayekh, S., Jeribi, A.: Fixed point theory in a new class of Banach algebras and application. Afr. Mat. **24**, 705–724 (2013)
6. El-Sayed, A.M.A., Hashem, H.H.G.: Integrable and continuous solutions of nonlinear quadratic integral equation. Electron. J. Qual. Theory Diff. Equ. **25**, 1–10 (2008)

Chapter 7
Two-Dimensional Value Problems for Differential Inclusions

In our final chapter, we discuss the existence of solutions for several nonlinear differential inclusions. Such problems are considered on compact intervals of the real line. We start the discussion with a quadratic initial value problems for differential inclusion. Next, we extend the results to the case of periodic boundary value problems of the first order (in short VPDI). In the second part of this chapter, we study several coupled systems of nonlinear functional integral inclusions and differential inclusions with bounded domains in Banach algebra satisfying the sequential condition ($\mathscr{P}$). This chapter is devoted to applications of the material in Chap. 3. For more details concerning the material of this chapter, we recommend [1–3]. Our novelty is mainly in the conditions we will impose on a nonlinearity multi-valued and in the method, we will use to prove the existence result which is, in fact, based upon continuous and Kakutani's fixed point theorem.

Existence of solutions for differential inclusions has been studied by many authors in the past half century with different application-directed motivations, including the issues of control and optimization, dynamical systems , and even biological sciences, see for example [4–11] and the references therein. In some earlier works, set-valued maps have been usually considered with convex values, see [12–14].

7.1 Nonlinear Differential Inclusions

A differential inclusion is a relation of the kind $x'(t) \in G(t, x(t))$, where G is a given multi-valued mapping, that we call an ordinary differential inclusion. In the first case, one seeks a function $x(\cdot)$ in general in the class of absolutely continuous functions, such that for almost every t in some interval, the derivative $x'(t)$ exists and contained in the set $G(t, x(t))$. In addition, initial or boundary conditions may be prescribed. This kind of problem appears in [15] in a form different, that is, on the left-hand side a multi-valued derivative. An approach to investigating the existence of solutions for

A. Jeribi and N. Kaddachi, *Topological Fixed-Point Theory in Suitable Banach Algebras with Applications*, Infosys Science Foundation Series,
https://doi.org/10.1007/978-981-96-5542-7_7

a differential inclusion is to reduce it to a problem for ordinary differential equations. In this setting, we are interested to know, under what conditions, the solution of the corresponding ordinary differential equation belongs to the right side of the given differential inclusion.

7.1.1 Multi-valued Initial Value Problems

We discuss on $C(J, \mathbb{R})$, where $J = [0, a]$, $a \in \mathbb{R}_+^*$, the existence of solutions of the following nonlinear differential inclusions of first order (IVP)

$$\begin{cases} \left(\dfrac{x(t) - k(t, x(t))}{f(t, y(t))}\right)' \in G(t, y(t)) \\ y(t) = \dfrac{1}{1 + b(t)|x(\theta(t))|} - p\left(t, \dfrac{1}{1 + b(t)|x(\theta(t))|}\right) + p(t, y(t)) \\ (x(0), y(0)) = (x_0, y_0) \in \mathbb{R}^2, \end{cases} \tag{7.1.1}$$

where $G : J \times \mathbb{R} \longrightarrow \mathscr{P}_{cp,cv}(\mathbb{R})$ and $x = x(t)$ and $y = y(t)$ are unknown functions, that satisfies, the following conditions.

(i) The partial $t \longrightarrow \dfrac{x(t) - k(t, x(t))}{f(t, y(t))}$ is differentiable, and

(ii) $\left(\dfrac{x(t) - k(t, x(t))}{f(t, y(t))}\right)' = v(t)$, for some $v \in L^1(J, \mathbb{R})$ such that $v(t) \in G(t, y(t))$.

Observe that the coupled system (7.1.1) may be transformed into the form

$$\begin{cases} x(t) \in k(t, x(t)) + f(t, y(t)) \cdot \left[\dfrac{x_0 - k(0, x_0)}{f(0, y_0)} + \displaystyle\int_0^t G(t, y(s))\, ds\right] \\ y(t) = \dfrac{1}{1 + b(t)|x(\theta(t))|} - p\left(t, \dfrac{1}{1 + b(t)|x(\theta(t))|}\right) + p(t, y(t)) \\ (x(0), y(0)) = (x_0, y_0) \in \mathbb{R}^2. \end{cases}$$

In what follows and for our subsequent analysis, we need the following assumptions.

(H_{47}) The mapping $k : J \times \mathbb{R} \longrightarrow \mathbb{R}$ is such that

(i) k is continuous,

(ii) k is generalized Lipschitzian with function l_1 with respect to the second variable, that is, there exists a continuous function $l_1 : J \longrightarrow \mathbb{R}$ with bound $\|l_1\|_\infty \leq \frac{1}{4}$ such that

$$|k(t, x) - k(t, y)| \leq l_1(t)|x - y|, \quad \text{for all } x,\ y \in \mathbb{R}.$$

(H_{48}) The mapping $f : J \times \mathbb{R} \longrightarrow \mathbb{R}\backslash\{0\}$ is such that

(i) f is continuous,

(ii) f is generalized Lipschitzian with function l_2 with respect to the second variable.

(H_{49}) The multi-valued mapping $G : J \times \mathbb{R} \longrightarrow \mathscr{P}_{cp,cv}(\mathbb{R})$ is $L^1_{\mathbb{R}}$-Carathéodory with a growth function h.

(H_{50}) The mapping $p : J \times \mathbb{R} \longrightarrow \mathbb{R}$ is such that
(i) the partial $x \longrightarrow p(t, x)$ is a contraction with a constant q,
(ii) the partial $t \longrightarrow p(t, x)$ is continuous,
(iii) $|p(t, 0)| = 0$, for all $t \in J$.

(H_{51}) The mapping $b : J \longrightarrow \mathbb{R}_+$ is continuous.

Theorem 7.1.1 *Suppose that the assumptions (H_{47})-(H_{51}) are satisfied. Moreover, if*

$$\begin{cases} 4\|b\|_\infty\|l_2\|\left(\left|\dfrac{x_0 - k(0, x_0)}{f(0, y_0)}\right| + \|h\|_{L^1}\right) \le 1 \\ q < \dfrac{1 - 4\|l_2\|\|b\|_\infty\left(\left|\dfrac{x_0 - k(0, x_0)}{f(0, y_0)}\right| + \|h\|_{L^1}\right)}{1 + 4\|l_2\|\left(\left|\dfrac{x_0 - k(0, x_0)}{f(0, y_0)}\right| + \|h\|_{L^1}\right)} \\ \max\left\{\|l_2\|\left(\left|\dfrac{x_0 - k(0, x_0)}{f(0, y_0)}\right| + \|h\|_{L^1}\right),\ K + \left(\left|\dfrac{x_0 - k(0, x_0)}{f(0, y_0)}\right| + \|h\|_{L^1}\right)F\right\} \le \dfrac{1}{4}, \end{cases} \tag{7.1.2}$$

where $K = \sup_{t \in J} |k(t, 0)|$ and $F = \sup_{t \in J} |f(t, 0)|$, then the system (7.1.1) has a solution in $C(J, \mathbb{R}) \times C(J, \mathbb{R})$.

Proof Let us consider, on the space $X = C(J, \mathbb{R})$, five operators A, B, C, D, and B' by putting

$$\begin{cases} Ax(t) = k(t, x(t)), \\ Bx(t) = f(t, x(t)), \\ Cx(t) = \dfrac{1}{1 + b(t)|x(\theta(t))|} - p\left(t, \dfrac{1}{1 + b(t)|x(\theta(t))|}\right), \\ Dx(t) = p(t, x(t)), \\ B'x = \left\{u \in X \text{ such that } u(t) = \dfrac{x_0 - k(0, x_0)}{f(0, y_0)} + \displaystyle\int_0^t v(s)\,ds,\ v \in S^1_G(x)\right\}, \end{cases}$$

where S^1_G is given by the formula

$$S^1_G(x) = \{v \in L^1(J, \mathbb{R}),\ v(t) \in G(t, x(t)),\ \text{for all } t \in J\}.$$

Then, the system IVP (7.1.1) can be transformed into the following fixed point problem

$$\begin{cases} x(t) \in Ax(t) + By(t) \cdot B'y(t) \\ y(t) = Cx(t) + Dy(t). \end{cases}$$

We prove that, these operators A, B, C, D, and B' satisfy all the assumptions of Theorem 3.3.4. Define the two sets S and S' of $C(J, \mathbb{R})$ by putting

$$S = \{y \in C(J, \mathbb{R}),\ \|y\|_\infty \leq 1 + q\}$$

and

$$S' = \left\{y \in C(J, \mathbb{R}),\ \|y\|_\infty \leq \frac{4}{1-q}\right\}.$$

Since $S^1_G(x) \neq \emptyset$, for each $x \in X$, it follows that B' is well defined.
Claim 1. We show that $B' : X \longrightarrow \mathscr{P}_{cp,cv}(X)$. At first, we prove that $B'x \in \mathscr{P}_{cp}(X)$, for all $x \in X$. After simple calculations, we can see that $B' = \mathscr{K} \circ S^1_G$, composition of two mapping on $L^1(J, \mathbb{R})$, where $\mathscr{K} : L^1(J, \mathbb{R}) \longrightarrow X$ is the continuous operator defined by

$$\mathscr{K} v(t) = \frac{x_0 - k(0, x_0)}{f(0, y_0)} + \int_0^t v(s)\, ds.$$

Let $x \in X$ be fixed and take a sequence $(v_n)_n$ contained in the set $\in S^1_G(x)$. By the definition of $S^1_G(x)$, we can verify that $v_n(t) \in G(t, x(t))$ and so, there exists a renamed convergent subsequence of $(v_n(t))_n$ in measure to some $v(t) \in G(t, x(t))$. Due to the fact that $\mathscr{K}$ is continuous, we infer that

$$\mathscr{K} v_n(t) \to \mathscr{K} v(t).$$

We prove now the equi-continuity of the sequence $(\mathscr{K} v_n)_n$ so as, to demonstrate the uniformity convergence. Let us fix $t_1, t_2 \in J$, then

$$|\mathscr{K} v_n(t_1) - \mathscr{K} v_n(t_2)| \leq \left| \int_{t_1}^{t_2} v_n(s)\, ds \right|.$$

Due to the fact that $v_n \in L^1(J, \mathbb{R})$ and the criterion of Cauchy, we can deduce that

$$\lim_{t_1 \to t_2} \int_{t_1}^{t_2} v_n(s)\, ds = 0.$$

This implies that, the sequence $(\mathscr{K} v_n)_n$ is equi-continuous. An application of Theorem 1.3.42 yields that, the sequence $(\mathscr{K} v_n)_n$ possesses a uniformly convergent subsequence. By composition, we obtain the mapping $\left(\mathscr{K} \circ S^1_G\right)(x)$ is compact, for any $x \in X$. Let u_1 and u_2 be two arbitrarily fixed elements of the set $B'x$. Then, there exist two elements $v_1, v_2 \in S^1_G(x)$ such that

$$\begin{cases} u_1(t) = \dfrac{x_0 - k(0, x_0)}{f(0, y_0)} + \displaystyle\int_0^t v_1(s)\, ds \\ u_2(t) = \dfrac{x_0 - k(0, x_0)}{f(0, y_0)} + \displaystyle\int_0^t v_2(s)\, ds. \end{cases}$$

Consider $\lambda \in [0, 1]$. Then,

$$\begin{aligned} \lambda u_1(t) + (1-\lambda)u_2(t) &= \frac{x_0 - k(0, x_0)}{f(0, y_0)} + \int_0^t (\lambda v_1(s) + (1-\lambda)v_2(s))\, ds \\ &= \frac{x_0 - k(0, x_0)}{f(0, y_0)} + \int_0^t v_3(s)\, ds, \end{aligned}$$

where $v_3(s) = \lambda v_1(s) + (1-\lambda)v_2(s) \in G(s, x(s))$. This implies that

$$\lambda u_1 + (1-\lambda)u_2 \in B'x.$$

In other words, we have $B'x \in \mathscr{P}_{cv}(X)$, for all $x \in X$ and the first claim is proved.
Claim 2. After simple calculations, we can see that

$$\|Ax - Ay\| \leq \|l_1\|_\infty \|x - y\|,$$

whenever $x, y \in X$. Similarly, we can verify that, the operators B and C are Lipschitzians with constants $\|l_2\|$ and $\|b\|_\infty + k$, respectively.
Claim 3. Let $x \in S$ be fixed and define the mapping $\varphi_x : C(J, \mathbb{R}) \longrightarrow C(J, \mathbb{R})$ by putting

$$\varphi_x(y)(t) = \frac{1}{1 + b(t)|x(\theta(t))|} - p\left(t, \frac{1}{1 + b(t)|x(\theta(t))|}\right) + p(t, y(t)).$$

After simple calculations, we can verify that φ_x is a contraction with a constant k, in view of assumption (H_{50}). It follows from Theorem 1.2.4, that φ_x has exactly one fixed point y in X. This implies that

$$C(S) \subset (I - D)(C(J, \mathbb{R})).$$

Since $y \in C(J, \mathbb{R})$, then there exists $t^* \in J$ such that

$$\begin{aligned} \|y\|_\infty &= |y(t^*)| \\ &= |Cx(t^*) + Dy(t^*)| \\ &\leq \frac{1}{1 + b(t^*)|x(\theta(t^*))|} + \left|p\left(t^*, \frac{1}{1 + b(t^*)|x(\theta(t^*))|}\right)\right| + |p(t^*, y(t^*))| \\ &\leq 1 + q + q|y(t^*)|. \end{aligned}$$

Thus,

$$|y(t^*)| \leq \frac{1+q}{1-q}.$$

This means that $C(S) \subset (I - D)(S')$ and the claim is proved.
Claim 4. Let $x \in S'$ and take $u \in B'(S')$. By composition, we have

$$u(t) = \frac{x_0 - k(0, x_0)}{f(0, y_0)} + \int_0^t v(s)\, ds,$$

for some $v \in S^1_G(x)$. Based on hypothesis (H_{49}), we have

$$\begin{aligned} |u(t)| &\leq \left| \frac{x_0 - k(0, x_0)}{f(0, y_0)} \right| + \int_0^t |v(s)|\, ds \\ &\leq \left| \frac{x_0 - k(0, x_0)}{f(0, y_0)} \right| + \int_0^t \|G(s, x(s))\|_{\mathcal{P}}\, ds \\ &\leq \left| \frac{x_0 - k(0, x_0)}{f(0, y_0)} \right| + \int_0^t h(s)\, ds \\ &\leq \left| \frac{x_0 - k(0, x_0)}{f(0, y_0)} \right| + \|h\|_{L^1}. \end{aligned}$$

Since $x \in S'$ is arbitrary, we get $B'(S')$ is a uniformly bounded subset of X. Similarly as in Claim 1, we can show the equi-continuity of the set $B'(S')$. Our next task is to show the upper semi-continuity of the multi-valued mapping B'. Let $(x_n)_n$ be a convergent sequence contained in the set S to a point x and take a convergent sequence $y_n \in B'x_n$ to a point y. Since $y_n \in B'x_n$, there exists a sequence $v_n \in S^1_G(x_n)$, that satisfies,

$$y_n(t) = \frac{x_0 - k(0, x_0)}{f(0, y_0)} + \int_0^t v_n(s)\, ds.$$

We want to show that, there exists a function $v \in S^1_G(x)$ satisfying

$$y(t) = \frac{x_0 - k(0, x_0)}{f(0, y_0)} + \int_0^t v(s)\, ds, \ t \in J.$$

Define the linear continuous mapping $\mathcal{K} : L^1(J, \mathbb{R}) \longrightarrow C(J, \mathbb{R})$ by putting

$$\mathcal{K} v(t) = \int_0^t v(s)\, ds, \ t \in J.$$

It should be noted, on account of Lemma 1.4.29, that the graph of $\mathcal{K} \circ S^1_G(x)$ is closed. Again, by using the definition of the operator $\mathcal{K}$, we get

$$y_n(t) - \frac{x_0 - k(0, x_0)}{f(0, y_0)} \in \mathcal{K} \circ S^1_G(x).$$

Due to the fact that $y_n \to y$, we infer that there exists a function $v \in S^1_G(x)$ satisfying

$$y(t) = \frac{x_0 - k(0, x_0)}{f(0, y_0)} + \int_0^t v(s)\, ds, \ t \in J.$$

This means that B' is upper semi-continuous and so, it is completely continuous. Based on assumption (H_{49}), it is shown that

$$\begin{aligned} M &= \|B'(I - D)^{-1}C(S)\|_{\mathscr{P}} \\ &= \sup\{\|B'(I - D)^{-1}Cx\|_{\mathscr{P}},\ x \in S\} \\ &\leq \sup\{\|B'x\|_{\mathscr{P}},\ x \in S'\} \\ &\leq \left|\frac{x_0 - k(0, x_0)}{f(0, y_0)}\right| + \|h\|_{L^1}. \end{aligned}$$

In order to simplify the calculations, it is assumed that $x_0 = k(0, x_0)$. Going back to estimate (7.1.2), we can prove that

$$q\left(1 + \frac{1}{4\|h\|_{L^1}\|l_2\|}\right) + \|b\|_\infty < \frac{1}{4\|h\|_{L^1}\|l_2\|}.$$

This implies that

$$\frac{\|l_2\|\,(\|b\|_\infty + q)}{1 - q}\|h\|_{L^1} < \frac{1}{4}.$$

Consequently,

$$\|l_1\| + \frac{\|l_2\|\,(\|b\|_\infty + q)}{1 - q}\|h\|_{L^1} < \frac{1}{2}.$$

Finally, let $x \in S$ be fixed and take $y \in X$ such that

$$y \in A(x) + B(I - D)^{-1}C(x) \cdot B'(I - D)^{-1}C(x).$$

In other words, we have

$$y(t) = k(t, x(t) + f(t, z(t)) \int_0^t w(s)\, ds, \ t \in J,$$

for some $z \in S'$ and $w \in S^1_G(z)$. Therefore, we have

$$\begin{aligned} |y(t)| &\leq |k(t, x(t))| + |f(t, z(t))| \int_0^t |w(s)|\, ds \\ &\leq |k(t, x(t)) - k(t, 0)| + |k(t, 0)| + (|f(t, z(t)) - f(t, 0)| + |f(t, 0)|)\, \|h\|_{L^1} \\ &\leq \|l_1\||x(t)| + K + (\|l_2\||z(t)| + F)\|h\|_{L^1} \\ &\leq (1 + q)(\|l_1\| + \|l_2\|\|h\|_{L^1}) + K + \|l_2\|\|h\|_{L^1} + F\|h\|_{L^1} \\ &\leq 1 + q. \end{aligned}$$

Hence, $y \in S$. So, all the conditions of Theorem 3.3.4 are satisfied and, then its application ensures that, the system (7.1.1) has, at least, one solution in $S \times S'$.

Q.E.D.

Remark 7.1.2 The system (7.1.1) is new in functional analysis. In the special case, where

$$x = y, \quad f(t, x) = 1 \quad \text{and} \quad p(t, y) = y,$$

then Theorem 7.1.1 reduces to the existence of solutions, established by J. P. Aubin and A. Cellina in [16], see also [17], of the following nonlinear initial value problem (IVP)

$$\begin{cases} x'(t) \in G(t, x(t)) \\ x(0) = x_0 \in \mathbb{R}. \end{cases}$$

Similarly, if we take $x = y$ and $p(t, x) = x$ and G is a single-valued mapping, then Theorem 7.1.1 reduces to the existence results, proved in [18, 19], of the following differential equation

$$\begin{cases} \dfrac{\partial}{\partial t}\left(\dfrac{x(t) - k(t, x(t))}{f(t, y(t))}\right) = g(t, y(t)) \\ x(0) = x_0 \in \mathbb{R}. \end{cases}$$

7.1.2 *Periodic Boundary Value Problems of First Order*

The results of this subsection can be found in [2]. Let us discuss, on the space $C(J, \mathbb{R}) \times C(J, \mathbb{R})$, where $J = [0, \mathscr{T}]$, $0 < \mathscr{T} < +\infty$, the existence of solutions of the following periodic boundary value problem, for the first order of ordinary differential inclusion

$$\begin{cases} \left(\dfrac{x(t) - k(t, x(t))}{f(t, y(t))}\right)' + h(t)\left(\dfrac{x(t) - k(t, x(t))}{f(t, y(t))}\right) \in G_h(t, x(t), y(t)) \\ y(t) = \dfrac{1}{1 + b(t)|x(\theta(t))|} - p\left(t, \dfrac{1}{1 + b(t)|x(\theta(t))|}\right) + p(t, y(t)) \\ (x(0), y(0)) = (x(\mathscr{T}), y(\mathscr{T})) \in \mathbb{R}^2, \end{cases} \tag{7.1.3}$$

where $G_h : J \times \mathbb{R} \times \mathbb{R} \longrightarrow \mathscr{P}_{cp,cv}(\mathbb{R})$ is given by

$$G_h(t, x, y) = G(t, y) + h(t)\left(\frac{x - k(t, x)}{f(t, y)}\right).$$

By a solution of problem (7.1.3), we mean two absolutely continuous functions x, $y : J \longrightarrow \mathbb{R}$ such that the following conditions are satisfied

(i) The partial $t \longrightarrow \dfrac{x(t) - k(t, x(t))}{f(t, y(t))}$ is differentiable, and

(ii) there exists $v \in L^1(J, \mathbb{R})$ with $v(t) \in G(t, y(t))$ such that

$$\left(\frac{x(t) - k(t, x(t))}{f(t, y(t))}\right)' = v(t).$$

The interest of the operators in the form that satisfies the above conditions lies in the following lemma, which can be found in [20].

Lemma 7.1.3 (Nieto) *Let $h \in L^1(J, \mathbb{R}^+)$ and $\sigma \in L^1(J, \mathbb{R})$ be two functions satisfying the following differential equation*

$$\begin{cases} x' + h(t)x = \sigma(t) \\ x(0) = x(\mathcal{T}). \end{cases} \tag{7.1.4}$$

Then, a function x is a solution of the equation (7.1.4) if, and only, if it is a solution of the following equation

$$x(t) = \int_0^{\mathcal{T}} g_h(t, s)\sigma(s)\, ds,$$

where

$$g_h(t, s) = \begin{cases} \dfrac{e^{H(s)-H(t)+H(\mathcal{T})}}{e^{H(\mathcal{T})} - 1} & \text{if } 0 \le s \le t \le \mathcal{T} \\ \dfrac{e^{H(s)-H(t)}}{e^{H(\mathcal{T})} - 1} & \text{if } 0 \le t < s \le \mathcal{T}, \end{cases} \tag{7.1.5}$$

here $H : J \longrightarrow \mathbb{R}$ is given by $H(t) = \int_0^t h(s)\, ds$.

Let us now introduce the following assumptions, in order to study the existence of solutions of the nonlinear system (7.1.3).

(H_{52}) $f, k : J \times \mathbb{R} \longrightarrow \mathbb{R}$ are periodic of period $\mathcal{T}$ with respect the first variable,

(H_{53}) The partial $(x, y) \longrightarrow \dfrac{x - k(0, x)}{f(0, y)}$ is one-to-one on $\mathbb{R}^2$.

Notice that, if the assumptions (H_{52}) and (H_{53}) are satisfied, then a point (x, y) is a solution of (7.1.3) if, and only, if (x, y) satisfies the following inclusion

$$\begin{cases} x(t) \in k(t, x(t)) + f(t, y(t)) \cdot \int_0^{\mathscr{T}} g_h(t,s) G_h(s, x(s), y(s))\, ds \\ y(t) = \dfrac{1}{1 + b(t)|x(\theta(t))|} - p\left(t, \dfrac{1}{1 + b(t)|x(\theta(t))|}\right) + p(t, y(t)) \\ (x(0), y(0)) = (x(\mathscr{T}), y(\mathscr{T})) \in \mathbb{R}^2, \end{cases}$$

for any bounded function $h \in L^1(J)$, where the function g_h is described in (7.1.5).

When dealing with the system (7.1.3), some difficulties arise, so we need the following assumptions.

(H_{54}) The mapping $k : J \times \mathbb{R} \longrightarrow \mathbb{R}$ is such that
(i) k is continuous,
(ii) The following estimate

$$|k(t, x) - k(t, y)| \leq l_1(t)|x - y|, \quad \text{for all } x,\ y \in \mathbb{R}$$

is satisfied, for some bounded function $l_1 : J \longrightarrow \mathbb{R}$ with bound $\|l_1\|_\infty \leq \frac{1}{6}$.

(H_{55}) The mapping $f : J \times \mathbb{R} \longrightarrow \mathbb{R}\backslash\{0\}$ is such that
(i) f is continuous,
(ii) The following estimate

$$|f(t, x) - f(t, y)| \leq l_2(t)|x - y|, \quad \text{for all } x,\ y \in \mathbb{R}$$

is satisfied, for some bounded function $l_2 : J \longrightarrow \mathbb{R}$.

(H_{56}) The mapping $p : J \times \mathbb{R} \longrightarrow \mathbb{R}$ is such that
(i) the partial $t \longrightarrow p(t, x)$ is Lipschitzian with a Lipschitz constant $q \in]0, \frac{1}{4}[$,
(ii) the partial $x \longrightarrow p(t, x)$ is continuous,
(iii) $|p(t, 0)| = 0$, for all $t \in J$.

(H_{57}) The mapping $b : J \longrightarrow \mathbb{R}_+$ is continuous such that $4\|b\|_\infty \leq 1$.

(H_{58}) $G : J \times \mathbb{R} \longrightarrow \mathscr{P}_{cp,cv}(\mathbb{R})$ is Carathéodory.

(H_{59}) There exists a continuous and increasing function $\varphi : \mathbb{R}_+ \longrightarrow \mathbb{R}_+$ and there exists a function $\rho \in L^1(J, \mathbb{R}_+^*)$ such that

$$\|G_h(t, x, y)\|_{\mathscr{P}} \leq \rho(t)\varphi(|y|),$$

for all x, $y \in \mathbb{R}$.

We are ready to state the main result of this section.

Theorem 7.1.4 *Assume that the assumptions (H_{52})-(H_{59}) are satisfied. Moreover, if there exists $r_0 > \frac{3}{2}$ satisfying the following conditions*

$$\begin{cases} \|l_1\|_\infty + \|l_2\|_\infty M_h \|\rho\|_{L^1} \varphi(r_0) < \dfrac{1}{2} \\ K + F M_h \|\rho\|_{L^1} \varphi(r_0) \leq \dfrac{r_0}{2}, \end{cases} \tag{7.1.6}$$

where $K = \sup_{t \in J}\{|k(t,0)|\}$ *and* $F = \sup_{t \in J}\{|f(t,0)|\}$, *then the system* (7.1.3) *has a solution in* $C(J, \mathbb{R}) \times C(J, \mathbb{R})$.

Proof Consider the mapping A, B, C, D, and B', on the space $X = C(J, \mathbb{R})$, by putting

$$\begin{cases} Ax(t) = k(t, x(t)), \\ Bx(t) = f(t, x(t)), \\ Cx(t) = \dfrac{1}{1 + b(t)|x(\theta(t))|} - p\left(t, \dfrac{1}{1 + b(t)|x(\theta(t))|}\right), \\ Dx(t) = p(t, x(t)), \\ B'x = \left\{u \in X \text{ such that } u(t) = \dfrac{x_0 - k(0, x_0)}{f(0, y_0)} + \displaystyle\int_0^t g_h(t,s) v(s)\, ds,\ v \in S^1_{G_h}(x)\right\}. \end{cases}$$

Then, problem (7.1.3) is equivalent to the following coupled system

$$\begin{cases} x(t) \in Ax(t) + By(t) \cdot B'y(t) \\ y(t) = Cx(t) + Dy(t). \end{cases}$$

By using Lemma 1.4.28, together with assumption (H_{59}), we can show that the set $S^1_{G_h}(x)$ is nonempty, for all $x \in \overline{\mathscr{B}}_{C(J,\mathbb{R})}(r_0)$. By the definition of B', we get

$$B' = \mathscr{K} \circ S^1_{G_h}.$$

We prove that $B' : \overline{\mathscr{B}}_{C(J,\mathbb{R})}(r_0) \longrightarrow \mathscr{P}_{cp,cv}(X)$. In fact, for all $u_1, u_2 \in B'x$, there are two functions $v_1, v_2 \in S^1_{G_h}(x)$ such that

$$\begin{cases} u_1(t) = \dfrac{x_0 - k(0, x_0)}{f(0, y_0)} + \displaystyle\int_0^t g_h(t,s) v_1(s)\, ds \\ u_2(t) = \dfrac{x_0 - k(0, x_0)}{f(0, y_0)} + \displaystyle\int_0^t g_h(t,s) v_2(s)\, ds. \end{cases}$$

Then,

$$\begin{aligned} u_3(t) &= \lambda u_1(t) + (1-\lambda)u_2(t) \\ &= \frac{x_0 - k(0,x_0)}{f(0,y_0)} + \int_0^t g_h(t,s)\left(\lambda v_1(s) + (1-\lambda)v_2(s)\right) ds \\ &= \frac{x_0 - k(0,x_0)}{f(0,y_0)} + \int_0^t v_3(s)\, ds, \end{aligned}$$

for each $\lambda \in [0,1]$, where

$$v_3(s) = \lambda v_1(s) + (1-\lambda)v_2(s) \in G_h(s, x(s), y(s)),$$

for all $s \in J$. This implies that $u_3 \in B'x$. Thus, in virtue of assumption (H_{59}), we get

$$B' : \overline{\mathscr{B}}_{C(J,\mathbb{R})}(r_0) \longrightarrow \mathscr{P}_{cp,cv}(C(J,\mathbb{R}))$$

is well defined. In the similar way as in Theorem 7.1.1, we can show that the mappings A, B, and C are Lipschitzians with constants $\|l_1\|_\infty$, $\|l_2\|_\infty$ and $\|b\|_\infty + q$, respectively. Based on assumption (H_{56}), it is shown that

$$C\left(\overline{\mathscr{B}}_{C(J,\mathbb{R})}(r_0)\right) \subset (I-D)\left(C(J,\mathbb{R})\right).$$

Since $y \in X$, then there exists a point $t^* \in J$ such that

$$\begin{aligned} \|y\|_\infty &= |Cx(t^*) + Dy(t^*)| \\ &\leq \frac{1+q}{1-q} \\ &\leq r_0. \end{aligned}$$

This means, in particular, that

$$C\left(\overline{\mathscr{B}}_{C(J,\mathbb{R})}(r_0)\right) \subset (I-D)\left(\overline{\mathscr{B}}_{C(J,\mathbb{R})}(r_0)\right).$$

Next, we demonstrate that the set $B'\left(\overline{\mathscr{B}}_{C(J,\mathbb{R})}(r_0)\right)$ is uniformly bounded and equicontinuous. To this end, let us fix arbitrarily $u \in B'\left(\overline{\mathscr{B}}_{C(J,\mathbb{R})}(r_0)\right)$. By the definition of B', we get there exists a function $v \in S^1_{G_h}(x)$, for some $x \in \overline{\mathscr{B}}_{C(J,\mathbb{R})}(r_0)$ such that

$$u(t) = \frac{x_0 - k(0,x_0)}{f(0,y_0)} + \int_0^t g_h(t,s)v(s)\, ds.$$

This implies that

$$\begin{aligned}
|u(t)| &\le \left|\frac{x_0 - k(0, x_0)}{f(0, y_0)}\right| + \int_0^t g_h(t, s)\, \|G_h(s, x(s), x(s))\|_{\mathscr{P}}\, ds \\
&\le \left|\frac{x_0 - k(0, x_0)}{f(0, y_0)}\right| + \int_0^t g_h(t, s)\rho(s)\varphi(|x(s)|)\, ds \\
&\le \left|\frac{x_0 - k(0, x_0)}{f(0, y_0)}\right| + \int_0^t g_h(t, s)\|\rho\|_{L^1}\varphi(r_0)\, ds \\
&\le \left|\frac{x_0 - k(0, x_0)}{f(0, y_0)}\right| + M_h\|\rho\|_{L^1}\varphi(r_0).
\end{aligned}$$

Thus, $B'\left(\overline{\mathscr{B}}_{C(J,\mathbb{R})}(r_0)\right)$ is uniformly bounded. Finally, it remains to be shown that the derivative u' is bounded on J, for all $u \in B'\left(\overline{\mathscr{B}}_{C(J,\mathbb{R})}(r_0)\right)$. In fact, for all $t \in J$, we get

$$\begin{aligned}
|u'(t)| &\le \int_0^t \frac{\partial}{\partial t} g_h(t, s)\, \|G_h(s, x(s), y(s))\|_{\mathscr{P}}\, ds \\
&\le \int_0^t (-h(t))\|\rho\|_{L^1}\varphi(r_0)\, ds \\
&\le \max_{t\in J}\{h(t)\} M_h\|\rho\|_{L^1}\varphi(r_0).
\end{aligned}$$

Then, for all $t_1, t_2 \in J$, we obtain

$$|B'x(t_1) - B'x(t_2)| \le \left(\max_{t\in J}(h(t)) M_h\|\rho\|_{L^1}\varphi(r_0)\right)|t_1 - t_2|.$$

This prove that the set $B'\left(\overline{\mathscr{B}}_{C(J,\mathbb{R})}(r_0)\right)$ is equi-continuous and so, it is totally bounded, in view of Theorem 1.3.42. Now, we show the upper semi-continuity of the multi-valued mapping B'. Indeed, consider $(x_n)_n$ as a convergent sequence contained in $C(J, \mathbb{R})$ to a point x and take a sequence $y_n \in B'x_n$ such that $y_n \to y$. Due to the fact $y_n \in B'x_n$, there exists a function $v_n \in S^1_G(x_n)$ satisfying

$$y_n(t) = \frac{x_0 - k(0, x_0)}{f(0, y_0)} + \int_0^t g_h(t, s)v_n(s)\, ds, \ t \in J.$$

Using an argument similar to the one above, we can find a function $v \in S^1_{G_h}(x)$ such that

$$y(t) = \frac{x_0 - k(0, x_0)}{f(0, y_0)} + \int_0^t g_h(t, s)v(s)\, ds,$$

which means that B' is upper semi-continuous. In our next step, we show that there is satisfied assumption (iv) of Corollary 3.3.4 on $\overline{\mathscr{B}}_{C(J,\mathbb{R})}(r_0)$. At first, observe that

$$\begin{aligned}
M &= \|B'(I - D)^{-1}C(S)\|_{\mathscr{P}} \\
&\le M_h\|\rho\|_{L^1}\varphi(r_0) + \left|\frac{x_0 - k(0, x_0)}{f(0, y_0)}\right|.
\end{aligned}$$

In order to simplify, we assume that $x_0 = k(0, x_0)$. Going back to (7.1.6), we get

$$\begin{aligned}\frac{\|l_2\|\,(\|b\|_\infty + q)}{1-q}M + \|l_1\|_\infty &< \frac{\|b\|_\infty + q}{1-q}\left(\frac{1}{2} - \|l_1\|_\infty\right) + \|l_1\|_\infty \\ &< \frac{1}{2}.\end{aligned}$$

Let $x \in S$ be fixed and let $y \in X$ such that

$$y \in A(x) + B(I-D)^{-1}C(x) \cdot B'(I-D)^{-1}C(x).$$

Then, there exist two functions $z \in \overline{\mathscr{B}}_{C(J,\mathbb{R})}(r_0)$ and $v \in S^1_{G_h}(z)$ such that

$$y(t) = k(t, x(t) + f(t, z(t)) \cdot \int_0^t g_h(t,s)v(s)\,ds, \ t \in J.$$

This yields that

$$\begin{aligned}|y(t)| &\le |k(t,x(t))| + |f(t,z(t))| \int_0^t g_h(t,s)|v(s)|\,ds \\ &\le |k(t,x(t)) - k(t,0)| + |k(t,0)| + (|f(t,z(t)) - f(t,0)| + |f(t,0)|)\, M_h \|\rho\|_{L^1}\varphi(r_0) \\ &\le \|l_1\||x(t)| + K + (\|l_2\|_\infty |z(t)| + F) M_h \|\rho\|_{L^1}\varphi(r_0) \\ &\le \left(\|l_1\|_\infty + \|l_2\|_\infty M_h \|\rho\|_{L^1}\varphi(r)\right) r_0 + K + F M_h \|\rho\|_{L^1}\varphi(r_0) \\ &< r_0.\end{aligned}$$

Hence, $y \in \overline{\mathscr{B}}_{C(J,\mathbb{R})}(r_0)$, for an arbitrary $x \in \overline{\mathscr{B}}_{C(J,\mathbb{R})}(r_0)$, and when we apply Corollary 3.3.4, we reach the desired conclusion. Q.E.D.

7.2 Biological Coupled System Under Weak Topology

Let X be a Banach algebra satisfying condition $(\mathscr{P})$. The aim of this section is to discuss the existence of results, on $C(J, X)$, where $J = [0, \mathscr{T}]$, $0 < \mathscr{T} < \infty$, of the following system of quadratic problems for ordinary differential inclusion arising in biology problems

$$\begin{cases} \dfrac{\partial}{\partial t}\left(\dfrac{x(t) - k(t, x(t))}{f(t, y(t))}\right) \in G(t, y(t)) \\ y(t) - f_1(t, y(t)) = \left(\lambda + \displaystyle\int_0^{\sigma(t)} p(s, x(\lambda s))\,ds\right) w, \\ (x(0), y(0)) = (x_0, y_0) \in \mathbb{R}^2 \end{cases} \tag{7.2.1}$$

for all $t \in J$, where $w \in X\backslash\{0\}$ is fixed with $\|w\|_\infty < 1$, $\lambda \in [\frac{1}{2}, 1[$, while x and y are unknown functions. Suppose that the functions involved in problem (7.2.1) verify the following conditions

(H_{60}) $f, k : J \times X \longrightarrow X$ are weakly sequentially continuous and such that
(i) the partial $x \longrightarrow f(t, x)$ is a contraction with a constant q_1,
(ii) the partial $x \longrightarrow k(t, x)$ is $\mathscr{D}$-Lipschitzian with $\mathscr{D}$-function ϕ_k.

(H_{61}) $p : J \times X \longrightarrow \mathbb{R}$ is such that
(i) $x \longrightarrow p(t, x)$ is weakly sequentially continuous,
(ii) $x \longrightarrow p(t, x)$ is $\mathscr{D}$-Lipschitzian with $\mathscr{D}$-function ϕ_p.

(H_{62}) $f_1 : J \times X \longrightarrow X$ is such that
(i) $x \longrightarrow f_1(t, x)$ is weakly sequentially continuous,
(ii) $x \longrightarrow f_1(t, x)$ is nonlinear contraction with $\mathscr{D}$-function Φ_{f_1},
(iii) $\|f_1(t, 0)\| = (1 - \lambda)\|w\|_\infty$, for all $t \in J$,
(iv) $\phi_{f_1}(r) < (1 - \lambda)r$, for all $r > 0$.

(H_{63}) $b : J \longrightarrow \mathbb{R}_+$ is continuous.

(H_{64}) $G : J \times X \longrightarrow \mathscr{P}_p(X)$ is such that
(i) there exists a measurable function v satisfying $v(t) \in G(t, x(t))$, for all $t \in J$, and for each $x \in C(J, X)$,
(ii) the partial $x \longrightarrow G(t, x)$ is weakly completely continuous,
(iii) for all $r > 0$, there exists a function $h_r \in L^1(J, \mathbb{R})$ such that

$$\|G(t, x)\|_{\mathscr{P}} \leq h_r(t), a.e \text{ for } t \in J,$$

for all $x \in \mathbb{R}$ with $|x| \leq r$.

Theorem 7.2.1 *Suppose that the assumptions* (H_{60})-(H_{64}) *are satisfied. Moreover, if there exists* $r_0 > \frac{2}{1-\mathscr{T}}$ *satisfying the following conditions*

$$\begin{cases} \phi_k(r_0) + K + F\|h_{r_0}\|_{L^1} + q_1 r_0 \|h_{r_0}\|_\infty \leq r_0, \\ \|p(t, x)\| \leq \lambda r_0, \text{ for } x \in C(J, X) \text{ with } \|x\|_\infty \leq r_0, \end{cases}$$

then problem (7.2.1) *has, at least, one solution in* $\overline{\mathscr{B}}_{C(J,X)}(r_0) \times \overline{\mathscr{B}}_{C(J,X)}(r_0)$.

Proof Let $S = \overline{\mathscr{B}}_{C(J,X)}(r_0)$ and define five operators A, B, C, D, and B' on $C(J, X)$ by putting

$$\begin{cases} Ax(t) = k(t, x(t)) \\ Bx(t) = f(t, x(t)) \\ Cx(t) = \left(\lambda + \int_0^{\sigma(t)} p(s, x(\lambda s))\, ds\right) w \\ Dx(t) = f_1(t, x(t)) \\ B'x(t) = \left\{ \dfrac{x_0 - k(0, x_0)}{f(0, y_0)} + \displaystyle\int_0^t v(s)\, ds,\ v \text{ is Pettis-integrable},\ v(t) \in G(t, x(t)) \right\}. \end{cases}$$

Then, problem (7.2.1) can be transformed into the following fixed point problem

$$\begin{cases} x(t) \in Ax(t) + By(t) \cdot B'y(t) \\ y(t) = Cx(t) + Dy(t). \end{cases}$$

We shall show that A, B, C, D, and B' satisfy all the conditions of Theorem 3.3.13. This will be achieved in a series of steps.

Step 1. The assumption (H_{60}) guaranties that

$$Ax,\ Bx \in C(J, X), \quad \text{for all } x \in S.$$

Let $(x_n)_n$ be a weakly convergent sequence contained in the set S to an element x. Then, on account of Theorem 1.3.50, we get

$$(t, x_n(t)) \rightharpoonup (t, x(t)) \ \text{ in } J \times X.$$

Based on assumption (H_{60}), we obtain

$$(A(x_n))(t) \rightharpoonup (Ax)(t) \ \text{ in } X. \tag{7.2.2}$$

Due to the fact that the set $\{A(x_n),\ n \in \mathbb{N}\}$ is bounded, and using (7.2.2), we infer that A is weakly sequentially continuous. Similarly, we can prove that B, C, and D are also three weakly sequentially continuous operators. Now, we show that

$$B'x \in \mathscr{P}_{cv}(S), \ \text{for all } \ x \in C(J, X).$$

Indeed, let $x \in C(J, X)$ be fixed and take $u_1, u_2 \in B'x$. Then, there exists two Pettis-integrable functions v_1 and v_2 with $v_1(s), v_2(s) \in G(s, x(s))$ such that

$$\begin{cases} u_1(t) = \dfrac{x_0 - k(0, x_0)}{f(0, y_0)} + \displaystyle\int_0^t v_1(s)\, ds \\ u_2(t) = \dfrac{x_0 - k(0, x_0)}{f(0, y_0)} + \displaystyle\int_0^t v_2(s)\, ds. \end{cases}$$

Consider $\gamma \in [0, 1]$. Then

$$\gamma u_1(t) + (1-\gamma)u_2(t) = \frac{x_0 - k(0, x_0)}{f(0, y_0)} + \int_0^t v(s)\, ds,$$

where $v(s) = \gamma v_1(s) + (1-\gamma)v_2(s) \in G(s, x(s))$, for any $s \in J$. This implies that

$$\gamma u_1 + (1-\gamma)u_2 \in B'x.$$

Also, for an arbitrary $u \in B'x$, there exists a Pettis-integrable function v such that $v(s) \in G(s, x(s))$ and

$$u(t) = \frac{x_0 - k(0, x_0)}{f(0, y_0)} + \int_0^t v(s)\, ds.$$

If we take $t, t' \in J$ with $u(t) \neq u(t')$, then there exists $\varphi \in (C(J, X))^*$ with $\|\varphi\| = 1$ such that

$$\|u(t) - u(t')\| = \varphi\left(u(t) - u(t')\right).$$

Using hypothesis (H_{64}), we obtain

$$\|u(t) - u(t')\| \leq \int_t^{t'} h_r(s)\, ds.$$

Thus, $B'x \in \mathscr{P}\left(C(J, X)\right)$.

Step 2. After simple calculations, we can prove that

$$\begin{aligned}
\|Cx - Cy\| &\leq \|w\|_\infty \sup_{t\in J} \int_0^{\sigma(t)} \|p(s, x(\lambda s)) - p(s, y(\lambda(s)))\|\, ds \\
&\leq \|w\|_\infty \sup_{t\in J} \int_0^{\sigma(t)} \phi_p\left(\|x(\lambda s) - y(\lambda s)\|\right) ds \\
&\leq \|w\|_\infty \sup_{t\in J} \int_0^{\sigma(t)} \phi_p\left(\|x - y\|_\infty\right) ds \\
&\leq \|w\|_\infty \mathscr{T} \phi_p\left(\|x - y\|_\infty\right),
\end{aligned}$$

whenever $x, y \in C(J, X)$. Then, C is $\mathscr{D}$-Lipschitzian with $\mathscr{D}$-function ψ, where

$$\psi(r) = \|w\|_\infty \mathscr{T} \phi_p(r).$$

Using an argument similar, we may demonstrate that A and B are $\mathscr{D}$-Lipschitzians with $\mathscr{D}$-functions $\phi_A = \phi_k$ and $\phi_B(r) = q_1 r$, respectively. Using Lemma 1.5.9, we get A, B, and C are $\mathscr{D}$-set-Lipschitzians with $\mathscr{D}$-functions ϕ_A, ϕ_B and ψ, respectively.
Step 3. The use of assumption (H_{62}) as well as Theorem 1.2.26 leads to $(I - D)^{-1}$ exists on $C(J, X)$. Let us take a point $(x, y) \in S \times C(J, X)$ such that the following equality

$$y(t) = \left(\lambda + \int_0^{\sigma(t)} p\,(t, x(\lambda s))\; ds\right) w + f_1\,(t, y(t))$$

is satisfied, for all $t \in J$. Since $y \in C(J, X)$, then $\|y\|_\infty = \|y(t^*)\|$, for some $t^* \in J$. Then,

$$\begin{aligned}\|y\|_\infty &\le \left|\lambda + \int_0^{\sigma(t^*)} p\left(t^*, x(\lambda s)\right) ds\right| \|w\|_\infty + \|f_1(t^*, y(t^*)) - f_1(t^*, 0)\| + \|f_1(t^*, 0\| \\ &\le (\lambda + \mathscr{T}\lambda r_0)\, \|w\|_\infty + \phi_{f_1}\left(\|y(t^*)\|\right) + (1-\lambda)\|w\|_\infty.\end{aligned}$$

This implies that

$$\|y\|_\infty \le \mathscr{T} r_0 \|w\|_\infty + 2.$$

This means that $(I - D)^{-1}C(S) \subseteq S$.

Step 4. Consider $(x_n)_n$ as a weakly convergent sequence contained in the set S to an element x and take $y_n \in B'(x_n)$ such that $y_n \rightharpoonup y$. Then, there exists a sequence of Pettis-integrable mapping $(v_n)_n$ with $v_n(s) \in G(s, x_n(s))$ such that

$$y_n(t) = \frac{x_0 - k(0, x_0)}{f(0, y_0)} + \int_0^t v_n(s)\, ds.$$

Using the weak completely continuity of the multi-valued mapping $G(t, \cdot)$, for all $t \in J$, we obtain the set $\{v_n(s),\ n \in \mathbb{N}\}$ as well as the set $\{v_n,\ n \in \mathbb{N}\}$ is relatively weakly compact. Then, the sequence $(v_n)_n$ possesses a weakly convergent subsequence, say $(v_{n_k})_k$, to a function v. Then, invoking assumption (H_{64}), we have v is Pettis-integrable, and

$$\frac{x_0 - k(0, x_0)}{f(0, y_0)} + \int_0^t v_n(s)\, ds \rightharpoonup \frac{x_0 - k(0, x_0)}{f(0, y_0)} + \int_0^t v(s)\, ds, \quad \text{for all } t \in J.$$

This implies that $v(s) \in G(s, x(s))$ and $y \in B'x$. Hence, B' has a weakly sequentially closed graph on S.

Step 5. We show that the set $B'(S)$ is relatively weakly compact. In fact, let $(x_n)_n$ be a sequence contained in the set S and take a sequence $y_n \in B'(S)$. Then, there exists a sequence of Pettis-integrable mapping $v_n(s) \in G(s, x_n(s))$ such that

$$y_n(t) = \frac{x_0 - k(0, x_0)}{f(0, y_0)} + \int_0^t v_n(s)\, ds.$$

Based on assumption (H_{64}), we get $(v_n)_n$ possesses a renamed weakly convergent subsequence to a function v. Then, on account of Theorem 1.3.50, we arrive at $B'(S)$ is relatively weakly compact. Let $x \in C(J, X)$ be fixed and take $z, z' \in X$ such that

$$z,\ z' \in A(x) + B(I - D)^{-1}C(x) \cdot B'(I - D)^{-1}C(x).$$

From the above discussion, we conclude that there exists exactly one point $y \in S$ such that

$$\begin{cases} z(t) = f(t, y(t)) \left[\dfrac{x_0 - k(0, x_0)}{f(0, y_0)} + \displaystyle\int_0^t v(s)\, ds \right] + k(t, x(t)), \quad \text{and} \\ z'(t) = f(t, y(t)) \left[\dfrac{x_0 - k(0, x_0)}{f(0, y_0)} + \displaystyle\int_0^t v'(s)\, ds \right] + k(t, x(t)), \end{cases}$$

where v, v' are two Pettis-integrable functions with $v(s), v'(s) \in G(s, y(s))$. Consider $\gamma \in [0, 1]$. Then,

$$\gamma z(t) + (1 - \gamma) z'(t) = f(t, y(t)) \int_0^t \big[\gamma v(s) + (1 - \gamma) v'(s)\big]\, ds + k(t, x(t)).$$

Using hypothesis (H_{64}), we obtain

$$\gamma z + (1 - \gamma) z' \in A(x) + B(I - D)^{-1} C(x) \cdot B'(I - D)^{-1} C(x).$$

This implies that

$$A(x) + B(I - D)^{-1} C(x) \cdot B'(I - D)^{-1} C(x) \in \mathscr{P}_{cv}(C(J, X)).$$

Second, if we fix a point $x \in S$ and take $z \in A(x) + B(I - D)^{-1} C(x) \cdot B'(I - D)^{-1} C(x)$, then there exists exactly one point $y \in S$ and there exists a Pettis-integrable function v with $v(s) \in G(s, y(s))$ such that

$$y(t) = k(t, x(t)) + f(t, y(t)) \cdot \int_0^t v(s)\, ds, \; t \in J.$$

After simple calculations, we can see that

$$\begin{aligned} |y(t)| &\leq |k(t, x(t))| + |f(t, y(t))| \int_0^t |v(s)|\, ds \\ &\leq |k(t, x(t)) - k(t, 0)| + |k(t, 0)| + (|f(t, y(t)) - f(t, 0)| + |f(t, 0)|)\, \|h_{r0}\|_{L^1} \\ &\leq \phi_k(|x(t)|) + K + [q_1 |y(t)| + F]\, \|h_{r0}\|_{L^1} \\ &\leq \phi_k(|x(t)|) + K + [q_1 |y(t)| + F]\, \|h_{r0}\|_{L^1}. \end{aligned}$$

This yields that

$$\|y\|_\infty \leq \frac{\phi_k(r_0) + K + F \|h_{r_0}\|_{L^1}}{1 - q_1 \|h_{r_0}\|_\infty}.$$

Hence, $A(x) + B(I - D)^{-1} C(x) \cdot B'(I - D)^{-1} C(x) \in \mathscr{P}_{cv}(S)$, for all $x \in S$, and when we apply Theorem 3.3.13, we reach the desired conclusion. Q.E.D.

7.3 Existence Result in Banach Space of a Nonlinear Integral Inclusions

Let X be a Banach space (not necessarily reflexive). The results of this section can be found in [1]. We are mainly concerned with the existence of solutions on the space $E = C([0, 1], X)$, for the following problem of nonlinear integral inclusion

$$\begin{cases} x(t) \in \displaystyle\int_0^{\sigma_1(t)} k(t, s)\, F(s, x(\tau(s)))\, ds + \int_0^{\sigma_2(t)} G(s, y(s))\, ds \\ y(t) = \Phi\left(t, \displaystyle\int_0^{\sigma_3(t)} f(s, x(\kappa(s)))\, ds\right) + g(t, y(t)), \end{cases} \tag{7.3.1}$$

for all $t \in J = [0, 1]$, where σ_1, τ, σ_2, σ_3, κ are continuous real functions on J, the real function k is continuous on $J \times J$ and $f, g, \Phi : J \times X \longrightarrow X$ are supposed to be weakly sequentially continuous with respect to the second variables. Here $x = x(t)$ and $y = y(t)$ are unknown functions. In our consideration, we need the following definition.

Definition 7.3.1 A mapping $h : [0, 1] \times X \longrightarrow X$ is called to be a weak Carathéodory if h is measurable with respect the first variable and it is weakly sequentially continuous with respect the second variable.

Let us recall some facts concerning the superposition operators required below.

Definition 7.3.2 Let $m(J, X)$ be the set of all measurable functions $\psi : J \longrightarrow X$. If h is a weak Carathéodory function, then we can define the mapping N_h on $m(J, X)$ by the formula

$$(N_h\psi)(t) = h(t, \psi(t)).$$

The mapping N_h is called to be the Nemytskii operator generated by h.

In what follows, suppose that

(H_{65}) $\sigma_1, \tau, \sigma_2, \sigma_3, \kappa : J \longrightarrow J$ and $k : J \times J \longrightarrow \mathbb{R}$ are continuous.

(H_{66}) $f, g : J \times X \longrightarrow X$ are such that
- (i) f, g are weakly sequentially continuous with respect to the second variable,
- (ii) the partial $x \longrightarrow f(t, x)$ is Lipschitzian with a Lipschitz constant $\lambda > 0$,
- (iii) the partial $x \longrightarrow g(t, x)$ is Lipschitzian with a Lipschitz constant $\gamma < 1$,
- (iv) the partial $x \longrightarrow f(t, x(t))$ is Pettis-integrable.

(H_{67}) The function $\Phi : J \times X \longrightarrow X$ is such that
- (i) the partial $x \longrightarrow \Phi(t, x)$ is weakly sequentially continuous,
- (ii) the partial $x \longrightarrow \Phi(t, x)$ is Lipschitzian with a Lipschitz constant $\frac{1-\gamma}{2\lambda}$,
- (iii) The partial $t \longrightarrow \Phi(t, x)$ is continuous uniformly, for $x \in X$.

(H_{68}) $F, G : J \times X \longrightarrow \mathscr{P}_{cv}(X)$ are such that

(i) for each $x \in E$, there exist two measurable functions $v, w : J \longrightarrow X$ such that $v(t) \in F(t, x(t))$ and $w(t) \in G(t, x(t))$,

(ii) there exist two functions $\alpha_F, \alpha_G \in L^1(J)$ satisfying

$$\|F(s, x)\|_{\mathscr{P}} \le \alpha_F(s)\|x\| \quad \text{and} \quad \|G(s, x)\|_{\mathscr{P}} \le \alpha_G(s)\|x\|,$$

(iii) there exists a nondecreasing continuous mapping $\varphi : \mathbb{R}^+ \longrightarrow \mathbb{R}^+$ satisfying $\varphi(r) < (1 - \delta)r$, for all $r > 0$ and $\varphi(0) = 0$ such that

$$\beta(F(J \times M)) \le \varphi(\beta(M)),$$

for any bounded subset M of X,

(iv) there exists $\delta > 0$ such that

$$\beta(G(J \times M(J))) \le \delta\, \beta(M(J)),$$

for all countably bounded subset M of E,

(v) F and G have weakly sequentially closed graphs with respect to the second variable,

(vi) the partial $t \longrightarrow F(t, x)$ has a convex graph, for all $x \in X$.

Theorem 7.3.3 *Assume that assumptions (H_{65})-(H_{68}) are satisfied. If there exists $r > 0$ such that*

$$\begin{cases} \dfrac{\|\alpha_G\|_{L^1}}{1 - M\|\alpha_F\|_{L^1}} < 1, \\ \dfrac{1}{1-\gamma}(\|g(\cdot, \theta)\|_\infty + \|\Phi(\cdot, \theta)\|_\infty) + \dfrac{1}{2\lambda}\|f(\cdot, \theta)\|_\infty < r, \end{cases}$$

where $M = \|k\|_\infty$, then problem (7.3.1) has, at least, one solution.

Proof Define on E four operators A, B, C and D by putting

$$\begin{cases} (Ax)(t) = \left\{ \displaystyle\int_0^{\sigma_1(t)} k(t, s)v(s)\, ds, \ v \text{ is Pettis Integrabe with } v(s) \in F(s, x(\tau(s))) \right\}, \\ (By)(t) = \left\{ \displaystyle\int_0^{\sigma_2(t)} v(s)\, ds, \ \ v \text{ is Pettis Integrabe with } v(s) \in G(s, y(s)) \text{ a.e. } s \in J \right\}, \\ (Cx)(t) = \Phi\left(t, \displaystyle\int_0^{\sigma_3(t)} f(s, x(\kappa(s)))\, ds\right), \\ (Dy)(t) = g(t, y(t)). \end{cases} \tag{7.3.2}$$

Let Γ_{eq} be a nonempty and equi-continuous subset of E and let us define the subsets Γ and Ω on E by

$$\Gamma = \bigcup_{\Gamma_{eq} \subset E} \Gamma_{eq}$$

and

$$\Omega = \overline{co}\left(\Gamma \bigcap \overline{\mathscr{B}}_E(r)\right).$$

We prove that these operators satisfy the assumptions of Theorem 3.3.41. At first, let us observe that assumption (H_{66}) guarantees that D maps E into itself. Due to the fact that C is equivalent to the composition $\mathscr{N}_\Phi \circ \mathscr{K}$ of two function on E, where $\mathscr{K}$ is the continuous operator defined on $\overline{\mathscr{B}}_E(r)$ by

$$(\mathscr{K}x)(t) = \int_0^{\sigma_3(t)} f(s, x(\kappa(s)))\, ds.$$

Let $x \in \overline{\mathscr{B}}_E(r)$ be fixed and take $t, t' \in [0, 1]$. Then,

$$\begin{aligned}\|Cx(t) - Cx(t')\| &= \|\mathscr{N}_\Phi\mathscr{K}(x)(t) - \mathscr{N}_\Phi\mathscr{K}(x)(t')\| \\ &\leq \left\|\mathscr{N}_\Phi\mathscr{K}(x)(t) - \Phi\left(t, \mathscr{K}(x)(t')\right)\right\| + \left\|\Phi\left(t, \mathscr{K}(x)(t')\right) - \mathscr{N}_\Phi\mathscr{K}(x)(t')\right\|.\end{aligned}$$

Based on assumption (H_{67}), we get

$$\|Cx(t) - Cx(t')\| \leq \frac{1-\gamma}{2\,\lambda}\left\|\int_{\sigma_3(t)}^{\sigma_3(t')} f(s, x(\kappa(s)))\, ds\right\| + \|\Phi(t, \mathscr{K}(x)(t')) - \Phi(t', \mathscr{K}(x)(t'))\|.$$

Hence, on account of Theorem 1.3.18, we have there exists $\varphi \in X^*$ with $\|\varphi\| = 1$ such that

$$\left\|\int_{\sigma_3(t)}^{\sigma_3(t')} f(s, x(\kappa(s)))\, ds\right\| = \varphi\left(\int_{\sigma_3(t)}^{\sigma_3(t')} f(s, x(\kappa(s)))\, ds\right).$$

From assumption (H_{66}), it follows that

$$\begin{aligned}\left\|\int_{\sigma_3(t)}^{\sigma_3(t')} f(s, x(\kappa(s)))\, ds\right\| &\leq \int_{\sigma_3(t)}^{\sigma_3(t')} \|f(s, x(\kappa(s))) - f(s, \theta)\|\, ds + \|f(\cdot, \theta)\|_\infty\, |\sigma_3(t) - \sigma_3(t')| \\ &\leq [\lambda\, r + \|f(\cdot, \theta)\|_\infty]\, |\sigma_3(t) - \sigma_3(t')|.\end{aligned}$$

So,

$$\|Cx(t) - Cx(t')\| \leq k_\lambda\, |\sigma_3(t) - \sigma_3(t')| + \|\Phi(t, \mathscr{K}(x)(t')) - \Phi(t', \mathscr{K}(x)(t'))\|,$$

where

$$k_\lambda = \frac{1-\gamma}{2\,\lambda}\,(\lambda\, r + \|f(\cdot, \theta)\|_\infty).$$

Since σ_3 and $\Phi(\cdot, x)$ are uniformly continuous on J, for all $x \in X$, then $C(x)$ is continuous. We prove that $A : \Omega \longrightarrow \mathscr{P}(E)$ and $B : E \longrightarrow \mathscr{P}(E)$ are well defined. Let us fix an element $x \in \Omega$. Then, in virtue of assumption (H_{68}), we get there exists a Pettis-integrable mapping $v : J \longrightarrow X$ with $v(s) \in F(s, x(\tau(s)))$ such that

$\varphi(v(s)) \in L^1(J)$, for any $\varphi \in E^*$. Hence, for all $t \in J$, the following mapping

$$\varphi(k(t,s)\, v(s)) = k(t,s)\, \varphi(v(s))$$

is Lebesgue integrable and so, the mapping $s \longrightarrow k(t,s)\, v(s)$ is Pettis-integrable. Let us fix an element $x \in \Omega$ and take $u \in A(x)$. Then, there exists a Pettis-integrable function v with $v(s) \in F(s, x(\tau(s))))$ such that

$$u(t) = \int_0^{\sigma_1(t)} k(t,s)\, v(s)\, ds.$$

Let $t, t' \in J$ such that

$$u(t) \neq u(t') \text{ and } \sigma_1(t) - \sigma_1(t') > 0.$$

Then, on account of Lemma 1.3.19, we get there exists a function $\varphi_1 \in E^*$ with

$$\|\varphi_1\| = 1 \text{ and } \|u(t) - u(t')\| = \varphi_1(u(t) - u(t')).$$

This implies that

$$\begin{aligned}
\|u(t) - u(t')\| &\leq \int_0^1 |k(t,s) - k(t',s)|\, \|\varphi_1(v(s))\|\, ds + \int_{\sigma_1(t')}^{\sigma_1(t)} |k(t',s)|\, \|\varphi_1(v(s))\|\, ds \\
&\leq \int_0^1 |k(t,s) - k(t',s)|\, \|v(s)\|\, ds + \int_{\sigma_1(t')}^{\sigma_1(t)} |k(t',s)|\, \|v(s)\|\, ds \\
&\leq \int_0^1 \alpha_F(s)\, |k(t,s) - k(t',s))|\, \|x(\tau(s))\|\, ds + M \int_{\sigma_1(t')}^{\sigma_1(t)} \alpha_F(s)\, \|x(\tau(s))\|\, ds.
\end{aligned}$$

Then,

$$\begin{aligned}
\|u(t) - u(t')\| \leq r &\left(\int_0^1 \alpha_F(s)\, |k(t,s) - k(t',s)|\, ds \right) \\
&+ r \left(M \int_{\sigma_1(t')}^{\sigma_1(t)} \alpha_F(s)\, ds \right).
\end{aligned} \tag{7.3.3}$$

Since the partial $t \longrightarrow k(t,s)$ is continuous on $[0, 1]$, it follows that $u \in E$. The claim regarding the multi-valued mapping B is immediate, from assumption (H_{68}).

We show that A is a $\mathcal{D}$-set-contraction with a weakly sequentially closed graph. Indeed, let $(x_n)_n$ be a weakly convergent sequence contained in the set Ω to an element x and take a weakly converging sequence $y_n \in A(x_n)$ to an element y. It follows from Theorem 1.3.50, that

$$x_n(t) \rightharpoonup x(t) \text{ and } y_n(t) \rightharpoonup y(t) \text{ in } X.$$

Since $y_n \in A(x_n)$, then there exists a sequence of Pettis-integrable functions $(v_n)_n$ with $v_n(s) \in F(s, x_n(\tau(s)))$ such that

$$y_n(t) = \int_0^{\sigma_1(t)} k(t,s)\, v_n(s)\, ds.$$

Based on assumption (H_{68}), we get

$$\beta(\{v_n(t),\ n \in \mathbb{N}\}) \leq \varphi(\beta(\{x_n(\tau(t)),\ n \in \mathbb{N}\})).$$

The use of Theorem 1.2.42 leads to the set

$$\{v_n,\ n \in \mathbb{N}\} = \prod_{s \in J} \{v_n(s),\ n \in \mathbb{N}\}$$

is relatively weakly compact and so, there exists a renamed convergent subsequence of $(v_n)_n$ to a point v. Taking into account the following inequality

$$\|v_n(s)\| \leq r\ \alpha_F(s), \quad \text{for all } s \in J,$$

together with the dominated convergence theorem for Pettis-integral, we obtain v is a Pettis-integrable function and

$$y(t) = \int_0^{\sigma_1(t)} k(t,s)\, v(s)\, ds.$$

Based on assumption (H_{68}), it is shown that $v(t) \in F(t, x(t))$ and so,

$$y(t) \in (Ax)(t).$$

This means that A has a weakly sequentially closed graph on Ω. Now, we prove that A is $\mathcal{D}$-set-contraction on Ω. Let $t \in J$ be fixed. Making use of assumption (H_{68}), together with Lemma 1.4.30 enables us to have

$$\begin{aligned}
\beta(A(\Omega)(t)) &\leq \beta\left(\int_0^{\sigma_1(t)} k(t,s)\, F(s, \Omega(s))\right) \\
&\leq \beta(\overline{co}(k(t,J)\, F(J \times \Omega(J)))) \\
&\leq M\ \beta(F(J \times \Omega(J))) \\
&\leq M\ \varphi(\beta(\Omega(J))).
\end{aligned}$$

It follows from Theorem 1.5.41, that

$$\beta(A(\Omega)(t)) \leq \varphi(\beta(\Omega)).$$

Going back to the inequality (7.3.3), we obtain $A(\Omega)$ is a equi-continuous subset of E and so,

$$\beta(A(\Omega)) \leq \varphi(\beta(\Omega)).$$

This means that A is nonlinear $\mathcal{D}$-set-contraction on Ω with $\mathcal{D}$-function φ. Our next task is to show that B is countably δ-set-contraction and possess a weakly sequentially closed graph. For this purpose, consider $(x_n)_n$ as a weakly convergent sequence contained in the set Ω to an element x and take a sequence $y_n \in B(x_n)$ such that $y_n \rightharpoonup y$. By using Theorem 1.3.50, we have

$$x_n(t) \rightharpoonup x(t) \quad \text{and} \quad y_n(t) \rightharpoonup y(t) \text{ in } X.$$

In the similar way, we can show that there exists a sequence of Pettis-integrable functions $w_n : J \longrightarrow X$ with $w_n(s) \in G(s, x_n(s))$ such that

$$y_n(t) = \int_0^{\sigma_2(t)} w_n(s)\, ds.$$

Going back to the following inclusion

$$\{w_n(t),\ n \in \mathbb{N}\} \subset G\left(t, \{x_n(t),\ n \in \mathbb{N}\}\right),$$

we get

$$\begin{aligned}\beta\left(\{w_n(t),\ n \in \mathbb{N}\}\right) &\leq \beta\left(G\left(t, \{x_n(t),\ n \in \mathbb{N}\}\right)\right)\\ &\leq \delta\, \beta\left(\{x_n(t),\ n \in \mathbb{N}\}\right).\end{aligned}$$

Then, the sequence $(w_n(t))_n$ has a weakly convergent subsequence, say $\left(w_{n_k}(t)\right)_k$, to a function $w(t)$ and so, $w(t) \in G\left(t, x(t)\right)$ in view of assumption (H_{68}). Arguing as before, we can show that

$$\{w_n,\ n \in \mathbb{N}\} = \prod_{s \in J} \{w_n(s),\ n \in \mathbb{N}\}$$

is relatively weakly compact. Then, the sequence $(w_n)_n$ possesses a renamed weakly convergent subsequence to a point w. It follows from assumption (H_{68}), that

$$\|w_n(s)\| \leq r\, \alpha_G(s).$$

Then, on account of the dominated convergence theorem for Pettis-integral, we get w as a Pettis-integrable function and

$$y(t) = \int_0^{\sigma_2(t)} w(s)\, ds,\ t \in J.$$

Let M be an arbitrary countably subset of Ω. By using of Lemma 1.4.30, we get

$$\begin{aligned}\beta(B(M)(t)) &\leq \beta\left(\bigcup\left\{\int_0^{\sigma_2(t)} G(s, x(s))\, ds : x \in M\right\}\right)\\ &\leq \beta\left(\int_0^{\sigma_2(t)} G(s, M(s))\, ds\right)\\ &\leq \beta\left(\overline{co}(G(J \times M(J)))\right)\\ &\leq \delta\, \beta((M(J))).\end{aligned}$$

Since Ω is equi-continuous, we obtain

$$\beta(B(M)(t)) \leq \delta\, \beta(M).$$

From the above discussion, we see that $B(M)$ is equi-continuous and, when we apply Theorem 1.5.41, we reach the result

$$\beta(B(M)) \leq \delta\beta(M).$$

We show that C is countably ρ-set-contraction and it is weakly sequentially continuous. Indeed, consider $(x_n)_n$ as a weakly converging sequence of the set Ω to an element x. Taking into account assumption (H_{66}), and using Theorem 1.3.50, we get

$$\lim_{n\to+\infty} \left|x^*(f(s, x_n(\kappa(s)))) - x^*(f(s, x(\kappa(s))))\right| = 0, \quad \text{for all } x^* \in X^*.$$

Due to the fact that $\|f(s, x_n(\kappa(s))\| \leq \lambda\, r + \|f(\cdot, \theta)\|_\infty$, we infer that

$$\int_0^{\sigma_3(t)} f(s, x_n(\kappa(s)))\, ds \rightharpoonup \int_0^{\sigma_3(t)} f(s, x(\kappa(s)))\, ds.$$

The assumption (H_{66}) guaranties that C is Lipschitzian with a constant $\frac{1-\gamma}{2}$. If we consider the fact that $C(\Omega)$ is bounded and deploy Theorem 1.3.50, we get

$$C(x_n) \rightharpoonup Cx.$$

Besides, on account of Lemma 1.5.9, we obtain C is countably ρ-set-contraction, where $\rho = \frac{1-\gamma}{2}$. After simple calculations, we can see that the operator D is nonlinear contraction with $\mathcal{D}$-function ϕ_D, where

$$\phi_D(r) < \frac{1+\gamma}{2}\, r.$$

Moreover, it follows from assumption (H_{68}), that $B(I - D)^{-1}C(y)$ is relatively weakly compact and the set $A(\Omega) + B(I - D)^{-1}C(y)$ is bounded, for all $y \in \Omega$. Due to the fact that A is nonlinear $\mathcal{D}$-set-contraction and Lemma 3.3.42, we can define the multi-valued mapping Q on Ω by the formula

$$Qx = (I - A)^{-1} B(I - D)^{-1} C(x).$$

Let us fix $x \in \Omega$ and take $y_1, y_2 \in Q(x)$. Then,

$$\begin{cases} y_1(t) = \int_0^{\sigma_1(t)} k(t,s)u_1(s)\,ds + \int_0^{\sigma_2(t)} v_1(s)\,ds, & \text{and} \\ y_2(t) = \int_0^{\sigma_1(t)} k(t,s)u_2(s)\,ds + \int_0^{\sigma_2(t)} v_2(s)\,ds, \end{cases}$$

for some Pettis-integrable functions u_1, u_2, v_1 and v_2 with $u_1(s), u_2(s) \in F(s, y_1(s))$ and $v_1(s), v_2(s) \in G\left(s, (I - D)^{-1} Cx(s)\right)$. If we consider $\tau \in (0, 1)$, we get

$$\begin{aligned} \tau y_1(t) + (1-\tau)y_2(t) &= \int_0^{\sigma_1(t)} k(t,s)(\tau u_1(s) + (1-\tau)u_2(s))\,ds \\ &\quad + \int_0^{\sigma_2(t)} (\tau v_1(s) + (1-\tau)v_2(s))\,ds \\ &= \int_0^{\sigma_1(t)} k(t,s)u_3(s)\,ds + \int_0^{\sigma_2(t)} v_3(s)\,ds, \end{aligned}$$

where

$$\begin{cases} u_3(s) = \tau u_1(s) + (1-\tau)u_2(s) \\ v_3(s) = \tau v_1(s) + (1-\tau)v_2(s). \end{cases}$$

Since G has a convex value, it follows that

$$v_3(t) \in G\left(t, (I - D)^{-1} Cx(t)\right).$$

Going back to $u_3(t) \in \tau F(t, y_1(t)) + (1-\tau)F(t, y_2(t))$, and using assumption (H_{68}), we get

$$u_3(t) \in F(t, \tau y_1(t) + (1-\tau)y_2(t)).$$

It is easy to see that u_3 and v_3 are Pettis-integrable functions and consequently, we have

$$\tau y_1 + (1-\tau)y_2 \in A(\tau y_1 + (1-\tau)y_2) + B(I - D)^{-1} Cx.$$

Next, we show that

$$Q\left(\overline{\mathscr{B}}_E(r)\right) \subset \overline{\mathscr{B}}_E(r).$$

Indeed, it is enough to proving that

$$y \in A(y) + B(I - D)^{-1} C(x),\ x \in \overline{\mathscr{B}}_E(r) \Rightarrow y \in \overline{\mathscr{B}}_E(r).$$

At first, we need to show the following inclusion

$$(I - D)^{-1}C(\overline{\mathscr{B}}_E(r)) \subset \overline{\mathscr{B}}_E(r).$$

Let y be an arbitrarily fixed element contained in the set $(I - D)^{-1}C\left(\overline{\mathscr{B}}_E(r)\right)$. Then,

$$y(t) = g(t, y(t)) + \Phi\left(t, \int_0^{\sigma_3(t)} f(s, x(\kappa(s)))\, ds\right),$$

for some $x \in E$ with $\|x\| \leq r$. Since $y \in C(J, X)$, then there exists a point $t^* \in J$ with $\|y\|_\infty = \|y(t^*)\|$ and so,

$$\begin{aligned}
\|y(t^*)\| &= \left\| g(t^*, y(t^*)) + \Phi\left(t^*, \int_0^{\sigma_3(t^*)} f(s, x(\kappa(s)))\, ds\right) \right\| \\
&\leq \|g(t^*, y(t^*))\| + \left\| \Phi\left(t^*, \int_0^{\sigma_3(t^*)} f(s, x(\kappa(s)))\, ds\right) \right\| \\
&\leq \|g(\cdot, \theta)\|_\infty + \gamma\ \|y(t^*)\| + \|\Phi(\cdot, \theta)\|_\infty + \frac{1-\gamma}{2\lambda} \left\| \int_0^{\sigma_3(t^*)} f(s, x(\kappa(s)))\, ds \right\|.
\end{aligned}$$

Thus, on account of Theorem 1.3.18, we get there exists a linear function $x^* \in E^*$ with $\|x^*\| = 1$ such that

$$\left\| \int_0^{\sigma_3(t^*)} f(s, x(\kappa(s)))\, ds \right\| = x^*\left(\int_0^{\sigma_3(t^*)} f(s, x(\kappa(s)))\, ds \right).$$

Thus,

$$\begin{aligned}
\|y(t^*)\| &\leq \frac{1}{1-\gamma}\left(\|g(\cdot, \theta)\|_\infty + \|\Phi(\cdot, \theta)\|_\infty\right) + \frac{1}{2\lambda}\, x^*\left(\int_0^{\sigma_3(t^*)} f(s, x(\kappa(s)))\, ds \right) \\
&\leq \frac{\|g(\cdot, \theta)\|_\infty + \|\Phi(\cdot, \theta)\|_\infty}{1-\gamma} + \frac{1}{2\lambda}\left(\int_0^{\sigma_3(t^*)} \|f(s, x(\kappa(s))) - f(s, \theta)\|\, ds + \|f(\cdot, \theta)\| \right) \\
&\leq \frac{\|g(\cdot, \theta)\|_\infty + \|\Phi(\cdot, \theta)\|_\infty}{1-\gamma} + \frac{1}{2\lambda}\|f(\cdot, \theta)\|_\infty + \frac{r}{2}.
\end{aligned}$$

Then,

$$\|y\|_\infty \leq \frac{1}{1-\gamma}\left(\|g(\cdot, \theta)\|_\infty + \|\Phi(\cdot, \theta)\|_\infty\right) + \frac{1}{2\lambda}\|f(\cdot, \theta)\|_\infty + \frac{r}{2}.$$

This means, in particular, that

$$(I - D)^{-1}C(\overline{\mathscr{B}}_E(r)) \subset \overline{\mathscr{B}}_E(r).$$

Let $x \in \overline{\mathscr{B}}_E(r)$ be fixed and take a function $y \in A(y) + B(I - D)^{-1}C(x)$. Then, there exist two Pettis-integrable functions v and w with $v(s) \in F(s, y(s))$ and $w(s) \in G(s, (I - D)^{-1}Cx(s))$ such that

$$y(t) = \int_0^{\sigma_1(t)} k(t, s)\, v(s)\, ds + \int_0^{\sigma_2(t)} w(s)\, ds, \ t \in J.$$

It follows from Theorem 1.3.18, that

$$\begin{aligned}
\|y(t)\| &\leq \int_0^{\sigma_1(t)} |k(t, s)|\, \|v(s)\|\, ds + \int_0^{\sigma_2(t)} \|w(s)\|\, ds \\
&\leq M \int_0^{\sigma_1(t)} \|F(s, y(s))\|\, ds + \int_0^{\sigma_2(t)} \left\|G\left(s, \left((I - D)^{-1}Cx\right)(s)\right)\right\| ds \\
&\leq M\, \|\alpha_F\|_{L^1}\, \|y\|_\infty + \|\alpha_G\|_{L^1}\, \left\|(I - D)^{-1}Cx\right\|_\infty \\
&\leq M\, \|\alpha_F\|_{L^1}\, \|y\|_\infty + \|\alpha_G\|_{L^1}\, r.
\end{aligned}$$

Then,

$$\|y\|_\infty \leq \frac{\|\alpha_G\|_{L^1}}{1 - M\, \|\alpha_F\|_{L^1}} r < r.$$

Now, it remains to prove the equi-continuity of the set $Q(\Omega)$. In other words, we only show that $(I - A)^{-1}B(\Omega_1)$, where $\Omega_1 = (I - D)^{-1}C(\Omega)$, is equi-continuous. In fact, let $x \in \Omega_1$ and take $y \in (I - A)^{-1}B(\Omega_1)$. Then,

$$y(t) = \int_0^{\sigma_1(t)} k(t, s)u(s)\, ds + \int_0^{\sigma_2(t)} v(s)\, ds, \ t \in J,$$

where u and v are two Pettis-integrable functions with $u(s) \in F(s, y(\tau(s)))$ and $v(s) \in G(s, x(s))$. Choose $t, t' \in J$ such that $\sigma_1(t) - \sigma_1(t') > 0$ and $\sigma_2(t) \neq \sigma_2(t')$. Then, on account of Lemma 1.3.19, we get there exists a function $x^* \in E^*$ with $\|x^*\| = 1$ such that

$$\|y(t) - y(t')\| = x^*(y(t) - y(t')).$$

This implies that

$$\begin{aligned}
\|y(t) - y(t')\| &\leq x^* \left(\int_0^{\sigma_1(t)} (k(t, s) - k(t', s))\, u(s)\, ds + \int_{\sigma_1(t)}^{\sigma_1(t')} k(t', s)\, u(s)\, ds \right) \\
&\quad + x^* \left(\int_{\sigma_2(t')}^{\sigma_2(t)} v(s)\, ds \right) \\
&\leq x^* \left(\int_0^{\sigma_1(t)} |k(t, s) - k(t', s)|\, \|F(s, y(\tau(s)))\|\, ds \right) \\
&\quad + \int_{\sigma_1(t')}^{\sigma_1(t)} |k(t', s)|\, \|F(s, y(\tau(s)))\|\, ds + \int_{\sigma_2(t')}^{\sigma_2(t)} \|G(s, x(s))\|\, ds.
\end{aligned}$$

From assumption (H_{68}), it follows that

$$\|y(t) - y(t')\| \leq r\left(\int_0^{\sigma_1(t)} |k(t,s) - k(t',s)|\, \alpha_F(s)\, ds + M \int_{\sigma_1(t')}^{\sigma_1(t)} \alpha_F(s)\, ds\right) + \|\Omega_1\| \int_{\sigma_2(t')}^{\sigma_2(t)} \alpha_G(s)\, ds.$$

This means, in particular, that $Q(\Omega)$ is equi-continuous. This achieves the claim and completes the proof. Q.E.D.

Example 7.3.1 Consider the following nonlinear differential integral inclusion

$$\begin{cases} \dfrac{\partial}{\partial t} x(t) \in k(t)\, F(t, x(t)) + G(t, y(t)), \\ y(t) = \left(u + \displaystyle\int_0^t f(s, x(\kappa(s)))\, ds\right) \cdot v + h(t, y(t)),\ t \in J, \\ (x(0), y(0)) = (x_0, y_0) \in X^2, \end{cases} \tag{7.3.4}$$

where $u, v \in X \backslash \{0\}$ with $\|v\| < \frac{1}{2\lambda}$, the functions $\kappa(\cdot)$ and $k(\cdot)$ are continuous on J, the single-valued maps $f, h : J \times X \longrightarrow X$ satisfy assumption (H_{66}) and the multi-valued maps $F, G : J \times X \longrightarrow \mathscr{P}(X)$ satisfy assumption (H_{68}). In addition, suppose that

$$\|k(\cdot)\|_\infty\, \|\alpha_F\|_{L^1} + \|\alpha_G\|_{L^1} < 1$$

and

$$\frac{1}{1-\gamma}\left(\|y_0 - u \cdot v\| + \|(u + x_0) \cdot v\|\right) + \frac{1}{2\lambda}\|f(\cdot, \theta)\|_\infty < r,$$

where λ and γ are as given in assumption (H_{66}), α_F and α_G are given in (H_{68}) and the constant r is given in Theorem 7.3.3. Then, problem (7.3.4) has a solution. In fact, the system (7.3.4) can be transformed into the following fixed point problem

$$\begin{cases} x(t) \in Ax(t) + By(t), \\ y(t) = Cx(t) + Dy(t), \\ (x(0), y(0)) = (x_0, y_0), \end{cases}$$

where A and B are given in (7.3.2), and the operators C and D defined by

$$\begin{cases} (Cx)(t) = \left(u + \displaystyle\int_0^t f(s, x(\kappa(s)))\, ds\right) \cdot v, \\ (Dy)(t) = h(t, y(t)) + y_0 - (u \cdot v + h(0, y_0)). \end{cases}$$

Observe that

$$(Cx)(t) = \Phi\left(t, \int_0^t f(s, x(\kappa(s)))\, ds\right)$$

and

$$(Dy)(t) = g(t, y(t)),$$

where $\Phi, g : J \times X \longrightarrow X$ are given by

$$\Phi(t, x) = (u + x) \cdot v$$

and

$$g(t, y) = h(t, y) + y_0 - (u \cdot v + h(0, y_0)).$$

Due to the fact that the right-hand multiplication operator $R_v(x) = x \cdot v$ is linear and continuous, we infer that R_v is weakly continuous. After simple calculations, we can verify that Φ is Lipschitzian with a Lipschitz constant $\|v\|$ and it is continuous uniformly with respect to the first variable, for all $x \in X$. This implies that Φ satisfies assumption (H_{67}). This completes the proof.

References

1. Ben Amara, K., Kaddachi, N., Jeribi, A.: New fixed point theorems for countably condensing maps with an application to functional integral inclusions. Math. Slovaca **71**, 1487–1510 (2021)
2. Jeribi, A., Kaddachi, N., Krichen, B.: Existence results for a coupled system of perturbed functional differential inclusions in Banach algebras. Bull. Malays. Math. Sci. Soc. **41**, 893–918 (2018)
3. Kaddachi, N.: Existence results of ordinary differential inclusions in Banach algebra under weak topology. J. Phys. Math. **9**, 123 (2018)
4. Barbu, V.: Nonlinear semigroups and differential equations in Banach spaces. Noordhoff International Publishing, Leyden (1976)
5. Cellina, A.: Multi-valued differential equations and ordinary differential equations. SIAM J. Appl. Math. **18**, 533–538 (1970)
6. Cellina, A.: A view on differential inclusions. Rend. Semin. Mat. Univ. Politec. Torino. **63**(3), 197–209 (2005)
7. Deimling, K.: Multi-valued Differential Equations. De Gruyter, Berlin (1998)
8. Djebali, S., Górniewicz, L., Ouahab, A.: Solution Sets for Differential Equations and Inclusions, De Gruyter Series in Nonlinear Analysis and Applications, vol. 18. Walter de Gruyter, Berlin (2013)
9. Filippov, A.F.: Classical solutions of differential equations with multi-valued right-hand side. SIAM J. Control **5**, 609–621 (1967)
10. Granas, A., Guenther, R.B., Lee, J.W.: Some general existence principles in the Carathéodory theory of nonlinear differential systems. J. Math. Pures Appl. **70**, 153–196 (1991)
11. Kamenskii, M., Obukhovski, V., Zecca, P.: Condensing multi-valued maps and semilinear differential inclusion in Banach spaces. De Gruyter Series in Nonlinear Analysis and Applications 7. Walter de Gruyter, Berlin-New York (2001)
12. Bressan, A.: On the qualitative theory of lower semi-continuous differential inclusions. J. Diff. Equ. **77**(2), 379–391 (1989)
13. Bressan, A.: Selections of Lipschitz multifunctions generating a continuous flow. Diff. Integral Equ. **4**(3), 483–490 (1991)
14. Conway, J.B.: A Course in Functional Analysis, 2nd edn. Springer, Berlin (1990)
15. Zaremba, S.C.: A propos des champs de demi-cônes convexes. Bull. Sci. Math. **64**, 5–12 (1940)

16. Aubin, J.P., Cellina, A.: Differential Inclusions. Springer (1984)
17. Hu, S., Papageorgiou, N.S.: Handbook of multi-valued analysis. Vol. I. Theory. Mathematics and its Applications, vol. 419, pp. 7923–4682. Kluwer Academic Publishers, Dordrecht (1997)
18. Dhage, B.C.: Some nonlinear alternatives in Banach algebras with applications I. Nonlinear Studies **12**, 271–283 (2005)
19. Dhage, B.C.: Some nonlinear alternatives in Banach algebras with applications II. Kyungpook Math. J. **45**, 281–292 (2005)
20. Nieto, J.J.: Periodic boundary value problems for first order impulsive ODEs. Nonlinear Anal. **51**, 1223–1232 (2002)

Index

A. Jeribi and N. Kaddachi, *Topological Fixed-Point Theory in Suitable Banach Algebras with Applications*, Infosys Science Foundation Series,
https://doi.org/10.1007/978-981-96-5542-7

The manufacturer's authorised representative in the EU is Springer Nature Customer Service Centre GmbH, Europaplatz 3, 69115 Heidelberg, Germany. If you have any concerns regarding our products, please contact ProductSafety@springernature.com

Printed and bound by CPI Group (UK) Ltd, Croydon, CR0 4YY

01/10/2026

02225909-0011